Groundwater Quality and Public Health

Groundwater Quality and Public Health

Editors

Jianhua Wu
Peiyue Li
Saurabh Shukla

MDPI • Basel • Beijing • Wuhan • Barcelona • Belgrade • Manchester • Tokyo • Cluj • Tianjin

Editors

Jianhua Wu
School of Water and Environment
Chang'an University
Xi'an
China

Peiyue Li
School of Water and Environment
Chang'an University
Xi'an
China

Saurabh Shukla
Faculty of Civil Engineering
Shri Ramswaroop Memorial University
Barabanki
India

Editorial Office
MDPI
St. Alban-Anlage 66
4052 Basel, Switzerland

This is a reprint of articles from the Special Issue published online in the open access journal *Water* (ISSN 2073-4441) (available at: www.mdpi.com/journal/water/special_issues/waterquality_publichealth).

For citation purposes, cite each article independently as indicated on the article page online and as indicated below:

LastName, A.A.; LastName, B.B.; LastName, C.C. Article Title. *Journal Name* **Year**, *Volume Number*, Page Range.

ISBN 978-3-0365-5836-3 (Hbk)
ISBN 978-3-0365-5835-6 (PDF)

© 2022 by the authors. Articles in this book are Open Access and distributed under the Creative Commons Attribution (CC BY) license, which allows users to download, copy and build upon published articles, as long as the author and publisher are properly credited, which ensures maximum dissemination and a wider impact of our publications.

The book as a whole is distributed by MDPI under the terms and conditions of the Creative Commons license CC BY-NC-ND.

Contents

About the Editors vii

Preface to "Groundwater Quality and Public Health" ix

Peiyue Li, Jianhua Wu and Saurabh Shukla
Achieving the One Health Goal: Highlighting Groundwater Quality and Public Health
Reprinted from: *Water* **2022**, *14*, 3540, doi:10.3390/w14213540 1

Ana Machado, Eva Amorim and Adriano A. Bordalo
Spatial and Seasonal Drinking Water Quality Assessment in a Sub-Saharan Country (Guinea-Bissau)
Reprinted from: *Water* **2022**, *14*, 1987, doi:10.3390/w14131987 9

Abel Nsabimana, Peiyue Li, Song He, Xiaodong He, S. M. Khorshed Alam and Misbah Fida
Health Risk of the Shallow Groundwater and Its Suitability for Drinking Purpose in Tongchuan, China
Reprinted from: *Water* **2021**, *13*, 3256, doi:10.3390/w13223256 25

Jiao Li, Congjian Sun, Wei Chen, Qifei Zhang, Sijie Zhou and Ruojing Lin et al.
Groundwater Quality and Associated Human Health Risk in a Typical Basin of the Eastern Chinese Loess Plateau
Reprinted from: *Water* **2022**, *14*, 1371, doi:10.3390/w14091371 45

Xueshan Bai, Xizhao Tian, Junfeng Li, Xinzhou Wang, Yi Li and Yahong Zhou
Assessment of the Hydrochemical Characteristics and Formation Mechanisms of Groundwater in A Typical Alluvial-Proluvial Plain in China: An Example from Western Yongqing County
Reprinted from: *Water* **2022**, *14*, 2395, doi:10.3390/w14152395 63

Jing Jin, Zihe Wang, Yiping Zhao, Huijun Ding and Jing Zhang
Delineation of Hydrochemical Characteristics and Tracing Nitrate Contamination of Groundwater Based on Hydrochemical Methods and Isotope Techniques in the Northern Huangqihai Basin, China
Reprinted from: *Water* **2022**, *14*, 3168, doi:10.3390/w14193168 81

Siyuan Liu, Xiao Han, Shaopeng Li, Wendi Xuan and Anlei Wei
Stimulating Nitrate Removal with Significant Conversion to Nitrogen Gas Using Biochar-Based Nanoscale Zerovalent Iron Composites
Reprinted from: *Water* **2022**, *14*, 2877, doi:10.3390/w14182877 107

Imen Ben Salem, Yousef Nazzal, Fares M. Howari, Manish Sharma, Jagadish Kumar Mogaraju and Cijo M. Xavier
Geospatial Assessment of Groundwater Quality with the Distinctive Portrayal of Heavy Metals in the United Arab Emirates
Reprinted from: *Water* **2022**, *14*, 879, doi:10.3390/w14060879 129

Zizhao Cai, Lingxia Liu, Wei Xu, Ping Wu and Chuan Lu
Hydrochemical Characteristics of Arsenic in Shallow Groundwater in Various Unconsolidated Sediment Aquifers: A Case Study in Hetao Basin in Inner Mongolia, China
Reprinted from: *Water* **2022**, *14*, 669, doi:10.3390/w14040669 147

Ruiping Liu, Fei Liu, Jiangang Jiao, Youning Xu, Ying Dong and El-Wardany R.M. et al.
Potential Toxic Impacts of Hg Migration in the Disjointed Hyporheic Zone in the Gold Mining Area Experiencing River Water Level Changes
Reprinted from: *Water* **2022**, *14*, 2950, doi:10.3390/w14192950 . . . . **165**

Chengcheng Liang, Wei Wang, Xianmin Ke, Anfeng Ou and Dahao Wang
Hydrochemical Characteristics and Formation Mechanism of Strontium-Rich Groundwater in Tianjiazhai, Fugu, China
Reprinted from: *Water* **2022**, *14*, 1874, doi:10.3390/w14121874 . . . . **181**

Zhiyuan Ma, Junfeng Li, Man Zhang, Di You, Yahong Zhou and Zhiqiang Gong
Groundwater Health Risk Assessment Based on Monte Carlo Model Sensitivity Analysis of Cr and As—A Case Study of Yinchuan City
Reprinted from: *Water* **2022**, *14*, 2419, doi:10.3390/w14152419 . . . . **197**

About the Editors

Jianhua Wu

Jianhua Wu is an Associate Professor of Hydrogeology and Environmental Sciences in the School of Water and Environment at the Chang'an University, China. She obtained her Ph.D. in Hydrology and Water Resources from Chang'an University (2015), and is interested in groundwater flow modeling, groundwater quality and soil moisture transport. She has published over 80 research articles in international journals on topics ranging from groundwater quality and health risk assessment to hydrological processes and contaminant transport modeling. She is at present Associate Editor of *Exposure and Health*, Guest Editor of *Human and Ecological Risk Assessment* and *Water*, and is Review Editor of *Frontiers in Environmental Science*.

Peiyue Li

Peiyue Li is a Professor of Hydrogeology and Environmental Sciences in the School of Water and Environment at the Chang'an University, China. Peiyue Li holds a Ph.D. in Groundwater Science and Engineering from Chang'an University (2014), and has wide experience in groundwater quality assessment, hydrogeochemistry, and groundwater modeling. He has published over 150 articles in refereed journals on topics that range from groundwater quality assessment and groundwater hydrochemistry to groundwater pumping tests and in situ tracer tests. He serves as Associate Editor for *Exposure and Health, Mine Water and the Environment, Human and Ecological Risk Assessment, Archives of Environmental Contamination and Toxicology, Discover Water* and *Frontiers in Environmental Science*, and is an editorial board member for *Environmental Monitoring and Assessment, Water* and some national key journals.

Saurabh Shukla

Saurabh Shukla is currently working as an Associate Professor at the Faculty of Civil Engineering at Shri Ramswaroop Memorial University (SRMU), India. His research interests include monitoring of surface and ground water resources, and water-soil interactions. He is also involved in exploring novel and sustainable techniques of wastewater treatment through biochar and nanotechnology. He has published more than 40 research articles including book chapters on topics like groundwater quality and health risk assessment, use of nanotechnology for wastewater remediation etc. He is serving *Environment, Development and Sustainability* as an Associate Editor. He is Guest Editor of *Environmental Monitoring and Assessment, Groundwater for Sustainable Development* and Review Editor of *Frontiers in Environmental Science*.

Preface to "Groundwater Quality and Public Health"

Groundwater is a primary source of water for nearly half of the world's population. Its quality has become as important as its quantity, as serious groundwater pollution has been identified in many regions and countries of the world. The contaminants in groundwater generally have two sources: one is geogenic origin and the other is human activities. The contaminants in groundwater can seriously affect human health, inducing a number of waterborne diseases. This book attempts to provide a platform for researchers, policy makers, and engineers to share their latest thoughts and findings on groundwater quality and public health, as well as novel methods dealing with groundwater pollution. The chapters published in this book include the latest research results by world-renowned researchers, whose findings can benefit researchers, engineers, policy makers, and government officials in future groundwater quality research and policy making. This book focuses on understanding the relationship between groundwater quality and public health, and the current state of knowledge on the links between geological/geochemical processes and human health across the world. The factors that accelerate or decelerate geological/geochemical processes, thus affecting human health, are also considered.

For their help in completing this book, we acknowledge the voluntary reviewers for their useful and critical comments that helped the contributing authors further improve the quality of their manuscripts. The authors whose manuscripts were accepted for publication in this book and those whose manuscripts were unfortunately rejected are also acknowledged for showing their interest in this book. Your participation in this topic makes this book unique and the world a better place.

Jianhua Wu, Peiyue Li, and Saurabh Shukla
Editors

Editorial

Achieving the One Health Goal: Highlighting Groundwater Quality and Public Health

Peiyue Li [1,2,*], Jianhua Wu [1,2] and Saurabh Shukla [3,*]

1. School of Water and Environment, Chang'an University, No. 126 Yanta Road, Xi'an 710054, China
2. Key Laboratory of Subsurface Hydrology and Ecological Effects in Arid Region of the Ministry of Education, Chang'an University, No. 126 Yanta Road, Xi'an 710054, China
3. Faculty of Civil Engineering, Shri Ramswaroop Memorial University, Barabanki 225003, India
* Correspondence: lipy2@163.com or peiyueli@chd.edu.cn (P.L.); saurabh.shukla2020@gmail.com (S.S.)

Abstract: In many regions of the world, groundwater is the main water source for multiple uses, including for drinking, irrigation, and industry. Groundwater quality, therefore, is closely related to human health, and the consumption of contaminated groundwater can induce various waterborne diseases. In the last ten years, the world has witnessed a rapid development in groundwater quality research and the assessment of associated health risks. This editorial introduced the foundation of the current Special Issue, *Groundwater Quality and Public Health*, briefly reviewed recent research advances in groundwater quality and public health research, summarized the main contribution of each published paper, and proposed future research directions that researchers should take into account to achieve the one health goal. It is suggested that groundwater quality protection should be further emphasized to achieve the one health goal and the UN's SDGs. Modern technologies should be continuously developed to remediate and control groundwater pollution, which is a major constrain in the development of a sustainable society.

Keywords: groundwater quality; groundwater pollution; contamination remediation; one health; public health; health risk assessment

1. Introduction

The United Nations (UN) established the Sustainable Development Goals (SDGs) to guide its member states' agendas and political policies through the next 15 years [1]. Among the 17 goals, SDG-6, *Ensure Access to Water and Sanitation for All*, requires improving water and sanitation management for all people, having urged every nation in the world to take the necessary actions to ensure its people's basic human right: gaining access to safe drinking water. Since the 21st century, with the acceleration of global integration, the population increase, continuous climate change, and rapid development of international trade, some public health events, including sudden infectious and foodborne/waterborne diseases, have occurred frequently, aggravating the complexity of health problems. For example, starting in late 2019, the COVID-19 pandemic has caused over six million deaths worldwide, still affecting daily life to this day. In order to find solutions to these problems, the concept of "One Health" was introduced, practiced, and applied in increasingly more international organizations and countries in the process of health governance [2,3].

Groundwater is one of the most valuable natural resources supporting the survival of human beings and the development of human societies [4–6]. However, serious groundwater pollution can have significant negative impacts on human health [7–11]. Contaminants in groundwater generally have two sources, one being of geogenic origin and the other human activities [12]. To reflect the most recent progress in groundwater quality research and associated public health issues, the Special Issue of the journal *Water*, entitled *Groundwater Quality and Public Health*, was developed. Its aim is to attempt to provide a platform for researchers, policy makers, and engineers to share their latest thoughts and findings on this topic, as well as novel methods dealing with groundwater pollution. The papers published in this Special Issue include the latest research results by world renowned researchers, whose findings could benefit researchers, engineers, policy makers, and government officials in future groundwater quality research and policy making.

2. Recent Research Advances in Groundwater Quality and Public Health

The importance of groundwater quality in maintaining human health has long been recognized [13–17]. As early as the 11th century AD, the Chinese population recognized the impacts of geological conditions on human health [18]; however, modern medical geological disciplines were not established to address the relationships between geoenvironmental elements and the health or occurrence of diseases in the environment until the 1930s, when Russian, Scandinavian and British geochemists established relationships between geochemistry and health in both humans and animals [18,19]. Since the 1980s, a number of books concerning highly interdisciplinary medical geology were edited and published [20–25], summarizing this field's contemporary advances. Among these, the book *Essentials of Medical Geology*, edited by Selinus et al. [20], is an award-winning work used worldwide as both a text and reference book. It involves environmental biological processes of elements, exposure pathways of elements, toxicology and pathology, and the techniques and tools in medical geological studies. Another book, *Introduction to Medical Geology: Focus on Tropical Environments*, edited by Dissanayake and Chandrajith [18], focuses on the impacts of medical geology on the health of millions of people in unique tropical lands. These valuable books provide comprehensive insights into the current developments and future prospects concerning medical geology.

In addition to books, there has been a large number of journal articles published over the past two decades, with the number of journals focusing on geological factors and human health also increasing. Some examples of these journals include *Exposure and Health*, *Human and Ecological Risk Assessment*, and *Environmental Geochemistry and Health*, presenting many papers reporting on the effects of geoenvironmental factors on human health [26–30]. In addition, some water- and geology-related journals also published a number of papers regarding water quality and public health [31–33]. Most recently, Fida et al. [34] reviewed the pollution status of water in Pakistan, where many people do not have access to safe and healthy drinking water, and summarized the significant health problems associated with the low-quality drinking water. Sathe et al. [35] conducted a comprehensive hydrogeochemical investigation in the north-eastern region of India to reveal the relationships between hydrogeological settings and groundwater with high arsenic and fluoride contents, assessing the health risks imposed due to exposure to these elements via drinking water intake. Alfeus et al. [36] assessed the human health risks caused with inhalation exposure to ambient PM2.5 and trace elements in Cape Town, South Africa. All the above studies showed that with the development in social economy, the public is seeking harmony between the rapid economic development and a sustainable environment, paying more attention to the health impacts of environmental pollution.

Particularly in China, on 11 September 2020, the security of public health was proposed to be one of the scientific and technological innovation targets when the Chinese president, Xi Jinping, chaired the symposium for scientists [37]. Since then, increasingly more research has been carried out seeking solutions to basic research questions behind this target. In 2021, the China Geological Survey implemented the Plan of Geological Survey to Support the Healthy China Strategy. The purposes of this were to accelerate the construction of technological systems, organizational structure systems, professional development systems, condition guarantee systems, and coordination and cooperation mechanisms for geological surveys to support the Healthy China Strategic, to systematically understand the status, health risks, and changing trend of major geological problems affecting human health in all of China, especially in key regions such as urban and periurban areas, and to fully reveal the mechanisms and laws behind how these geoenvironmental factors have affected human health. Based on these geological survey projects, medical geology in China is developing at a fast pace. In the past, medical geology mainly focused on the health problems caused by toxic elements and associated processes. However, particular attention has also been paid to the essential elements in the novel geological survey projects. For example, years ago, the main research focus was how toxic elements could cause diseases, such as fluorosis and arsenicosis [38–41], but, now, increasingly more research focuses on the effects of essential elements on human health [42–45]. This shift indicates that people's attitudes to health has changed from knowing how to avoid diseases to knowing how to maintain physical and mental health. This is a big step from traditional medical geology to health geology, and is more supportive towards fulfilling the One Health goal.

3. Papers Published in This Special Issue

The Special Issue *Groundwater Quality and Public Health* in the journal *Water* attracted 23 submissions, and after a rigorous peer review, 11 research papers were published. The topics of these papers ranged from regional groundwater quality and human health to specific elements or pollutants in groundwater (Table 1). Specifically, the 11 published articles could be classified into three topical clusters. The first topical cluster was regional groundwater quality and human health, and included four research articles. The second cluster was nitrate pollution, including two articles; the final cluster centered on trace elements, consisting of five research articles. In addition, as shown in the word cloud map generated using the titles and abstracts of papers in this Special Issue (Figure 1), the most frequently used words or terms in the papers comprising this Special Issue were groundwater, water, risk nitrate, health, hydrochemical, quality, drinking, basin, and HCO_3^-. This indicated that the main research objective among these papers was groundwater, with nitrate being a very common groundwater contaminant. In addition, groundwater quality studies are usually associated with studies on groundwater hydrochemistry represented by major ions such as HCO_3^-, and drinking water intake is the most significant exposure pathway causing human health risks.

Figure 1. Word cloud generated with titles and abstracts of the Special Issue papers.

Table 1. Information of research papers in the Special Issue.

Topic Clusters	Authors	Titles	DOIs
Regional groundwater quality and human health	Machado et al.	Spatial and seasonal drinking water quality assessment in a sub-Saharan country (Guinea-Bissau)	10.3390/w14131987
	Nsabimana et al.	Health risk of the shallow groundwater and its suitability for drinking purpose in Tongchuan, China	10.3390/w13223256
	Li et al.	Groundwater Quality and Associated Human Health Risk in a Typical Basin of the Eastern Chinese Loess Plateau	10.3390/w14091371
	Bai et al.	Assessment of the hydrochemical characteristics and formation mechanisms of groundwater in a typical alluvial-proluvial plain in China: an example from western Yongqing County	10.3390/w14152395
Nitrate pollution	Jin et al.	Delineation of hydrochemical characteristics and tracing nitrate contamination of groundwater based on hydrochemical methods and isotope techniques in the northern Huangqihai Basin, China	10.3390/w14193168
	Liu et al.	Stimulating nitrate removal with significant conversion to nitrogen gas using biochar-based nanoscale zerovalent iron composites	10.3390/w14182877
Trace elements	Salem et al.	Geospatial assessment of groundwater quality with the distinctive portrayal of heavy metals in the United Arab Emirates	10.3390/w14060879
	Cai et al.	Hydrochemical characteristics of arsenic in shallow groundwater in various unconsolidated sediment aquifers: a case study in Hetao Basin in Inner Mongolia, China	10.3390/w14040669
	Liu et al.	Potential toxic impacts of Hg migration in the disjointed hyporheic zone in the gold mining area experiencing river water level changes	10.3390/w14192950
	Liang et al.	Hydrochemical characteristics and formation mechanism of strontium-rich groundwater in Tianjiazhai, Fugu, China	10.3390/w14121874
	Ma et al.	Groundwater health risk assessment based on Monte Carlo model sensitivity analysis of Cr and As—a case study of Yinchuan City	10.3390/w14152419

In the first topical cluster, the research paper by Machado et al. [46] investigated the seasonal and spatial dynamics of drinking water quality across Guinea-Bissau, an endemic cholera sub-Saharan country, to fully understand the impacts of drinking water quality on public health. Serious fecal contamination was discovered in the water resources in this research, and some short-term sustainable measures were proposed for mitigating the associated health risks. To evaluate the quality and potential health risks of groundwater in the Tongchuan area, China, Nsabimana et al. [47] conducted a water quality and health risk assessment. The main contribution of this research was that it combined a carcinogenic risk assessment and noncarcinogenic risk assessment, and proved that traditional water quality assessments must be supplemented with health risk assessments to obtain completeness and comprehensiveness of the assessment. Similarly, Li et al. [48] also assessed the quality and potential health risks associated with the Linfen basin of the eastern Chinese Loess Plateau, and concluded that F^-, Pb, and Cr^{6+} were major contaminants responsible for inducing noncarcinogenic health risks in their study area. Bai et al. [49] focused on interpreting the hydrochemical characteristics and formation mechanisms of unconfined groundwater in a local area in the North China Plain using multiple approaches. This research could provide significant guidance for further groundwater quality protection and management in overexploited groundwater regions.

Worldwide, nitrate is very commonly found in groundwater, especially in agricultural regions [50–53]. The two articles in the second topical cluster reported on research concerning the identification of sources of nitrate and its removal. Specifically, the research by Jin et al. [54] revealed that the hydrochemical evolution of groundwater regulated rock weathering and cation exchange, and adopted stable nitrogen isotopes to trace sources of groundwater nitrate pollution, revealing the main sources of nitrate to be manure, sewage, and NH_4 fertilizers. Liu et al. [55] prepared biochar-based nanoscale zerovalent iron composites for nitrate removal from synthetic groundwater. This experimental research provides the necessary basis for nitrate removal with high efficiency.

The five papers in the third topical cluster involved trace elements, such as arsenic, mercury, strontium, chromium, and other metals. Trace elements in water and soil mainly originated from rock crusts, but also from anthropogenic activities [56,57]. Some are essential for human health when digested in trace amounts, but quickly become toxic when consumed in large quantities [58]. Some trace elements are toxic even in trace amounts [59,60], thus, requiring particular attention from different stakeholders. Salem et al. [61] reported on the spatial variability of heavy metals, such as Al, Ba, Cr, Cu, Pb, Mn, Ni, and Zn, in groundwater in the Liwa area of the United Arab Emirates the using principal component analysis and geographic information systems, while Cai et al. [62] focused on arsenic in shallow groundwater in the Hetao Basin in Inner Mongolia, China. Liu et al. [63] investigated Hg migration via in situ testing in the disjointed hyporheic zone in the gold mining area where river water level changes were detected, and Liang et al. [64] conducted research on the formation mechanisms of Sr-rich groundwater in the Shimachuan River basin, China. Ma et al. [65] assessed the health risk associated with As and Cr in dry and wet seasons using the Monte Carlo model, and quantified the possible ionic forms of As. These articles investigating trace elements could be essential for setting up future groundwater quality monitoring systems and groundwater pollution remediation measures.

4. Future Prospects

Medical geology is truly a multidisciplinary research field, and requires collaboration among researchers, policy makers, and the general public, with groundwater quality research being a fundamental field of this discipline. Hence, for its further promotion, we propose some suggestions that may be fundamental and significant to guiding this discipline:

- Groundwater is influenced by multiple factors, increasing the complexity of groundwater quality research [66,67]. Therefore, continuous research should be promoted, focusing on the mutual interactions among different elements and substances, as well as their effects on the toxicity of newly formed species through their interactions [19]. Particularly, a number of studies [68–70] indicated that land use/land cover, though not capable of altering groundwater quality directly, can significantly alter elemental levels in groundwater, thus, affecting the suitability of groundwater for drinking. Therefore, research on these indirect influencing factors should be highlighted.
- The impacts of human activities on groundwater quality are increasing, and conditions are becoming more complex. Therefore, more monitoring data are required for groundwater quality research. However, obtaining them is not easy, thus, requiring the help of nonprofessional communities. The importance of citizen science in big data accumulation and analysis has been well recognized by the science community [71–73]. However, criticism also exists in the science community, such as concerning the lack of data reliability. Citizen science should, therefore, be further promoted to facilitate groundwater quality and public health research.
- The concept and theoretical basis for medical geology should be further updated. As mentioned previously in this editorial, people's attitude toward health has changed due to the spread of knowledge on how to avoid diseases to knowing how to maintain physical and mental health, which is a broader concept than just knowing how to

avoid diseases. Maintaining health does not only include avoiding diseases, but also keeping healthy via an appropriate intake of necessary essential elements through water, food, and other media. Therefore, the geology of health could be a possible replacement for medical geology in the future.

Author Contributions: Conceptualization, P.L.; writing—original draft preparation, P.L.; writing—review and editing, J.W. and S.S.; funding acquisition, P.L. and J.W. All authors have read and agreed to the published version of the manuscript.

Funding: This research was funded by the National Natural Science Foundation of China (42072286 and 42272302), the Qinchuangyuan "Scientist + Engineer" Team Development Program of the Shaanxi Provincial Department of Science and Technology (2022KXJ-005), the Fok Ying Tong Education Foundation (161098), and the Ten Thousand Talents Program (W03070125).

Acknowledgments: We are grateful for the support of the entire *Water* editorial team. Without their commitment to the entire editorial process, the Special Issue would have been impossible. For completing this Special Issue, we also acknowledge the voluntary reviewers for their useful and critical comments that helped the authors further improve the quality of their manuscripts. The authors whose manuscripts were accepted for publication in this Special Issue and those whose manuscripts were unfortunately rejected are also acknowledged for showing their interest in this Special Issue. Your contributions make this Special Issue unique and the world a better place.

Conflicts of Interest: The authors declare no conflict of interest.

References

1. United Nations. Sustainable Development Goals. Available online: https://www.unodc.org/roseap//en/sustainable-development-goals.html (accessed on 10 October 2022).
2. Zinsstag, J.; Schelling, E.; Waltner-Toews, D.; Tanner, M. From "one medicine" to "one health" and systemic approaches to health and well-being. *Prev. Vet. Med.* **2011**, *101*, 148–156. [CrossRef] [PubMed]
3. Gibbs, E.P.J. The evolution of One Health: A decade of progress and challenges for the future. *Vet. Rec.* **2014**, *174*, 85–91. [CrossRef] [PubMed]
4. Wang, D.; Wu, J.; Wang, Y.; Ji, Y. Finding high-quality groundwater resources to reduce the hydatidosis incidence in the Shiqu County of Sichuan Province, China: Analysis, assessment, and management. *Expo. Health* **2020**, *12*, 307–322. [CrossRef]
5. Li, P.; Wang, D.; Li, W.; Liu, L. Sustainable water resources development and management in large river basins: An introduction. *Environ. Earth Sci.* **2022**, *81*, 179. [CrossRef]
6. Zhang, Q.; Li, P.; Lyu, Q.; Ren, X.; He, S. Groundwater contamination risk assessment using a modified DRATICL model and pollution loading: A case study in the Guanzhong Basin of China. *Chemosphere* **2022**, *291*, 132695. [CrossRef]
7. Wei, M.; Wu, J.; Li, W.; Zhang, Q.; Su, F.; Wang, Y. Groundwater geochemistry and its impacts on groundwater arsenic enrichment, variation, and health risks in Yongning County, Yinchuan Plain of northwest China. *Expo. Health* **2022**, *14*, 219–238. [CrossRef]
8. Ji, Y.; Wu, J.; Wang, Y.; Elumalai, V.; Subramani, T. Seasonal variation of drinking water quality and human health risk assessment in Hancheng City of Guanzhong Plain, China. *Expo. Health* **2020**, *12*, 469–485. [CrossRef]
9. Wu, J.; Zhou, H.; He, S.; Zhang, Y. Comprehensive understanding of groundwater quality for domestic and agricultural purposes in terms of health risks in a coal mine area of the Ordos basin, north of the Chinese Loess Plateau. *Environ. Earth Sci.* **2019**, *78*, 446. [CrossRef]
10. Liu, L.; Wu, J.; He, S.; Wang, L. Occurrence and distribution of groundwater fluoride and manganese in the Weining Plain (China) and their probabilistic health risk quantification. *Expo. Health* **2022**, *14*, 263–279. [CrossRef]
11. Wang, Y.; Li, P. Appraisal of shallow groundwater quality with human health risk assessment in different seasons in rural areas of the Guanzhong Plain (China). *Environ. Res.* **2022**, *207*, 112210. [CrossRef]
12. Li, P.; Karunanidhi, D.; Subramani, T.; Srinivasamoorthy, K. Sources and consequences of groundwater contamination. *Arch. Environ. Contam. Toxicol.* **2021**, *80*, 1–10. [CrossRef] [PubMed]
13. Mthembu, P.P.; Elumalai, V.; Li, P.; Uthandi, S.; Rajmohan, N.; Chidambaram, S. Integration of heavy metal pollution indices and health risk assessment of groundwater in semi-arid coastal aquifers, South Africa. *Expo. Health* **2022**, *14*, 487–502. [CrossRef]
14. He, X.; Li, P.; Ji, Y.; Wang, Y.; Su, Z.; Elumalai, V. Groundwater arsenic and fluoride and associated arsenicosis and fluorosis in China: Occurrence, distribution and management. *Expo. Health* **2020**, *12*, 355–368. [CrossRef]
15. Duan, L.; Wang, W.; Sun, Y.; Zhang, C.; Sun, Y. Hydrogeochemical Characteristics and Health Effects of Iodine in Groundwater in Wei River Basin. *Expo. Health* **2020**, *12*, 369–383. [CrossRef]
16. Rasool, A.; Muhammad, S.; Shafeeque, M.; Ahmad, I.; Al-Misned, F.A.; El-Serehy, H.A.; Ali, S.; Murtaza, B.; Sarwar, A. Evaluation of arsenic contamination and potential health risk through water intake in urban and rural areas. *Hum. Ecol. Risk Assess.* **2021**, *27*, 1655–1670. [CrossRef]

17. Ahmed, J.; Wong, L.P.; Channa, N.; Ahmed, W.; Chua, Y.P.; Shaikh, M.Z. Arsenic contamination and potential health risk to primary school children through drinking water sources. *Hum. Ecol. Risk Assess.* **2022**, *early access*. [CrossRef]
18. Dissanayake, C.B.; Chandrajith, R. *Introduction to Medical Geology: Focus on Tropical Environments*; Springer: Berlin/Heidelberg, Germany, 2009; pp. 1–8.
19. Li, P.; Wu, J. Medical geology and medical geochemistry: An editorial introduction. *Expo. Health* **2022**, *14*, 217–218. [CrossRef]
20. Selinus, O.; Alloway, B.; Centeno, J.A.; Finkelman, R.B.; Fuge, R.; Lindh, U.; Smedley, P. *Essentials of Medical Geology, Revised ed.*; Springer: Dordrecht, The Netherlands, 2013; 805p.
21. Komatina, M.M. *Medical Geology: Effects of Geological Environments on Human Health*; Elsevier: Amsterdam, The Netherlands, 2004; 488p.
22. Centeno, J.A.; Finkelman, R.B.; Selinus, O. *Medical Geology: Impacts of the Natural Environment on Public Health*; MDPI: Basel, Switzerland, 2016; 238p.
23. Siegel, M.; Selinus, O.; Finkelman, R. *Practical Applications of Medical Geology*; Springer Nature: Berlin, Germany, 2021; 932p.
24. Ibaraki, M.; Mori, H. *Progress in Medical Geology*; Cambridge Scholars Publishing: Newcastle, UK, 2017; 329p.
25. Censi, P.; Darrah, T.H.; Erel, Y. *Medical Geochemistry: Geological Materials and Health*; Springer: Berlin/Heidelberg, Germany, 2013; 194p.
26. Guo, Y.; Li, P.; He, X.; Wang, L. Groundwater quality in and around a landfill in northwest China: Characteristic pollutant identification, health risk assessment, and controlling factor analysis. *Expo. Health* **2022**, *early access*. [CrossRef]
27. Subba Rao, N.; Ravindra, B.; Wu, J. Geochemical and health risk evaluation of fluoride rich groundwater in Sattenapalle Region, Guntur district, Andhra Pradesh, India. *Hum. Ecol. Risk Assess.* **2020**, *26*, 2316–2348. [CrossRef]
28. Lan, T.; Wang, F.; Bao, S.; Miao, J.; Bai, Y.; Jia, S.; Cao, Y. The human health risk assessment and countermeasures study of groundwater quality. *Environ. Geochem. Health* **2022**, *early access*. [CrossRef]
29. Wang, L.; Li, P.; Duan, R.; He, X. Occurrence, Controlling Factors and Health Risks of Cr^{6+} in Groundwater in the Guanzhong Basin of China. *Expo. Health* **2022**, *14*, 239–251. [CrossRef]
30. He, X.; Li, P.; Wu, J.; Wei, M.; Ren, X.; Wang, D. Poor groundwater quality and high potential health risks in the Datong Basin, northern China: Research from published data. *Environ. Geochem. Health* **2021**, *43*, 791–812. [CrossRef] [PubMed]
31. Kumar, P.J.S. Groundwater fluoride contamination in Coimbatore district: A geochemical characterization, multivariate analysis, and human health risk perspective. *Environ. Earth Sci.* **2021**, *80*, 232. [CrossRef]
32. Mishra, D.; Sen, K.; Mondal, A.; Kundu, S.; Mondal, N.K. Geochemical appraisal of groundwater arsenic contamination and human health risk assessment in the Gangetic Basin in Murshidabad District of West Bengal, India. *Environ. Earth Sci.* **2022**, *81*, 157. [CrossRef]
33. Kumar, A.; Roy, S.S.; Singh, C.K. Geochemistry and associated human health risk through potential harmful elements (PHEs) in groundwater of the Indus basin, India. *Environ. Earth Sci.* **2020**, *79*, 86. [CrossRef]
34. Fida, M.; Li, P.; Wang, Y.; Alam, S.M.K.; Nsabimana, A. Water Contamination and Human Health Risks in Pakistan: A Review. *Expo. Health* **2022**, *early access*. [CrossRef]
35. Sathe, S.S.; Mahanta, C.; Subbiah, S. Hydrogeochemical evaluation of intermittent alluvial aquifers controlling arsenic and fluoride contamination and corresponding health risk assessment. *Expo. Health* **2021**, *13*, 661–680. [CrossRef]
36. Alfeus, A.; Molnar, P.; Boman, J.; Shirinde, J.; Wichmann, J. Inhalation health risk assessment of ambient PM2.5 and associated trace elements in Cape Town, South Africa. *Hum. Ecol. Risk Assess.* **2022**, *28*, 917–929. [CrossRef]
37. China Daily. Available online: http://china.chinadaily.com.cn/a/202009/22/WS5f69b418a3101e7ce9725dfc.html?from=groupmessage&isappinstalled=0 (accessed on 13 October 2022).
38. Xu, B.; Zhang, Y.; Wang, J. Hydrogeochemistry and human health risks of groundwater fluoride in Jinhuiqu irrigation district of Wei river basin, China. *Hum. Ecol. Risk Assess.* **2019**, *25*, 230–249. [CrossRef]
39. He, S.; Wu, J.; Wang, D.; He, X. Predictive modeling of groundwater nitrate pollution and evaluating its main impact factors using random forest. *Chemosphere* **2022**, *290*, 133388. [CrossRef]
40. Li, P.; He, X.; Guo, W. Spatial groundwater quality and potential health risks due to nitrate ingestion through drinking water: A case study in Yan'an City on the Loess Plateau of northwest China. *Hum. Ecol. Risk Assess.* **2019**, *25*, 11–31. [CrossRef]
41. Li, P.; He, X.; Li, Y.; Xiang, G. Occurrence and health implication of fluoride in groundwater of loess aquifer in the Chinese Loess Plateau: A case study of Tongchuan, northwest China. *Expo. Health* **2019**, *11*, 95–107. [CrossRef]
42. Hao, L.; Zhang, S.; Luo, K. Content of selenium and other elements, water quality, health risks and utilization prospect in natural water of Southern Qinling-Daba Mountains, Southern Shaanxi, China. *Expo. Health* **2022**, *14*, 29–47. [CrossRef]
43. Deng, B.; Tian, S.; Li, S.; Guo, M.; Liu, H.; Li, Y.; Wang, Q.; Zhao, X. A simple, rapid and efficient method for essential element supplementation based on seed germination. *Food Chemistry* **2020**, *325*, 126827. [CrossRef] [PubMed]
44. Wu, J.; Wang, D.; Yan, L.; Jia, M.; Zhang, J.; Han, S.; Han, J.; Wang, J.; Chen, X.; Zhang, R. Associations of essential element serum concentrations with autism spectrum disorder. *Environ. Sci. Pollut. Res.* **2022**, *early access*. [CrossRef]
45. Peng, G.; Sun, J.; Peng, B.; Tan, Y.; Wu, Y.; Bai, X. Assessment of essential element accumulation in red swamp crayfish (*Procambarus clarkii*) and the highly efficient selenium enrichment in freshwater animals. *J. Food Compos. Anal.* **2021**, *101*, 103953. [CrossRef]
46. Machado, A.; Amorim, E.; Bordalo, A.A. Spatial and Seasonal Drinking Water Quality Assessment in a Sub-Saharan Country (Guinea-Bissau). *Water* **2022**, *14*, 1987. [CrossRef]
47. Nsabimana, A.; Li, P.; He, S.; He, X.; Alam, S.M.K.; Fida, M. Health Risk of the Shallow Groundwater and Its Suitability for Drinking Purpose in Tongchuan, China. *Water* **2021**, *13*, 3256. [CrossRef]

48. Li, J.; Sun, C.; Chen, W.; Zhang, Q.; Zhou, S.; Lin, R.; Wang, Y. Groundwater Quality and Associated Human Health Risk in a Typical Basin of the Eastern Chinese Loess Plateau. *Water* **2022**, *14*, 1371. [CrossRef]
49. Bai, X.; Tian, X.; Li, J.; Wang, X.; Li, Y.; Zhou, Y. Assessment of the Hydrochemical Characteristics and Formation Mechanisms of Groundwater in A Typical Alluvial-Proluvial Plain in China: An Example from Western Yongqing County. *Water* **2022**, *14*, 2395. [CrossRef]
50. Zhang, Y.; Wu, J.; Xu, B. Human health risk assessment of groundwater nitrogen pollution in Jinghui canal irrigation area of the loess region, northwest China. *Environ. Earth Sci.* **2018**, *77*, 273. [CrossRef]
51. Su, F.; Wu, J.; Wang, D.; Zhao, H.; Wang, Y.; He, X. Moisture movement, soil salt migration, and nitrogen transformation under different irrigation conditions: Field experimental research. *Chemosphere* **2022**, *300*, 134569. [CrossRef] [PubMed]
52. He, S.; Li, P.; Su, F.; Wang, D.; Ren, X. Identification and apportionment of shallow groundwater nitrate pollution in Weining Plain, northwest China, using hydrochemical indices, nitrate stable isotopes, and the new Bayesian stable isotope mixing model (MixSIAR). *Environ. Pollut.* **2022**, *298*, 118852. [CrossRef] [PubMed]
53. Ceballos, E.; Dubny, S.; Othax, N.; Zabala, M.E.; Peluso, F. Assessment of Human Health Risk of Chromium and Nitrate Pollution in Groundwater and Soil of the Matanza-Riachuelo River Basin, Argentina. *Expo. Health* **2021**, *13*, 323–336. [CrossRef]
54. Jin, J.; Wang, Z.; Zhao, Y.; Ding, H.; Zhang, J. Delineation of Hydrochemical Characteristics and Tracing Nitrate Contamination of Groundwater Based on Hydrochemical Methods and Isotope Techniques in the Northern Huangqihai Basin, China. *Water* **2022**, *14*, 3168. [CrossRef]
55. Liu, S.; Han, X.; Li, S.; Xuan, W.; Wei, A. Stimulating Nitrate Removal with Significant Conversion to Nitrogen Gas Using Biochar-Based Nanoscale Zerovalent Iron Composites. *Water* **2022**, *14*, 2877. [CrossRef]
56. Li, Y.; Li, P.; Liu, L. Source Identification and Potential Ecological Risk Assessment of Heavy Metals in the Topsoil of the Weining Plain (Northwest China). *Expo. Health* **2022**, *14*, 281–294. [CrossRef]
57. Snousy, M.G.; Li, P.; Ismail, E. Trace elements speciation and sources characterization in the main watercourses, middle-upper Egypt. *Hum. Ecol. Risk Assess.* **2021**, *27*, 1764–1785. [CrossRef]
58. Gaur, S.; Agnihotri, R. Health Effects of Trace Metals in Electronic Cigarette Aerosols—A Systematic Review. *Biol. Trace Elem. Res.* **2019**, *188*, 295–315. [CrossRef]
59. El-Kady, A.A.; Abdel-Wahhab, M.A. Occurrence of trace metals in foodstuffs and their health impact. *Trends Food Sci. Technol.* **2018**, *75*, 36–45. [CrossRef]
60. Logan, T.J.; Traina, S.J. Trace Metals in Agricultural Soils. In *Metals in Groundwater*, 1st ed.; Allen, H.E., Perdue, E.M., Brown, D.S., Eds.; CRC Press: Boca Raton, FL, USA, 1993; 39p.
61. Salem, I.B.; Nazzal, Y.; Howari, F.M.; Sharma, M.; Mogaraju, J.K.; Xavier, C.M. Geospatial Assessment of Groundwater Quality with the Distinctive Portrayal of Heavy Metals in the United Arab Emirates. *Water* **2022**, *14*, 879. [CrossRef]
62. Cai, Z.; Liu, L.; Xu, W.; Wu, P.; Lu, C. Hydrochemical Characteristics of Arsenic in Shallow Groundwater in Various Unconsolidated Sediment Aquifers: A Case Study in Hetao Basin in Inner Mongolia, China. *Water* **2022**, *14*, 669. [CrossRef]
63. Liu, R.; Liu, F.; Jiao, J.; Xu, Y.; Dong, Y.; RM, E.-W.; Zhang, X.; Chen, H. Potential Toxic Impacts of Hg Migration in the Disjointed Hyporheic Zone in the Gold Mining Area Experiencing River Water Level Changes. *Water* **2022**, *14*, 2950. [CrossRef]
64. Liang, C.; Wang, W.; Ke, X.; Ou, A.; Wang, D. Hydrochemical Characteristics and Formation Mechanism of Strontium-Rich Groundwater in Tianjiazhai, Fugu, China. *Water* **2022**, *14*, 1874. [CrossRef]
65. Ma, Z.; Li, J.; Zhang, M.; You, D.; Zhou, Y.; Gong, Z. Groundwater Health Risk Assessment Based on Monte Carlo Model Sensitivity Analysis of Cr and As—A Case Study of Yinchuan City. *Water* **2022**, *14*, 2419. [CrossRef]
66. Li, P. Groundwater Quality in Western China: Challenges and Paths Forward for Groundwater Quality Research in Western China. *Expo. Health* **2016**, *8*, 305–310. [CrossRef]
67. Li, P.; Tian, R.; Xue, C.; Wu, J. Progress, opportunities and key fields for groundwater quality research under the impacts of human activities in China with a special focus on western China. *Environ. Sci. Pollut. Res.* **2017**, *24*, 13224–13234. [CrossRef] [PubMed]
68. Xu, F.; Li, P.; Chen, W.; He, S.; Li, F.; Mu, D.; Elumalai, V. Impacts of land use/land cover patterns on groundwater quality in the Guanzhong Basin of northwest China. *Geocarto. Int.* **2022**, early access. [CrossRef]
69. Xu, F.; Li, P.; Du, Q.; Yang, Y.; Yue, B. Seasonal hydrochemical characteristics, geochemical evolution, and pollution sources of Lake Sha in an arid and semiarid region of northwest China. *Expo. Health* **2022**, early access. [CrossRef]
70. He, S.; Li, P.; Wu, J.; Elumalai, V.; Adimalla, N. Groundwater quality under land use/land cover changes: A temporal study from 2005 to 2015 in Xi'an, northwest China. *Hum. Ecol. Risk Assess.* **2020**, *26*, 2771–2797. [CrossRef]
71. Farnham, D.J.; Gibson, R.A.; Hsueh, D.Y.; McGillis, W.R.; Culligan, P.J.; Zain, N.; Buchanan, R. Citizen science-based water quality monitoring: Constructing a large database to characterize the impacts of combined sewer ovrflow in New York City. *Sci. Total Environ.* **2017**, *580*, 168–177. [CrossRef] [PubMed]
72. Jollymore, A.; Haines, M.J.; Satterfeld, T.; Johnson, M.S. Citizen science for water quality monitoring: Data implications of citizen perspectives. *J. Environ. Manag.* **2017**, *200*, 456–467. [CrossRef] [PubMed]
73. McKinley, D.C.; Miller-Rushing, A.J.; Ballard, H.L.; Bonney, R.; Brown, H.; Cook-Patton, S.C.; Evans, D.M.; French, R.A.; Parrish, J.K.; Phillips, T.B.; et al. Citizen science can improve conservation science, natural resource management, and environmental protection. *Biol. Conserv.* **2017**, *208*, 15–28. [CrossRef]

Article

Spatial and Seasonal Drinking Water Quality Assessment in a Sub-Saharan Country (Guinea-Bissau)

Ana Machado [1,2,*], Eva Amorim [1,2] and Adriano A. Bordalo [1,2]

1 Laboratory of Hydrobiology and Ecology, Instituto de Ciências Biomédicas Abel Salazar (ICBAS—UP), University of Porto, Rua Jorge Viterbo Ferreira 228, 4050-313 Porto, Portugal; ecamorim@icbas.up.pt (E.A.); bordalo@icbas.up.pt (A.A.B.)
2 Interdisciplinary Centre of Marine and Environmental Research (CIIMAR—UP), University of Porto, Novo Edifício do Terminal de Cruzeiros do Porto de Leixões, Avenida General Norton de Matos, S/N, 4450-208 Matosinhos, Portugal
* Correspondence: ammachado@icbas.up.pt

Citation: Machado, A.; Amorim, E.; Bordalo, A.A. Spatial and Seasonal Drinking Water Quality Assessment in a Sub-Saharan Country (Guinea-Bissau). *Water* 2022, 14, 1987. https://doi.org/10.3390/w14131987

Academic Editors: Saurabh Shukla, Jianhua Wu and Peiyue Li

Received: 23 May 2022
Accepted: 20 June 2022
Published: 21 June 2022

Publisher's Note: MDPI stays neutral with regard to jurisdictional claims in published maps and institutional affiliations.

Copyright: © 2022 by the authors. Licensee MDPI, Basel, Switzerland. This article is an open access article distributed under the terms and conditions of the Creative Commons Attribution (CC BY) license (https://creativecommons.org/licenses/by/4.0/).

Abstract: The United Nations Sustainable Development Goal target 6.1 calls for universal and equitable access to safe and affordable drinking water. Worldwide, about 2.2 billion people live without access to safe water, and millions of people suffer from waterborne pathogens each year, representing the most pressing situation in developing countries. The aim of this study was to investigate the drinking water quality dynamics across an endemic cholera sub-Saharan country (Guinea-Bissau), and understand its implications for public health. Microbiological and physical–chemical quality parameters of 252 major water sources spread all over the country were seasonally surveyed. These comprised hand-dug shallow wells and boreholes, fitted with a bucket or a pump to retrieve water. The results showed that the majority of water sources available to the population were grossly polluted with faecal material (80%), being unsuitable for consumption, with significantly ($p < 0.05$) higher levels during the wet season. Hand-dug wells revealed the highest contamination levels. The chemical contamination was less relevant, although 83% of the water sources were acidic (pH < 6.5). This study highlights the potential health risk associated with the lack of potable drinking water, reinforcing the evidence for water monitoring, and the need to improve WASH (water, sanitation, and hygiene) infrastructure and water management in West African countries. In addition, the authors suggest easy-to-implement interventions that can have a dramatic impact in the water quality, assisting to reduce the associated waterborne diseases rise.

Keywords: water quality; drinking water; WASH; waterborne diseases; One Health; Guinea-Bissau; sub-Saharan Africa

1. Introduction

Water is essential for human life, and the access to safe and sufficient water and sanitation are recognized as basic human rights [1]. Moreover, the United Nations Sustainable Development Goal (SDG) target 6.1 calls for universal and equitable access to safe and affordable drinking water [2]. However, about 2.2 billion people still live without access to safe water, the majority of which are in low- and middle-income countries [3], mainly in sub-Saharan Africa and Asia. In sub-Saharan Africa countries, about 319 million people live without access to improved reliable drinking water sources, and 695 million lack improved sanitation facilities [4].

Each year, 829,000 deaths are attributed to diarrhoeal diseases linked to inadequate WASH, including 297,000 in children under five [3]. Over the coming decades, with a growth rate of about 2.5%, the sub-Saharan region will account for most of the global population growth [5]. This near-future scenario, with densely populated urban and peri-urban areas, entails an increment in water demand and sanitation, pressuring the already compromised WASH infrastructure. The sub-Saharan country of Guinea-Bissau is one of the poorest countries in the world, ranked 175 out of 189 countries on the Human Development Index in 2019. Presently, with a total population of 1,920,922 inhabitants, the life expectancy at birth in Guinea-Bissau is 57 and 62 years for males and females, respectively. Moreover, the under-five mortality rate is 78.47 [6]. According to the WHO [7], in Guinea-Bissau, each person has only 21 L of water for daily personal needs. This value is well under the 50 L minimum known water requirement for human domestic use [8]. The latest data (2016 and 2017) reported that "improved" water sources (not necessarily meaning potable water, but rather a refurbished facility) were accessible to 73% of the population, whereas only 20.5% of the population has access to proper sanitation [9]. The health system is fragile and not universally available, with a very high health burden from malaria, diarrhoea and respiratory diseases, HIV, and malnutrition. Indeed, diarrhoeal diseases are the third leading cause of death, with 702,974 cases reported in Guinea-Bissau in 2019 [10]. Cholera is endemic in the country, being responsible for 71,307 cases and 1638 deaths between 1996 and 2017 [11].

Although a few studies concerning water quality in Guinea-Bissau are available [12–14], these were carried out in limited geographic areas. To the best of our knowledge, no study with broad range has been performed to understand the problem at the national level. However, to design and implement robust measures to ensure safe water for the population, it is pivotal to understand the national status of the water quality, as well as the regional differences. The purpose of this study was to investigate the water quality dynamics across this endemic cholera sub-Saharan country (Guinea-Bissau), and relate it to environmental constrains (including seasonality) and associated WASH components, in order to understand the implications for public health.

2. Materials and Methods

2.1. Study Area

The primary source of water to support the daily needs of the majority of the population in Guinea-Bissau, including drinking water, are open/unprotected, shallow hand-dug wells (<15 m). Previous studies reported that about 80% of those water sources were contaminated with faecal materials and had an acidic pH [12,14]. Guinea-Bissau has a tropical climate (hot and humid), with a dry season (DS) spanning from November to May, and a wet season (WS) with high precipitation from June to October. The most representative soil groups in the country are Ferrallisols, Plinthosols, Gleysols, Fluvisols, and Arenosols [15].

2.2. Sample Collection and Analytical Procedures

The data used in this study resulted from several surveys carried out throughout the country over a time span of 13 years (2006–2019). Sampling sites within different administrative regions were chosen, taking into account the number of people served and the accessibility (Figure 1). The water sources comprised hand-dug shallow wells ($n = 216$) and boreholes ($n = 35$), fitted with a bucket or a pump (manual, solar, or electric) to collect water (Figure S1). Out of the 252 water sources surveyed, 47 were examined in the dry season, whereas 83 were studied in the wet season. The remainder (122) water sources were assessed in both seasons. Moreover, throughout the study period, each sampling location was surveyed between 1 and 17 times.

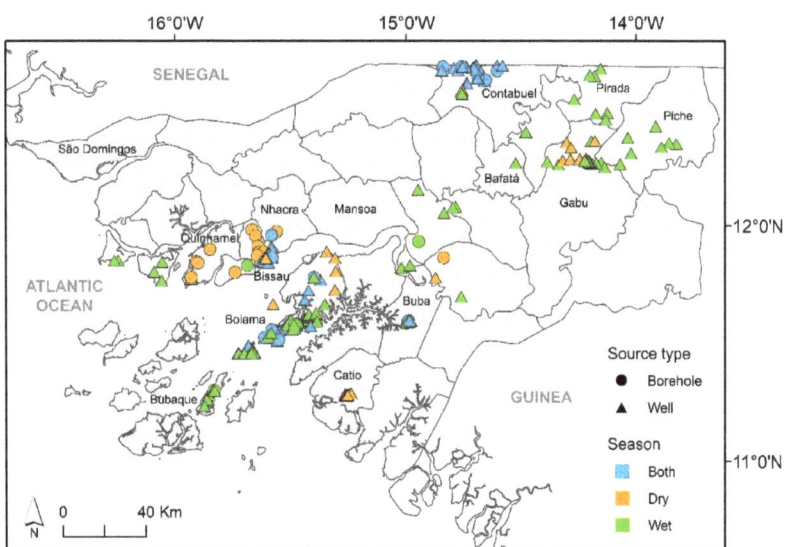

Figure 1. Location of surveyed water sources across Guinea-Bissau (source of second-level administrative divisions of Guinea-Bissau: Hjmans et al. [16]).

Water samples were collected using 500 mL plastic sterile flasks. All samples were kept in the dark in refrigerated ice chests and processed within 4 h of collection at a field laboratory similar to the one described in [13]. Water temperature, conductivity, dissolved oxygen, oxygen saturation, and pH were measured *in situ* using a Hanna Instruments 9828 portable meter. The exact position of each water source was obtained by means of GPS (Magellan 600).

Monthly precipitation data were extracted from the historical GHCN gridded V2 dataset provided by NOAA/OAR/ESRL PSD, available on their website (http://www.esrl.noaa.gov/psd/, accessed on 4 June 2020). Resolution was 2.5 degrees for the available grid of pixels covering the Guinea-Bissau country area (11–13.5° N, 18–14° W).

Samples for water colour, nitrate, nitrite, ammonium, aluminium, arsenic, copper, chromium, cyanide, and iron and were assayed in a 12 V multiparameter Hanna HI83200 photometer, according to standard methods supplied by the manufacturer(www.hannacom.pt, accessed on 1 June 2019). A Hanna HI-93102 Multi Range Portable Turbidity Meter for water analysis was used for turbidity assessment.

Samples for faecal indicators evaluation were filtered onto sterile gridded cellulose nitrate membranes (0.45 µm pore size, 47 mm diameter, Whatman, Maidstone, UK), and placed on mFC-agar (Difco, Le Pont de Claix, France) and Slanetz–Bartley agar (Oxoid, Hants, UK) plates, for faecal coliforms (FC) and intestinal enterococci (IE) enumeration, respectively. Incubation was performed at 44.5 °C for 24 h (FC) or 48 h (IE) [17] using solar-generated electricity in the absence of an electrical supply grid. Typical colonies were counted and results expressed as colony-forming units (CFU)/100 mL.

Guinea-Bissau does not have guidelines concerning drinking water quality; therefore, the parameters assayed were compared with the WHO [18], EU [19] (1998), and UK [20] guidelines to establish whether the quality of the water was fit for human consumption.

2.3. Statistical Analysis

Faecal indicators concentrations were Log (n + 1) transformed prior to analysis. The Spearman's rank correlation coefficient was used to assess the relationship of environmental factors and microbiological indicators. Spatial and seasonal statistically significant differences among samples were evaluated through analysis of variance (one-way ANOVA), followed by a post hoc Tukey honestly significant difference (HSD) multi-comparison test. The significance level used for all tests was 0.05.

Boosted regression trees (BRTs) were used to assess the relationship between environmental factors and microbiological indicators in the hand-dug wells only, because these represented the primary water source in Guinea-Bissau and revealed the highest contamination levels. BRTs are tree-based ensemble methods that combine the algorithms of regression trees and boosting (which build and combine a collection of models). The method works by iteratively fitting simple tree models using a forward stage-wise procedure, which progressively fits trees to the residuals of the previously fitted trees [21,22]. Some of the wells were surveyed several times; therefore, one sample per season (wet and dry) was selected at each location. Data were selected from the years when more samples were collected (2010 and 2009), and, when not available, from the closest years. When more than one sample was collected per season and year, the sample collected in the month closest to the middle of the season was chosen. Two BRT models were built, one for FCs and other for IE, using a Gaussian error distribution. Models were built in R software, version 4.0.5 [23], using the packages "dismo" and "gbm" [22,24]. Combinations of several settings were fitted before finding the optimal final setting: tree complexity (tc) of 1, learning rate of 0.001, bag fraction of 0.5 and k-fold cross validation of 10. The full models were then simplified by removing non-informative variables based on the decrease in variance. Final models were chosen based on their statistical performance, evaluated by the explained cross-validated deviance, i.e., the cross-validated correlation between training and testing data.

3. Results

Microbiological and physical–chemical quality parameters of 252 water sources spread all over Guinea-Bissau were evaluated. All the studied wells were hand-dug, shallow, without proper wall isolation, often located at the vicinity of latrines and waste dumps (<30 m), and most of them lacked any well cover or fence to prevent contamination. Regarding the method of water collection, the majority of the wells were fitted with a bucket (n = 122), 72 with a manual pump, 6 with a solar/electric pump, and 13 with a mixed version combining pump and bucket. Of the analysed boreholes, 4 were fitted with a distribution system with faucets, 23 with a solar/electric pump, 4 with manual pumps, and 4 with a mixed version combining pump and bucket (Table S1).

Microbiological and physical–chemical analysis results are summarized in Table 1 and Figure S2. Water temperature averaged 29.1 °C (22.9–35.3 °C) year-round, with only two wells sampled in each season below the 25 °C recommended maximum temperature for drinking water, according to UK standards [20]. The water conductivity was low to moderate (median 163 µS/cm), with oxygen concentrations averaging 5.5 mg/L (average oxygen saturation 59%). The water was acidic to very acidic, averaging pH 5.3. Averages were similar between seasons, being higher in boreholes (pH 6.6 vs. pH 5.2 in wells). Overall, 83% of the water sources surveyed exhibited pH below the EU parametric value for drinking water, representing 89% and 49% of the wells and boreholes studied, respectively. Acceptable standard values for colour and turbidity were exceeded in 56% and 40% of the sampled water sources, respectively. As expected, extremely high values of these parameters were recorded in the wet season in the shallow wells (Table 1). Additionally, higher values were obtained when a bucket was used to withdraw the water (Figure S2). The nitrate, nitrite, and ammonium concentrations were below the EU and WHO parametric values for drinking water in most wells, although a value slightly above the acceptable parametric threshold was registered in the wet season. The highest concentrations of nitrate and ammonium were observed in wells and associated with the use of buckets (Figure S2). Overall, heavy metal (Al, As, Cr, Cu, and Fe), and cyanide concentrations were under the parametric value for drinking water. However, a higher number of surveyed water sources revealed non-compliance with the standard parametric values related to the wet season, wells, and when using a bucket to collect water (Table 1 and Figure S2).

Table 1. Minimum and maximum values for water quality parameters seasonally assayed, according to water source type. In italics are the percentage of sites above the parametric values for drinking water, and the respective number of water sources surveyed. EU—European Union parametric values for drinking water [19], WHO—Word Health Organization guidelines values for drinking water [18].

Parameter	Unit	Season		Water Source		EU	WHO
		Wet	Dry	Well	Borehole		
Temperature	°C	24.4–35.3 ($n = 650$) 99% ($n = 204$)	22.9–34.5 ($n = 351$) 99% ($n = 204$)	22.9–35.3 ($n = 903$) 98% ($n = 216$)	25.3–32.8 ($n = 98$) 100% ($n = 35$)	-	-
Conductivity	µS/cm	12.0–2550 ($n = 648$) 0.5% ($n = 204$)	12.2–3125 ($n = 350$) 0.6% ($n = 165$)	12.0–2550 ($n = 900$) 0.5% ($n = 216$)	19.1–3125 ($n = 98$) 3% ($n = 35$)	2500	-
DO saturation	%	3.4–130 ($n = 595$)	7.3–97.6 ($n = 347$)	3.4–107.9 ($n = 847$)	12.5–130 ($n = 95$)	-	-
pH	-	3.37–8.62 ($n = 647$) 85% ($n = 204$)	3.36–8.33 ($n = 352$) 88% ($n = 165$)	3.36–8.62 ($n = 900$) 89% ($n = 216$)	3.88–8.57 ($n = 99$) 49% ($n = 35$)	≥ 6.5 and ≤ 9.5	-
Colour	PtCo	1–1210 ($n = 651$) 54% ($n = 204$)	1–520 ($n = 351$) 58% ($n = 163$)	1–1210 ($n = 902$) 59% ($n = 212$)	1–720 ($n = 100$) 40% ($n = 35$)	a	a
Turbidity	NTU	0.1–407 ($n = 610$) 42% ($n = 191$)	0.1–82 ($n = 339$) 25% ($n = 165$)	0.1–407 ($n = 856$) 45% ($n = 206$)	0.1–45 ($n = 93$) 13% ($n = 31$)	a	-
Ammonium	mg NH$_4$/L	0.01–20.70 ($n = 650$) 22% ($n = 204$)	0.01–14.50 ($n = 353$) 12% ($n = 165$)	0.01–20.70 ($n = 903$) 22% ($n = 216$)	0.01–1.22 ($n = 100$) 23% ($n = 35$)	0.5	-
Nitrate	mg NO$_3$/L	0.1–450 ($n = 649$) 19% ($n = 204$)	0.1–473 ($n = 354$) 14% ($n = 166$)	0.1–473 ($n = 903$) 20% ($n = 216$)	0.1–88 ($n = 100$) 3% ($n = 35$)	50	50
Nitrite	mg NO$_2$/L	0.01–4.50 ($n = 649$) 4% ($n = 204$)	0.01–0.90 ($n = 352$) 1% ($n = 166$)	0.01–4.50 ($n = 901$) 5% ($n = 216$)	0.01–0.24 ($n = 100$) 0% ($n = 35$)	0.5	0.2
Aluminium	mg/L	0.01–1.30 ($n = 573$) 37% ($n = 200$)	0.01–1.42 ($n = 166$) 36% ($n = 166$)	0.01–1.42 ($n = 818$) 42% ($n = 215$)	0.01–0.54 ($n = 90$) 18% ($n = 34$)	0.2	0.2
Arsenic	µg/L	5–250 ($n = 174$) 27% ($n = 51$)	5–22 ($n = 75$) 39% ($n = 44$)	5–290 ($n = 241$) 80% ($n = 30$)	5–10 ($n = 8$) 25% ($n = 4$)	10	10
Chromium	µg/L	1–300 ($n = 482$) 8% ($n = 190$)	1–226 ($n = 243$) 4% ($n = 136$)	1–300 ($n = 673$) 10% ($n = 199$)	1–46 ($n = 52$) 0% ($n = 30$)	50	50
Copper	µg/L	1–1000 ($n = 506$) 0% ($n = 171$)	1–1330 ($n = 243$) 0% ($n = 136$)	1–330 ($n = 702$) 0% ($n = 202$)	1–629 ($n = 47$) 0% ($n = 19$)	2000	2000
Cyanide	µg/L	1–250 ($n = 468$) 4% ($n = 167$)	1–47 ($n = 197$) 0% ($n = 103$)	1–250 ($n = 631$) 3% ($n = 199$)	1–19 ($n = 34$) 0% ($n = 18$)	50	70
Iron	µg/L	1–4770 ($n = 649$) 23% ($n = 204$)	1–2560 ($n = 352$) 20% ($n = 166$)	1–4770 ($n = 901$) 26% ($n = 216$)	2–1050 ($n = 100$) 29% ($n = 35$)	200	-
Faecal coliforms	CFU/100 mL	0–376,850 ($n = 636$) 84% ($n = 204$)	0–30,000 ($n = 354$) 83% ($n = 166$)	0–376,850 ($n = 891$) 65% ($n = 216$)	0–1520 ($n = 99$) 66% ($n = 35$)	0	0
Intestinal enterococci	CFU/100 mL	0–51,300 ($n = 625$) 83% ($n = 204$)	0–91,800 ($n = 354$) 73% ($n = 166$)	0–91,800 ($n = 881$) 81% ($n = 216$)	0–4320 ($n = 98$) 54% ($n = 35$)	0	0

a—Acceptable for consumption (<5 PtCo).

The obtained results showed that the drinking water available to the population was grossly polluted with faecal material. The majority of the water sources sampled (83% and 77% for FC and IE, respectively), failed to meet the microbiological quality standards for drinking water, as recommended by the WHO and the EU. Values averaged 4217 and 800 CFU/100 mL for FC and IE, respectively (Table 1). Overall, faecal contamination levels were significantly higher during the wet season (Tukey's HSD test, $p < 0.05$), with values as high as 376, 850 CFU/100 mL for FC, and 33,000 CFU/100 mL for IE. Additionally, significantly higher (Tukey's HSD test, $p < 0.05$) faecal contamination levels were associated with wells and the use of a bucket for water collection. The lowest levels of contamination were observed in water samples collected from boreholes associated with electric pumps with small water distribution systems fitted with faucets (Figure 2).

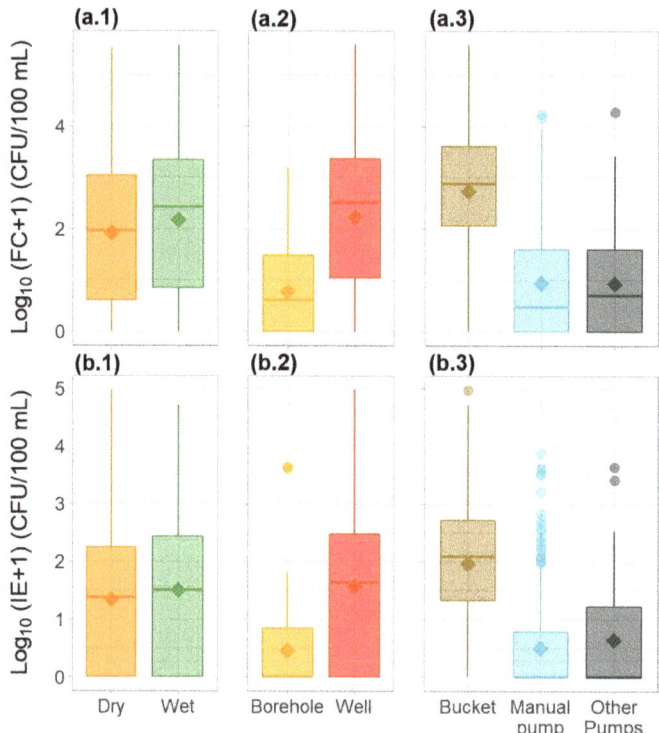

Figure 2. Variation in the microbiological water quality parameters—(**a**) faecal coliforms and (**b**) intestinal enterococci—in the water sources across Guinea-Bissau between (1) seasons, (2) source type, and (3) method of water collection. Diamond shapes represent the mean. Circles represent outliers.

FC and IE were significantly correlated (r = 0.85, $p < 0.05$), as expected. Significant positive correlations ($p < 0.05$) were observed between both faecal indicators and colour (r_{FC} = 0.32, r_{IE} = 0.30), turbidity (r_{FC} = 0.50, r_{IE} = 0.49), ammonium (r_{FC} = 0.32, r_{IE} = 0.30), nitrate (r_{FC} = 0.34, r_{IE} = 0.35), nitrite (r_{FC} = 0.33, r_{IE} = 0.30), aluminium (r_{FC} = 0.12, r_{IE} = 0.12), chromium (r_{FC} = 0.20, r_{IE} = 0.14), copper (r_{FC} = 0.15, r_{IE} = 0.20), cyanide (r_{FC} = 0.10). and iron (r_{FC} = 0.22, r_{IE} = 0.26). Turbidity and colour, associated with downpours during the wet season, showed positive significant correlations ($p < 0.05$) with the nitrogen species, and the majority of metal concentrations (Figure S3).

Hand-dug wells represented the majority of the water sources sampled, with the highest contamination levels; therefore, the relationship between environmental factors and microbiological indicators was further explored using a regression analysis approach. As a result, the fitted BRT models for the microbiological parameters FC and IE were similar in performance, although the FC model performed slightly better, with an explained deviance of 49% and a cross-validated correlation of 0.7 (Table 2).

Table 2. Predictive performance of the final models developed for the microbiological indicators quantified in wells in Guinea-Bissau. Final settings: bag fraction—0.5; tree complexity—1; no. folds—10; learning rate—0.001.

	Faecal Coliforms	Intestinal Enterococci
Number of trees	7650	6950
Deviance explained (%)	49	45
CV correlation	0.7	0.68

The partial responses for FC and IE for each predictor variable and variable contributions are shown in Figures 3 and 4, respectively. Both models retained the same eight predictor variables. For both models, the two most influential variables were turbidity and the method used to collect water, accounting for 57.9% and 55% of the explained deviance for FC and IE, respectively. The levels of faecal contamination increased with higher turbidity and the use of a bucket to collect water. The contributions of the remaining six variables differed between models, but the effects on the response were similar. The concentrations of faecal indicators were higher in the wet season, although precipitation was a more influential variable for the FC model. Latitude was also an influential variable in both models, with contamination increasing with increasing latitudes (towards northern regions). The levels of contamination also increased with higher concentrations of nitrate and pH. Faecal indicators concentration decreased as temperature rose. The effects of copper were different for the two models: FC concentration showed a slight decrease with higher levels of the metal, whereas IE concentration peaked and then decreased as metal levels increased.

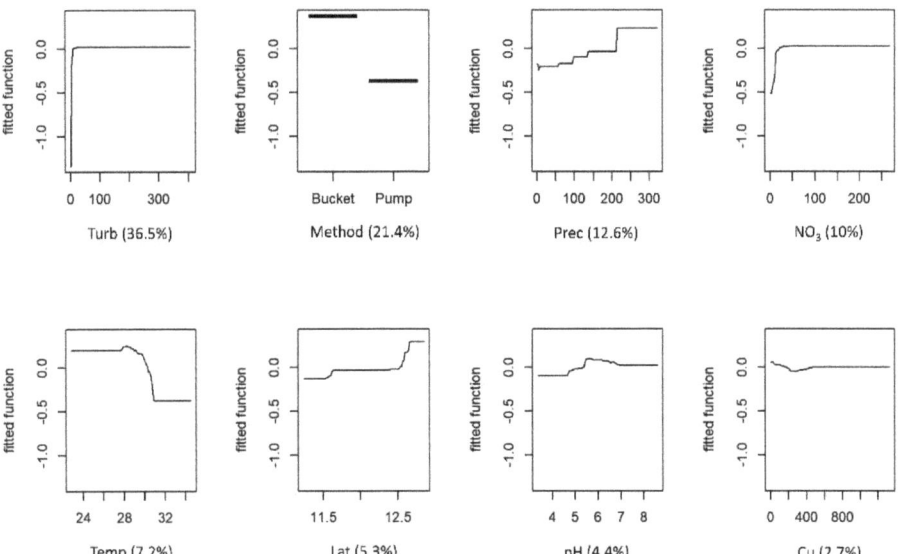

Figure 3. Boosted regression tree (BRT) partial dependence plots showing the effect of each predictor variable on faecal coliforms (FC) of the water of wells in Guinea-Bissau: Turb—turbidity; Method—method used for water collection; Prec—precipitation; NO_3—nitrate; Temp—temperature; Lat—latitude; Cu—copper. At each plot, the fitted function shows the relationship between the response variable (y-axis), and the predictor variable (x-axis), holding the values of all other variables at their mean. The relative contribution (%) of each predictor variable for the BRT model is shown in brackets.

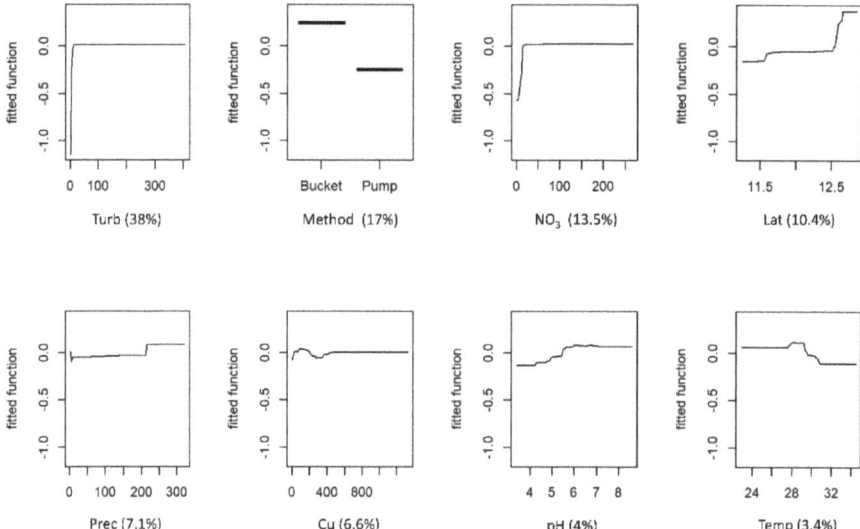

Figure 4. Boosted regression tree (BRT) partial dependence plots showing the effect of each predictor variable on intestinal enterococci (IE) of the water of wells in Guinea-Bissau: Turb—turbidity; Method—method used for water collection; NO_3—nitrate; Lat—latitude; Prec—precipitation; Cu—copper; Temp—temperature. At each plot, the fitted function shows the relationship between the response variable (y-axis), and the predictor variable (x-axis), holding the values of all other variables at their mean. The relative contribution (%) of each predictor variable for the BRT model is shown in brackets.

4. Discussion

The water quality assessment in Guinea-Bissau revealed that the water sources used for drinking purposes countrywide were acidic to very acidic, and heavily contaminated with faecal material. Higher levels of contamination were found during the wet season, associated with wells and the use of a bucket to withdraw water. This trend has previously been identified in regional studies performed in the country [12,25]. In Guinea-Bissau, the majority of the population retrieve water from shallow (<15 m), hand-dug wells, without any wall isolation or well cover protection. Typically, each well was fitted with a bucket and rope to withdraw water, although some had a manual pump associated. This proportion was visible during the survey, with about one borehole analysed for every six wells. Only 20.5% of the population has access to proper sanitation; thus, communal latrines are the standard, although in rural areas open defecation is still common [9]. Nevertheless, most are single pit latrines, with a basic thatched or galvanized corrugated sheet walls and no doors.

The high faecal contamination observed throughout the year is essentially a consequence of the proximity between wells and latrines (<30 m), the presence of freely wandering domestic animals (including cattle), and the contact of the bucket and rope with contaminated soil, as reported by several previous studies in Guinea-Bissau [12–14,26] and perceived by the authors in the field.

The influence of pit latrines on groundwater quality was previously described, being recognized as a major source of water contamination depending on the surrounding environment, particularly hydrological and soil conditions [27–29]. Rainfall promotes the mobilization of soil particles through infiltration and percolation that eventually reach the subsurface groundwater, conveying associated bacteria and viruses. Sediments are well known reservoirs for microorganisms, typically revealing higher bacterial levels than the water column [30]. Indeed, in accordance with other studies [13,26], high levels of contamination could be observed in the wet season, associated with a simultaneous increase in turbidity. Turbidity in water is caused by suspended particles or colloidal matter, and may be caused by inorganic or organic matter, or a combination of the two [31]. Due to the high superficial area and metabolic substrate concentration, high turbidity can provide an adequate environment for microorganism growth and persistence. Indeed, drinking water turbidity has been associated with the incidence of gastrointestinal diseases [32]. In Guinea-Bissau, the reported diarrhoea episodes systematically peak at the onset of the wet season [33,34], and the main cholera outbreaks have been reported during the wet season [11]. Additionally, the persistent high water temperature (average 29.1 °C) observed in wells year-round can also contribute to foster microbial growth [35].

Furthermore, the majority of wells were fitted with buckets and ropes that are often placed on the ground due to the lack of a basic overhead frame to hang them when not in use. This may allow contact with animals and human excreta; moreover, the tools may not exclusively be used to retrieve water. The aggravated situation was confirmed by the results obtained from the models, indicating that the method used to collect water was one of the most influential predictors. The use of a bucket considerably increased the contamination levels. On the other hand, boreholes (>20 m) were drilled mechanically and fitted with a pump (solar, electric), and were thus less prone to contamination.

The relationship between faecal indicators and the nitrogen species strengthen the hypothesis stated above; that microbial contamination enters groundwater due to infiltration and percolation from the surface. The presence of high concentrations of ammonium, nitrate, or nitrite in drinking water are recognized as indicators of possible bacterial, sewage, and animal waste pollution or agricultural runoff [36]. Moreover, there is an increased risk of methemoglobinemia development in bottle-fed infants associated with high concentrations of N species in drinking water, which can further be complicated by the concurrent presence of microbial contamination [36,37]. Additionally, nitrate appears to competitively inhibit iodine uptake, and the long-term exposure by drinking water intake can contribute to the iodine deficiency problem felt in countries such as Guinea-Bissau [36,38].

The faecal contamination was revealed to be related with latitude, with an increment towards the northern part of the country, that can be potentially explained by additional faecal contamination associated with the livestock presence. Although recent data are not available, according to the national livestock census in Guinea-Bissau conducted in 2009, it was estimated that the total number of cattle was approximately 1,325,412, mainly concentrated in the northern areas (Gabu, Bafata, and Oio regions) [39], owned by Fulani, Mandinga, and Balanta ethnic groups. The practice of extensive cattle ranching still dominates in Guinea-Bissau, with a marked transhumance period during the dry season for pasturage and water [40]. Cattle lairage is an unused practice in Guinea-Bissau and livestock roam free in the fields, representing a potential pollution source for unprotected shallow wells. It could be argued that the lairage of cattle could somehow contribute to a diffuse contamination decrease; however, the solution is unworkable due to the lack of water and fodder, technical resources, and infrastructure support.

Overall, similarly to other studies [12–14], the chemical contamination was less relevant, with metal concentrations below the acceptable limits for the majority of the studied water sources. Nonetheless, it is important to note that high metal levels were observed in several wells, and the consumption of elevated levels of metals through drinking water has been associated with the development of health problems, including cancer [31]. Again, the positive correlation found between metal concentration and turbidity indicated the potential role of soil particles as a natural contamination source. During the wet season, the soil particles and associated metal-rich leachates can easily be mobilized by percolation and infiltration, reaching the subsurface groundwater that feeds the shallow wells. In the dry season, the water–soil contact is promoted by the low water level and higher residence time of the water, with increased contact with the earth walls and bottom sediments, fostering high water turbidity.

The majority of the water sources were outside the suitable pH range for drinking water according to EU standards, in the acidic to very acidic interval, a consequence of red, sulphur-rich, sandy-clay soil characteristics [41]. The consumption of acidic water can have a direct impact on population health, particularly in dental erosion [42,43] and through the potential mobilization of heavy metals [44]. Currently, very limited working water networks are available in the country; therefore, the risk of metal contamination through pipes does not seem to be problematic, although such low pH values have to be taken into account in the construction of future infrastructure or rehabilitation initiatives, in order to choose the most suitable materials. Additionally, the corrosion of the materials used in constructing well and distribution systems can enable the feed and growth of microorganisms from adjacent areas in the water used for drinking. Bacteria can adapt to environmental conditions such as low pH, potentially reducing the stomach acidic barrier efficiency, and consequently, the required infectious dose to cause disease [45].

The access to safe drinking water, in addition to being a basic human right, is associated with population health, and consequently, with poverty. Waterborne diseases linked to inadequate WASH are a key public health concern [46–48], particularly in middle- and low-income countries, which needs to be addressed for country development. Furthermore, the expected population growth and the climate change scenario will pose an additional threat to WASH infrastructure and services [49].

This study highlighted the urgent need to improve the access to safe drinking water in sub-Saharan countries, such as Guinea-Bissau. For a long-term solution, a robust, country-wide intervention on WASH infrastructure is pivotal. New construction or rehabilitation of the limited water and sanitation networks should be performed, taking into account the local and regional conditions (such as the acidic groundwater), and forearm periodic maintenance and water quality monitoring, which are presently inexistent. To ensure water quality safety, government administration and non-governmental organizations must favour the construction of deep boreholes, even if it means cooperation and co-ordination between agencies, as well as constructing less infrastructure, due to the higher associated costs. Water quality should be regularly monitored, and if needed, promptly treated using the disinfection treatments techniques currently available and applied in countries with limited resources [50,51]. Emphasis should also be placed on the collection and treatment of effluent from latrines, including the construction of sealed tanks—septic tanks—and the disposal of waste into small collective treatment plants with biological beds. Furthermore, scientific and technical advice, prior and during construction, is essential for correct guidance on the establishment of local infrastructure and maintenance.

The traditional or settled way of living can be a hindrance to development; therefore, the implementation of behaviour change programs, with structured formative and educational interventions to raise awareness at the different societal levels (individual, household, community, and institutional) will be decisive to prevent disease transmission. The integration of local stakeholders and the community, throughout the implementation programs, will help to ensure the sustainable use and management of water.

On other hand, considering the urgency overt in the results of this study, the authors propose the short-term implementation of simple, sustainable measures, which could drastically improve the access to safe drinking water, mitigating the associated health risk:

- Remove potential contamination sources, such as latrines and garbage dumps from the vicinity of the water sources (>30 m). Promote latrine disinfection with quicklime and the controlled disposal of waste. Quicklime is easily available and can be used to disinfect faecal solids [52], with <90% efficiency in removing bacterial and viral pathogens [53].
- Whenever possible, favour the supply of water from boreholes over shallow wells. When using wells, favour those fitted with pumps, over buckets, proven to endure the harsh country conditions.

- Build and maintain proper security perimeters around water sources, deterring wandering animals by fencing and waterproofing the nearby terrain to prevent mud accumulation and avoiding the stagnation of water, particularly in the wet season.
- Fit the wells with covers and support systems for buckets and ropes in order to avoid contact with the soil.
- Promote the household storage of water in narrow-mouth containers fitted with faucets, avoiding contact between drinking water and hands or small containers used to collect water. Therefore, secondary contamination may be prevented. Promote the efficient and correct time frames to perform the disinfection of these containers.
- Promote household-level disinfection of drinking water to decrease the microbial load by filtration, chlorination, boiling, solar water disinfection (SODIS), or using plant extracts. Favouring household-level disinfection will prevent secondary contamination. Filtration through naturally occurring materials is a cost-effective and efficient treatment. For instance, slow sand filtration, ceramic filters, and biochar are widely used as disinfection techniques [50]. Chlorination with commercial bleach can also successfully decrease the bacterial load [50,54]. Although boiling water can be an efficient method of disinfection, in most situations is an unpractical solution, because it implies a time and economic expend. Moreover, boiling water can enhance metal concentrations, generating an additional problem. SODIS is a simple-to-use and inexpensive technique capable of microbial inactivation [55]. Plant extracts, such *Moringa oleifera*, can also be used as water treatment strategy, due to their antimicrobial and coagulant properties [56]. However, caution should be taken when water presents high turbidity values (>1 NTU), because this can interfere with the disinfection kinetics and efficiency, by providing protection and subtract for organisms [31,57,58]. Thus, resorting to multiple methods may be necessary to reduce particulate matter before disinfection. Filtration through easily available cotton cloth has been shown to be effective in the reduction of particulate matter and associated microbial loads [59]. The addition of *Moringa oleifera* has also been evaluated as a pretreatment for SODIS, reducing turbidity [60].
- Employ easily available oyster shells to increase water pH. Oyster shells, rich in calcium carbonate, can be used as a natural neutralizer to raise the pH in acidic waters [13]. Moreover, oyster shells can contribute to the heavy metal immobilization from drinking water [61,62].

5. Conclusions

The majority of the population in Guinea-Bissau still retrieve water for daily needs, including drinking water, from shallow wells. The water was acidic to very acidic and heavily contaminated with faecal matter, and thus unfit for human consumption. This situation was present throughout the year, but deteriorated further in the wet season, associated with high water turbidity. The highest levels of contamination were associated with shallow wells and bucket water retrieval. Although the chemical contamination was less relevant, and no overall trend could be found, several of the surveyed water sources revealed high values of nitrogen species and heavy metals.

A national concert intervention on WASH infrastructure is essential to provide safe drinking water to the population, as recognised in the United Nations Sustainable Development Goals. In the short-term, the authors suggest the implementation of simple measures to improve water potability and reduce the disease burden associated with waterborne pathogens.

The results of this research fill the pressing need for scientific background knowledge concerning the water quality in Guinea-Bissau, decisive to help the design of sustainable mitigation strategies. Improving water quality could have a dramatic impact on the population health status, and consequently, on the development of this low-income country.

Supplementary Materials: The following are available online at https://www.mdpi.com/article/10.3390/w14131987/s1, Figure S1: Example of typical water sources surveyed across Guinea-Bissau with the respective method used to retrieve water, Figure S2: Variation of key water quality parameters according to source type and method of water collection, Figure S3: Matrix of Spearman correlations between the physical–chemical quality parameters of the water sources across Guinea-Bissau, Table S1: Number of water sources surveyed according to the water collection system.

Author Contributions: Conceptualization, A.M. and A.A.B.; methodology, A.M., E.A. and A.A.B.; software, E.A.; validation, A.M., E.A. and A.A.B.; formal analysis A.M. and E.A.; investigation, A.M. and A.A.B.; resources, A.M. and A.A.B.; data curation A.M. and A.A.B.; writing—original draft preparation, A.M. and E.A.; writing—review and editing, A.M., E.A. and A.A.B.; visualization, A.M., E.A. and A.A.B.; supervision, A.A.B.; project administration, A.A.B.; funding acquisition, A.M. and A.A.B. All authors have read and agreed to the published version of the manuscript.

Funding: This study was partially funded through a fellowship to A. Machado (SFRH/BD/46146/2008), co-financed by POPH/FSE. This research was also partially supported by the Project BeachSafe (PTDC/SAU-PUB/31291/2017), co-financed by COMPETE 2020, Portugal 2020 and the European Union through the ERDF, and by FCT through national funds.

Institutional Review Board Statement: Not applicable.

Informed Consent Statement: Not applicable.

Data Availability Statement: All data generated or analysed during this study are included in this published article (and its supplementary information files). Notwithstanding that, the datasets used and/or analysed during the current study are available from the corresponding author upon reasonable request.

Acknowledgments: The authors wish to acknowledge the on-site logistic support of AIDA (Ayuda, Intercambio y Desarrollo), AMI (International Medical Assistance) and Cruz Roja Española. We thank Alfa and Samba, who assisted with sample collection and facilitated contact with the local population.

Conflicts of Interest: The authors declare no conflict of interest. The funders had no role in the design of the study; in the collection, analyses, or interpretation of data; in the writing of the manuscript, or in the decision to publish the results.

References

1. UN, United Nations. *The Human Right to Water and Sanitation [A/RES/64/292]*; United Nations General Assembly; UN, United Nations: New York, NY, USA, 2010. Available online: www.un.org/ga/search/view_doc.asp?symbol=A/RES/64/292 (accessed on 7 July 2021).
2. UN, United Nations. *Transforming Our World: The 2030 Agenda for Sustainable Development [A/RES/70/1]*; United Nations General Assembly. UN, United Nations: New York, NY, USA, 2015. Available online: https://www.un.org/en/development/desa/population/migration/generalassembly/docs/globalcompact/A_RES_70_1_E.pdf (accessed on 3 July 2021).
3. WHO, World Health Organization. *Results of Round II of the WHO International Scheme to Evaluate Household Water Treatment Technologies*; WHO: Geneva Switzerland, 2019. Available online: https://apps.who.int/iris/handle/10665/325896 (accessed on 2 July 2021).
4. WHO, World Health Organization. *Key Facts from 2015 JMP Report*; WHO, World Health Organization: Geneva, Switzerland, 2015. Available online: http://www.who.int/water_sanitation_health/publications/JMP-2015-keyfacts-en-rev.pdf?ua=1 (accessed on 3 July 2021).
5. UN, United Nations. *World Population Prospects 2019: Highlights [ST/ESA/SER.A/423]*; United Nations General Assembly, Department of Economic and Social Affairs, Population Division. UN, United Nations: New York, NY, USA, 2019. Available online: https://www.un.org/development/desa/publications/world-population-prospects-2019-highlights.html (accessed on 3 July 2021).
6. WHO, World Health Organization. *Guinea-Bissau Data. Global Health Observatory Data Repository*; WHO, World Health Organization: Geneva, Switzerland, 2021. Available online: https://apps.who.int/gho/data/view.main.SUBREGchildmortality-GNB (accessed on 3 July 2021).
7. WHO, World Health Organization. *Global Supply and Sanitation Assessment 2000 Report. Part II: Water Resources Development*; Joint Monitoring Programme for Water Supply and Sanitation; WHO: Geneva Switzerland, 2000. Available online: https://www.who.int/publications/i/item/9241562021 (accessed on 4 July 2021).
8. Gleick, P.H. The human right to water. *Water Policy* **1998**, *1*, 487–503. [CrossRef]
9. UN, United Nations. Global SDG Indicators Database—Guinea-Bissau. United Nations Statistics Division. 2021. Available online: https://www.sdg6data.org/country-or-area/Guinea-Bissau#anchor_6.1.1 (accessed on 12 October 2021).

10. INASA, National Institute of Public Health. *Relatório Nacional Sobre Casos de Diarreias em 2018 e 2019*; Ministry of Health: Bissau, Guinea-Bissau, 2019.
11. UNICEF, United Nations Children's Fund. *Cholera Factsheet Guinea-Bissau*; WHO, UNICEF West and Central Africa Regional Office: Geneva, Switzerland, 2017. Available online: https://www.unicef.org/cholera/files/UNICEF-Factsheet-Guinea-Bissau-EN-FINAL.pdf (accessed on 5 July 2021).
12. Bancessi, A.; Catarino, L.; José Silva, M.; Ferreira, A.; Duarte, E.; Nazareth, T. Quality Assessment of Three Types of Drinking Water Sources in Guinea-Bissau. *Int. J. Environ. Res. Public Health* **2020**, *17*, 7254. [CrossRef] [PubMed]
13. Bordalo, A.A.; Savva-Bordalo, J. The quest for safe drinking water: An example from Guinea-Bissau (West Africa). *Water Res.* **2007**, *41*, 2978–2986. [CrossRef] [PubMed]
14. Ferrante, M.; Signorelli, S.S.; Ferlito, S.L.; Grasso, A.; Dimartino, A.; Copat, C. Groundwater-based water wells characterization from Guinea Bissau (Western Africa): A risk evaluation for the local population. *Sci. Total Environ.* **2018**, *619*, 916–926. [CrossRef] [PubMed]
15. FAO, Food and Agriculture Organization of the United Nations. *World Soil Resources Report 84*; Food and Agriculture Organization of the United Nations; International Society of Soil Science: Rome, Italy, 1998. Available online: https://edepot.wur.nl/493579 (accessed on 10 July 2021).
16. Hijmans, R.; University of California. Second-Level Administrative Divisions, Guinea-Bissau, 2015. UC Berkeley, Museum of Vertebrate Zoology. 2015. Available online: http://purl.stanford.edu/hy559qv799. (accessed on 1 July 2021).
17. Bordalo, A.A. Microbiological water quality in urban coastal beaches: The influence of water dynamics and optimization of the sampling strategy. *Water Res.* **2003**, *37*, 3233–3241. [CrossRef]
18. WHO, World Health Organization. *Guidelines for Drinking-Water Quality, Incorporating 1st and 2nd Addenda, Recommendations*, 3rd ed.; WHO: Geneva, Switzerland, 2008; Volume 1. Available online: http://www.who.int/water_sanitation_health/dwq/fulltext.pdf?ua=1 (accessed on 5 July 2021).
19. EU, European Union. *Council Directive 98/83/EC of 3 November on the Quality of Water Intended for Human Consumption*; European Union: Brussels, Belgium, 1998. Available online: https://eur-lex.europa.eu/legal-content/EN/TXT/PDF/?uri=CELEX:31998L0083&from=EN (accessed on 5 July 2021).
20. Tebutt, T.H.Y. *Principles of Water Quality Control*, 5th ed.; Butter Worth-Heinemann: Oxford, UK, 1998.
21. Leathwick, J.; Elith, J.; Francis, M.; Hastie, T.; Taylor, P. Variation in demersal fish species richness in the oceans surrounding New Zealand: An analysis using boosted regression trees. *Mar. Ecol. Prog. Ser.* **2006**, *321*, 267–281. [CrossRef]
22. Elith, J.; Leathwick, J.R.; Hastie, T. A working guide to boosted regression trees. *J. Anim. Ecol.* **2008**, *77*, 802–813. [CrossRef]
23. R Core Team. *R: A Language and Environment for Statistical Computing*; R Foundation for Statistical Computing: Vienna, Austria, 2021. Available online: https://www.R-project.org/ (accessed on 13 November 2021).
24. Ridgeway, G. Generalized Boosted Regression Models. Documentation on the R Package 'gbm'. Version 1.6-3. 2007. Available online: https://cran.r-project.org/web/packages/gbm/gbm.pdf (accessed on 13 November 2021).
25. Machado, A.; Bordalo, A.A. Analysis of the bacterial community composition in acidic well water used for drinking in Guinea-Bissau, West Africa. *J. Environ. Sci.* **2014**, *26*, 1605–1614. [CrossRef]
26. Colombatti, R.; Vieira, C.S.; Bassani, F.; Cristofoli, R.; Coin, A.; Bertinato, L.; Riccardi, F. Contamination of drinking water sources during the rainy season in an urban post-conflict community in Guinea Bissau: Implications for sanitation priority. *Afr. J. Med. Med. Sci.* **2009**, *38*, 155–161.
27. Graham, J.P.; Polizzotto, M.L. Pit latrines and their impacts on groundwater quality: A systematic review. *Environ. Health Perspect.* **2013**, *121*, 521–530. [CrossRef]
28. Escamilla, V.; Knappett, P.S.K.; Yunus, M.; Streatfield, P.K.; Emch, M. Influence of Latrine Proximity and Type on Tubewell Water Quality and Diarrheal Disease in Bangladesh. *Ann. Assoc. Am. Geogr.* **2013**, *103*, 299–308. [CrossRef]
29. Martínez-Santos, P.; Martín-Loeches, M.; García-Castro, N.; Solera, D.; Díaz-Alcaide, S.; Montero, E.; García-Rincón, J. A survey of domestic wells and pit latrines in rural settlements of Mali: Implications of on-site sanitation on the quality of water supplies. *Int. J. Hyg. Environ. Health* **2017**, *220*, 1179–1189. [CrossRef] [PubMed]
30. Sander, B.C.; Kalff, J. Factors controlling bacterial production in marine and freshwater sediments. *Microb. Ecol.* **1993**, *26*, 79–99. [CrossRef] [PubMed]
31. WHO, World Health Organization. *Guidelines for Drinking-Water Quality*; Fourth Edition Incorporating 1st Addendum; WHO: Geneva, Switzerland, 2017. Available online: https://www.who.int/publications/i/item/9789241549950 (accessed on 8 August 2021).
32. Muoio, R.; Caretti, C.; Rossi, L.; Santianni, D.; Lubello, C. Water safety plans and risk assessment: A novel procedure applied to treated water turbidity and gastrointestinal diseases. *Int. J. Hyg. Environ. Health* **2020**, *223*, 281–288. [CrossRef] [PubMed]
33. Mølbak, K.; Jensen, H.; Lngholt, L.; Aaby, P. Risk Factors for Diarrheal Disease Incidence in Early Childhood: A Community Cohort Study from Guinea-Bissau. *Am. J. Epidemiol.* **1997**, *146*, 273–282. [CrossRef] [PubMed]
34. Alves, S. Water Diseases: Dynamic of Malaria and Gastrointestinal Diseases in the Tropical Guinea-Bissau (West Africa). Master's Thesis, University of Porto, Porto, Portugal, 2018. Available online: https://repositorio-aberto.up.pt/bitstream/10216/119094/2/316303.pdf (accessed on 12 October 2021).
35. Shiah, F.K.; Ducklow, H.W. Bacterioplankton growth responses to temperature and chlorophyll variations in estuaries measured by thymidine: Leucine incorporation ratio. *Aquatic Microb. Ecol.* **1997**, *13*, 151–159. [CrossRef]

36. WHO, World Health Organization. *Nitrate and Nitrite in Drinking-Water. Background Document for Preparation of WHO Guidelines for Drinking-Water Quality*; [WHO/HSE/AMR/07.01/16/Rev/1]; WHO: Geneva, Switzerland, 2011. Available online: https://www.who.int/water_sanitation_health/dwq/chemicals/nitratenitrite2ndadd.pdf. (accessed on 8 August 2021).
37. Sadeq, M.; Moe, C.L.; Attarassi, B.; Cherkaoui, I.; Elaouad, R.; Idrissi, L. Drinking water nitrate and prevalence of methemoglobinemia among infants and children aged 1–7 years in Moroccan areas. *Int. J. Hyg. Environ. Health* **2008**, *211*, 546–554. [CrossRef]
38. Carvalho, A.C.; Machado, A.; Embalo, A.R.; Bordalo, A.A. Endemic goiter and iodine deficiency status among Guinea-Bissau school-age children. *Eur. J. Clin. Nutr.* **2018**, *72*, 1576–1582. [CrossRef]
39. Republic of Guinea-Bissau. *Third National Communication: Report to the United Nations Framework Convention on Climate Change*; Government of the Republic of Guinea-Bissau: Bissau, Guinea-Bissau, 2018. Available online: https://unfccc.int/sites/default/files/resource/TCN_Guinea_Bissau.pdf (accessed on 29 October 2021).
40. BOAD, Banque Ouest Africaine de Développement. Enhancing Livestock Resilience to Drought in Guinea Bissau. Project Concept Note. Green Climate Fund. 2018. Available online: https://www.greenclimate.fund/sites/default/files/document/19670-enhancing-livestock-resilience-drought-guinea-bissau.pdf (accessed on 29 October 2021).
41. Jalloh, A.; Roy-Macauley, H.; Sereme, P. Major agro-ecosystems of west and central Africa: Brief description, species richness, management, environmental limitations and concerns. *Agric. Ecosyst. Environ.* **2012**, *157*, 5–16. [CrossRef]
42. O'Sullivan, E.; Milosevic, A. UK national clinical guidelines in paediatric dentistry: Diagnosis, prevention and management of dental erosion. *Int. J. Paediatr. Dent.* **2008**, *18*, 29–38. [CrossRef]
43. West, N.X.; Hughes, J.A.; Addy, M. The effect of pH on the erosion of dentine and enamel by dietary acids in vitro. *J. Oral Rehabil.* **2008**, *28*, 860–864. [CrossRef]
44. Calmano, W.; Hong, J.; Förstner, U. Binding and Mobilization of Heavy Metals in Contaminated Sediments Affected by pH and Redox Potential. *Water Sci. Technol.* **1993**, *28*, 223–235. [CrossRef]
45. Merrel, D.S.; Camilli, A. The *cadA* gene of *Vibrio cholerae* is induced during infection and plays a role in acid tolerance. *Mol. Microbiol.* **1999**, *34*, 836–949. [CrossRef] [PubMed]
46. WHO, World Health Organization. *2019. Drinking Water [Fact Sheet 391]*; WHO: Geneva, Switzerland, 2017. Available online: https://www.who.int/news-room/fact-sheets/detail/drinking-water (accessed on 19 October 2021).
47. Prüss, A.; Kay, D.; Fewtrell, L.; Bartram, J. Estimating the burden of disease from water, sanitation, and hygiene at a global level. *Environ. Health Perspect.* **2002**, *110*, 537–542. [CrossRef]
48. Ashbolt, N.J. Microbial contamination of drinking water and disease outcomes in developing regions. *Toxicology* **2004**, *198*, 229–238. [CrossRef] [PubMed]
49. Alhassan, S.; Hadwen, W.L. Challenges and Opportunities for Mainstreaming Climate Change Adaptation into WaSH Development Planning in Ghana. *Int. J. Environ. Res. Public Health* **2017**, *14*, 749. [CrossRef] [PubMed]
50. Pandit, A.B.; Kumar, J.K. Clean water for developing countries. *Annu. Rev. Chem. Biomol. Eng.* **2015**, *6*, 217–246. [CrossRef]
51. Thomas, B.; Vinka, C.; Pawan, L.; David, S. Sustainable groundwater treatment technologies for underserved rural communities in emerging economies. *Sci. Total Environ.* **2022**, *813*, 152633. [CrossRef]
52. Meckes, M.C.; Rhodes, E.R. Evaluation of bacteriological indicators of disinfection for alkaline treated biosolids. *J. Environ. Engin. Sci.* **2004**, *3*, 231–236. [CrossRef]
53. EPA, Environmental Protection Agency. *White House Document: Environmental Regulations and Technology: Control of Pathogens and Vector Attraction in Sewage Sludge*; EPA/625/R-92/013 [Revised in 2003]; US Environmental Protection Agency: Washington, DC, USA, 1995. Available online: https://www.epa.gov/sites/production/files/2015-07/documents/epa-625-r-92-013.pdf. (accessed on 19 October 2021).
54. Quick, R.E.; Kimura, A.; Thevos, A.; Tembo, M.; Shamputa, I.; Hutwagner, L.; Mintz, E. Diarrhea prevention through household-level water disinfection and safe storage in Zambia. *Am. J. Trop. Med. Hyg.* **2002**, *66*, 584–589. [CrossRef]
55. McGuigan, K.G.; Conroy, R.M.; Mosler, H.J.; du Preez, M.; Ubomba-Jaswa, E.; Fernandez-Ibañez, P. Solar water disinfection (SODIS): A review from bench-top to roof-top. *J. Hazard. Mater.* **2012**, *235-236*, 29–46. [CrossRef] [PubMed]
56. Bancessi, A.; Pinto, M.M.F.; Duarte, E.; Catarino, L.; Nazareth, T. The antimicrobial properties of *Moringa oleifera* Lam. for water treatment: A systematic review. *SN Appl. Sci.* **2020**, *2*, 323. [CrossRef]
57. Rowe, A.R. Chlorinating well water with liquid bleach was not an effective water disinfection strategy in Guinea-Bissau. *Int. J. Environ. Health Res.* **1998**, *8*, 339–340. [CrossRef]
58. Sichel, C.; Blanco, J.; Malato, S.; Fernández-Ibáñez, P. Effects of experimental conditions on E. coli survival during solar photocatalytic water disinfection. *J. Photochem. Photobiol. A* **2007**, *189*, 239–246. [CrossRef]
59. Huq, A.; Yunus, M.; Sohel, S.S.; Bhuiya, A.; Emch, M.; Luby, S.P.; Russek-Cohen, E.; Nair, G.B.; Sack, R.B.; Colwell, R.R. Simple sari cloth filtration of water is sustainable and continues to protect villagers from cholera in Matlab, Bangladesh. *MBio* **2010**, *1*, e00034-10. [CrossRef] [PubMed]
60. Keogh, M.B.; Elmusharaf, K.; Borde, P.; McGuigan, K.G. Evaluation of the natural coagulant *Moringa oleifera* as a pretreatment for SODIS in contaminated turbid water. *Sol. Energy* **2017**, *158*, 448–454. [CrossRef]

61. Tudor, H.E.A.; Gryte, C.C.; Harris, C.C. Seashells: Detoxifying Agents for Metal-Contaminated Waters. *Water Air Soil Pollut.* **2006**, *173*, 209–242. [CrossRef]
62. Xu, Z.; Valeo, C.; Chu, A.; Zhao, Y. The Efficacy of Whole Oyster Shells for Removing Copper, Zinc, Chromium, and Cadmium Heavy Metal Ions from Stormwater. *Sustainability* **2021**, *13*, 4184. [CrossRef]

Article

Health Risk of the Shallow Groundwater and Its Suitability for Drinking Purpose in Tongchuan, China

Abel Nsabimana [1,2], Peiyue Li [1,2,3,*], Song He [1,2], Xiaodong He [1,2], S. M. Khorshed Alam [1,2] and Misbah Fida [1,2]

1. School of Water and Environment, Chang'an University, No. 126 Yanta Road, Xi'an 710054, China; nsabby41@gmail.com (A.N.); hesong_chd@163.com (S.H.); hexiaod3@163.com (X.H.); khorshed11_31@yahoo.com (S.M.K.A.); misbahfida20@gmail.com (M.F.)
2. Key Laboratory of Subsurface Hydrology and Ecological Effects in Arid Region of the Ministry of Education, Chang'an University, No. 126 Yanta Road, Xi'an 710054, China
3. School of Water Resources and Environment, Hebei GEO University, No. 136 East Huai'an Road, Shijiazhuang 050031, China
* Correspondence: lipy2@163.com or peiyueli@chd.edu.cn

Abstract: Studying the quality and health risks of groundwater is of great significance for sustainable water resources utilization, especially in arid and semi-arid areas around the world. The current study is carried out to evaluate the quality and potential health risks of groundwater in the Tongchuan area on the Loess Plateau, northwest China. Water quality index (WQI) and hydrochemical correlation analysis were implemented to understand the status of groundwater quality. Daily average exposure dosages through the oral and dermal contact exposure pathways were taken into consideration to calculate the health risks to the human body. Additionally, graphical approaches such as Piper diagram, Durov diagram and GIS mapping were used to help better understand the results of this study. The WQI approach showed that 77.1% of the samples were of excellent quality. The most significant parameters affecting water quality were NO_3^-, F^-, and Cr^{6+}. The health risk assessment results showed that 27.1% and 54.2% of the samples lead to non-carcinogenic risks through oral intake for adults and children, respectively. In contrast, 12.5% of the groundwater samples would result in carcinogenic risks to the residents. This study showed that the WQI method needs to be supplemented by a health risk evaluation to obtain comprehensive results for groundwater quality protection and management in the Tongchuan area.

Keywords: water quality index; health risk assessment; Tongchuan city; Loess Plateau

1. Introduction

Groundwater is an important source for drinking and other various purposes for the majority of the population around the world, especially in arid and semiarid regions where precipitation and runoff are rare [1–6]. In addition to drinking, groundwater is useful for domestic, industrial and agricultural purposes. Due to the increased demand for groundwater, the groundwater table is subject to fluctuations, and aquifers are becoming contaminated in the context of climate change, rapid population growth, industrial development and urban expansions [7–12]. This situation is also aggravated where natural phenomena are controlling the physicochemical parameters of groundwater, such as rock influences, volcanic eruption or marine salt intrusions [13].

There is a critical increase in freshwater demand correlated with the rapid growth of the population all over the world [14] and intensive agriculture activities [15,16]. The increment of the population also leads to the expansion of cities and municipal waste that affect the groundwater quality through organic and inorganic contaminants [17–20]. Furthermore, industrialization is one of the most significant factors affecting groundwater quality through the effluents released into the nature [21–24]. Papazotos et al. [25]

investigated the impact of water–rock and agricultural activities in the Psachna Basin (Greece) on groundwater quality and found that groundwater was strongly affected by the ultramafic geological environment with anthropogenic activities as revealed by high concentrations of Cr, Cr^{6+}, and NO_3^-. Water-ultramafic rock processes can also increase the concentration of Cr in groundwater as investigated by Vasileiou et al. [26] in their study on hydrogeochemical processes and natural background levels of chromium in an ultramafic environment in Macedonia (Greece), and they found a high concentration of Cr^{6+} ranging from 0.5 to 131 µg/L in groundwater of western Vermino Mountain. In addition, Chen et al. [27] found rock dissolution and precipitation of Ca-As and CaF_2, which controlled a high concentration of As and CaF_2 in northwest China. In terms of groundwater pollution by marine intrusion, Zissimos et al. [28] tested the occurrence and distribution of Cr in groundwater and surface water in Cyprus and found that the highest Cr^{6+} concentration observed in the Troodos area was 26 µg/L. However, the abnormal concentrations of Cr^{6+} (460 µg/L) and As (15 µg/L) were detected in groundwater along the coastline in the Schinos area (Greece) due to seawater intrusion [29].

Given the importance of groundwater for humanity and considering its vulnerability facing pollution issues as aforementioned, numerous studies have been conducted to evaluate groundwater quality to ensure the health of consumers. As a result, governments and states implemented controlling structures for water quality in order to preserve the population health [14]. In this regard, many groundwater quality investigations have been conducted based on the guidelines set by governments and organizations such as the World Health Organization (WHO) and the Ministry of Environmental Protection of the P.R. China [30]. Based on the aforementioned guidelines, serious drinking groundwater contamination was reported by many scholars all over the world [15,16,18,31–36]. However, few of them were associated with groundwater pollution and health risk assessment. To obtain the results, many approaches were used by the researchers. Ni et al. [37] used the geostatistical spatial analysis function of ArcGIS to map the evaluated carcinogenic and non-carcinogenic risks in the Sichuan Basin, China. Their study showed that total cancerous and non-cancerous risks were found in 5% and 8% of samples, respectively. Using a comprehensive water quality index assessment, Wu and Sun [38] found that 60% of sampled water was unsuitable for drinking in the alluvial plain located in mid-west China. Chen and her colleagues [27] used a triangular fuzzy numbers approach to assess health risk by As and CaF_2 in groundwater and found that their concentrations were higher in the shallow groundwater, which exceeded the acceptable limit (1×10^{-6}) set by the Ministry of Environmental Protection of the P.R. China for Cr^{6+} and As [30].

Studies performed in the northwest of China reported high nitrate concentrations representing health risk concerns for the population [38] due to anthropogenic activities, especially fertilizers used in agriculture [39]. N-bearing and P-bearing fertilizers can cause the oxidation of geogenic Cr, which results in elevated Cr^{6+} [19]. Wei et al. [34] also reported that nitrate pollution was a major environmental geological problem in the groundwater in part of China. In addition, Li et al. [21] reported a severe water stress in the Chinese Loess Plateau aggravated by the high fluoride concentration in drinking water.

The Tongchuan region is situated in the middle edges of the Loess Plateau and is adjacent to the Weihe River Valley and Guanzhong Basin, and the main water supply aquifer in this area is a phreatic aquifer with thickness ranging from 25 to 60 m [34,39]. The main objective of the present study is to enhance the understanding of the association between water quality and health risk assessment. Specifically, this study aims to characterize the major pollutants in shallow groundwater, to check their concentration based on the depth of wells, to determine the water quality index and make its distribution map, and to assess the water's potential risks to human health. To understand the status of groundwater quality, the water quality index (WQI), hydrochemical correlation analysis, and graphical approaches were used. The health risk assessment was performed considering daily average exposure dosage through oral pathway per unit weight (mg/(kg.d)) for drinking water intake; and for dermal contact, the exposure dosage of every single event in mg/cm^2 and the skin surface (cm^2) were taken into consideration. Geographical information system approaches helped to better understand the results of this study.

2. Materials and Methods

2.1. Study Area

Tongchuan City is 70 km away from Xi'an City, the capital city of Shaanxi Province (Figure 1). It belongs to the Chinese Loess Plateau, with longitude between 108°35′44″ E and 109°29′22″ E and latitude between 34°48′27″ N and 35°35′23″ N. The altitude of Tongchuan City ranges from 900 to 1350 m above mean sea level [39]. The study area is situated in the middle edges of Loess Plateau and adjacent to the Weihe River Valley and Guanzhong Basin [34,40]. Tongchuan lies in the transition zone of semi-humid and semi-arid climate with annual mean rainfall and evaporation of around 540 and 1964 mm, respectively. The annual temperature of Tongchuan City is 8.9–12 °C [34,39]. Precipitation, reservoir leakage and irrigation are the main recharges of groundwater, whereas discharge to some rivers such as the Beiluo River and Juhe River, evaporation and artificial extraction [34] are the main discharge pathways of groundwater. Li et al. [39] estimated the groundwater recharge at 52.8% from precipitation and 40.1% from irrigation infiltration, whereas 37.4% and 44.9% of groundwater were discharged by artificial extraction and the lateral outflow, respectively. Geologically, the study area is dominated by Quaternary loess divided into three landforms, including loess tableland, loess gully and alluvial terrace. Furthermore, this area has several layers from top down [39]: Holocene loess layer and upper Pleistocene loess layer, which are unsaturated. The middle Pleistocene layer is composed of silty clay, which separates the phreatic aquifer and the confined aquifer partially formed by the lower Pleistocene loess layer, alluvial, sand and gravel layers. The phreatic aquifer with a thickness of 25 to 60 m is the main water supply aquifer in this area.

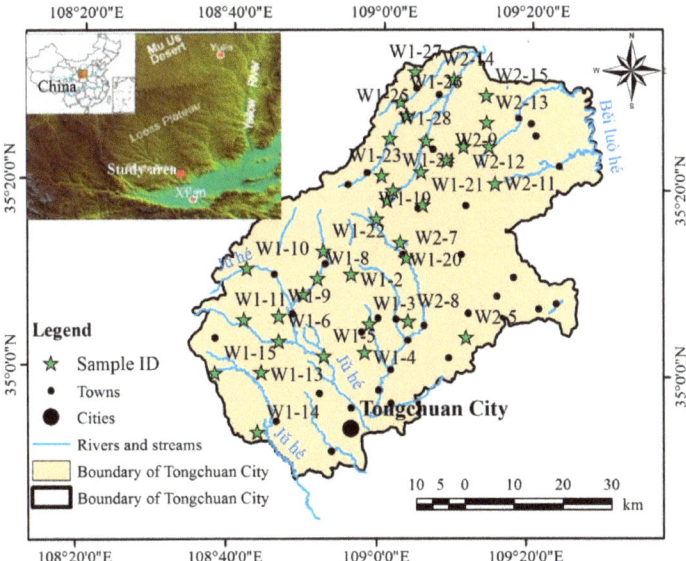

Figure 1. Study area and samples distribution.

2.2. Groundwater Samples

For this study, 48 groundwater samples were collected from the wells and boreholes distributed in the study area. The criteria for the selection of water samples were based on the depth of wells, water purposes and the zone of collection. The sampling locations were recorded by coordinates using a portable GPS device and are shown as Figure 1. Samples were collected in pre-cleaned plastic polyethylene bottles for physicochemical analysis after the wells were pumped for 10 min. Before sampling, all the containers were washed and rinsed thoroughly with the groundwater to be sampled. Water was filtered through 0.45 μm filter during sampling. Sample collection, handling, and preservation complied with the standard procedures recommended by Standard Examination Methods for Drinking Water [30] to ensure data quality and consistency. The water samples were analyzed in the Soil and Water Testing Center of Shaanxi Institute of Engineering Investigation, China.

2.3. Chemical Analysis and Data Processing

The samples were analyzed for physical and chemical parameters, including temperature, pH, electrical conductivity (EC), total hardness (TH), total dissolved solids (TDS), major ions (Na^+, K^+, Ca^{2+}, Mg^{2+}, Cl^-, SO_4^{2-}, HCO_3^-, NO_3^-, NO_2^- and F^-), and Cr^{6+}. Some parameters such as pH, EC and temperature were recorded on the field by portable multi-parameter devices. Drying and weighing approach was used to measure TDS. Na^+ and K^+ were determined using flame atomic absorption spectrometer and TH, Ca^{2+}, and Mg^{2+} were analyzed using EDTA titrimetric methods. Spectrophotometer and ion chromatography were used to determine the enrichment of NO_2^-, NO_3^-, and SO_4^{2-}, respectively. Standard titration method using $AgNO_3$ as a reactant solution was used to determine the concentration of Cl^-. Traditional titrimetric and ion selective electrode methods were used to determine HCO_3^-, and F^-, respectively. Ion chromatographic-colorimetric analytical principle was used to determine Cr^{6+}.

The evaluation of water suitability for drinking purposes was based on the concentrations of physical and hydrochemical characteristics of the considered samples compared to the limits of physicochemical parameters recommended by the WHO [14,41,42]. The groundwater quality standards set by the Ministry of Health of the People's Republic of China, and the Standardization Administration of the People's Republic of China [43] were also considered in this study.

2.4. Statistical Analysis and Computing

In this study, statistical analysis was conducted by using SPSS 25 for Pearson's correlation. Pearson's correlation coefficient (r) helps to quantify the significance of a relationship between two parameters and was widely used in groundwater quality assessment because it gives a quick correlation value. Its mathematical formula is expressed as follows [44]:

$$r_{xy} = \frac{i = \sum_{i=1}^{n}(x_i - \bar{x})(y_i - \bar{y})}{\sqrt{\sum_{i=1}^{n}(x_i - \bar{x})^2 \sum_{i=1}^{n}(y_i - \bar{y})^2}} \quad (1)$$

where, r_{xy} represents the correlation coefficient between the parameters x and y, n denotes the sample size, x_i is the individual value of the parameter x, $\bar{x}$ is the mean value of the parameter x, y_i stands for the individual value of the parameter y, and $\bar{y}$ denotes the mean value of the parameter y.

The values of correlation coefficient can be classified as very strong for $r \geq 0.80$, strong for $0.60 \leq r < 0.80$, moderate for $0.40 \leq r < 0.60$, weak for $0.20 \leq r < 0.40$, and very weak for $r < 0.20$. In addition, the correlation coefficient is evaluated on the basis of p value. The correlation coefficient is statistically considered as highly significant when $p < 0.01$, marginally significant when $p < 0.05$, and not significant when $p > 0.10$ [44].

For various computing and plots, Microsoft Office 2016 (Excel and Word), Origin 2018, and Grapher 12 were used. Parameter analysis, Piper [45] and Durov [46] diagrams plots were executed using AqQA software. Finally, for mapping, ArcMap 10.3 software was used to locate samples and make a water quality distribution map. This map was obtained using Bayesian Kriging method, which is an automatic Geo-statistical interpolation package incorporated in ArcGIS software. The general Kriging equation can be described as follows [47]:

$$Z^*(x_p) = \sum_{i=1}^{n} \lambda_i Z(x_i) \text{ with } \sum_{i=1}^{n} \lambda_i = 1 \qquad (2)$$

where λ_i is the Kriging weight; $Z^*(x_p)$ estimates the unknown true value.

2.4.1. Water Quality Index (WQI)

To evaluate groundwater quality status in the study area, method of water quality index (WQI) was used to integrate comprehensive information through the analysis of physicochemical parameters [31,48–51]. In other words, WQI is a single numerical value obtained by combining a large water quality data [52,53]. First, each chemical parameter is assigned with a weight (w_i), which is determined by affecting the degree of the parameters to groundwater quality. The relative weight (W_i) is computed as:

$$W_i = \frac{w_i}{\sum_{i=1}^{n} w_i} \qquad (3)$$

where, W_i is the relative weight, w_i is the assigned weight of each parameter, n is the number of parameters. The value of w_i ranges from 1 to 5 according to the impact of the contaminant on human health.

Then, the quality rating scale (q_i) can be computed by:

$$q_i = \frac{C_i}{S_i} \times 100 \qquad (4)$$

where, q_i is the quality rating scale, C_i is the concentration of each chemical parameter in each water sample in mg/L, and S_i is the standard for each chemical parameter.

To calculate the WQI, SI_i has to be determined with the following equations:

$$SI_i = W_i \times q_i \qquad (5)$$

$$WQI = \sum SI_i \qquad (6)$$

where, SI_i is the sub-index of the ith parameter and WQI is the water quality index.

The computed WQI values are classified into five categories [15,31,48,54]: excellent water (<50), good water (50–100), poor water (100–200), very poor water (200–300), and unsuitable water (>300).

2.4.2. Human Health Risk Assessment

The evaluation of drinking water quality needs to be completed by a health risk assessment as polluted water may cause adverse effects on the human body through water intake and dermal contact [1,38,42]. In this study, the potential risks through dermal contact pathway were neglected for non-carcinogenic risk because it is usually low [27,38,39], and water contamination in the study area was not considerably high as listed in Table 1. The risk assessment parameters selected for this study are NH_4^+, NO_3^-, NO_2^-, F^- and Cr^{6+}, using the models recommended by the Ministry of Environmental Protection of the P.R. China [30], which are also based on the model recommended by the United States Environmental Protection Agency [29,39].

Table 1. Statistical analysis of physicochemical indices for water samples collected in Tongchuan.

Indices	Sample Size	Min	Max	Mean	Median	Standard Deviation	Chinese Standards	WHO Guidelines	Detection Limits	% Exceeding Standards
pH	48	7.05	8.39	7.77	7.79	0.30	6.5–8.5	6.5–8.5	0.01	0 [1,2]
TH	48	175	731	350	340	115	450	500	1	17 [1], 10.4 [2]
TDS	48	252	1224	540	512	216	1000	1000	5	4.2 [1,2]
EC	48	519	1501	870	824	352	/	/	0.01	/
Na^+	48	4.8	282.0	51.8	29.6	65.7	200	200	2	8.3 [1,2]
K^+	48	0.88	73.10	3.99	2.04	10.36	/	/	0.01	0 [1,2]
Ca^{2+}	48	4.8	282.0	51.8	29.6	65.7	/	/	0.5	36 [2]
Cr^{6+}	48	BDL	0.071	0.027	0.010	0.030	0.05	0.05	0.0002	6.2 [1,2]
Mg^{2+}	48	2.4	57.1	26.4	26.1	11.3	/	/	0.5	4.2 [2]
Cl^-	48	2.0	144.0	37.5	18.0	40.0	250	250	0.5	0 [1,2]
SO_4^{2-}	48	4.80	572.00	79.19	48.00	93.76	250	500	0.5	2 [1,2]
HCO_3^-	48	201	604	389	384	91	/	/	1	/
NO_3^--N	48	BDL	262.00	32.66	16.41	49.25	20	50	0.009	45.8 [1], 18.5 [2]
NH_4^+	48	BDL	0.13	0.00	0.00	0.02	0.50	1.5	0.025	0 [1,2]
NO_2^--N	48	BDL	0.46	0.07	0.01	0.13	1	3	0.013	0 [1,2]
F^-	48	0.18	2.34	0.47	0.42	0.33	1	1.5	0.01	4.2 [1,2]

[1] Percentage of samples exceeding the P.R. China national standards, [2] percentage of samples exceeding WHO standards. BDL, below detection limit. All units for all parameter indices are in mg/L, except for pH (non-dimensional) and EC (μS/cm).

According to the references mentioned above, the non-carcinogenic risk through the oral intake pathway is calculated as follows:

$$Intake_{oral} = \frac{C \times IR \times EF \times ED}{BW \times AT} \quad (7)$$

$$HQ_{oral} = \frac{Intake_{oral}}{RfD_{oral}} \quad (8)$$

where $Intake_{oral}$ denotes the daily average exposure dosage through oral pathway per unit weight (mg/(kg·d)), C is the concentration of the parameter in water (mg/L), and IR represents the ingestion rate of water through drinking (L/d). EF and ED represent the exposure frequency (d/a) and exposure duration (a), respectively. BW and AT are the average body weight (kg) and average time of non-carcinogenic effects (d), respectively.

For this study, the ingestion rate of water used was based on statistical investigation that considers 1.5 L per day for adults and 0.7 L per day for children under 12 years old [38]. For non-carcinogenic risk assessment, EF is 365 days per year for both adults and children. ED is 30 years for adults and 12 years for children. BW is 15.9 kg for children, 56.8 kg for adults [30]. The average time (AT) for non-carcinogenic effects on children is 4380 days, whereas it is 10,950 days for female and male adults. HQ_{oral} and RfD_{oral} represent the hazard quotient and the reference dosage for non-carcinogenic pollutants through the oral exposure pathway (mg/(kg.d)), respectively. This study considered the RfD_{oral} values for NH_4^+, NO_3^-, NO_2^-, F^- and Cr^{6+} as 0.97, 1.6, 0.1, 0.04 and 0.003 mg/(kg.d), respectively [1,30,38]. HQ with a value exceeding 1 indicates a high potential health risk [21]. In addition, Cr^{6+} can also cause carcinogenic risks through drinking water intake and dermal contact. The total carcinogenic risk is the sum of calculated cancer risk through drinking pathway and that of dermal contact and is calculated as follows [1,30]:

$$CR_{oral} = Intake_{oral} \times SF_{oral} \quad (9)$$

$$CR_{dermal} = Intake_{dermal} \times SF_{dermal} \quad (10)$$

$$SF_{dermal} = \frac{SF_{oral}}{ABS_{gi}} \quad (11)$$

$$CR_{total} = CR_{oral} + CR_{dermal} \tag{12}$$

where CR_{oral} represents the carcinogenic risk through the oral exposure pathway. The CR limit is set as 1×10^{-6}. $Intake_{oral}$ denotes daily average exposure dosage through oral pathway per unit weight (mg/(kg·d)), SF_{oral} is the slope factor for the carcinogenic pollutants (mg/(kg·d))$^{-1}$. The SF_{oral} value of Cr^{6+} is set as 0.5 (mg/(kg·d))$^{-1}$ by the Ministry of Environmental Protection of the P.R. China [30]. ABS_{gi} is the gastrointestinal absorption factor, and its value is 1 for all contaminants except for Cr^{6+}, with ABS_{gi} equals 0.025 [1,30,55].

The $Intake_{dermal}$ is calculated as [1,30,38] as in Equations (13)–(15):

$$Intake_{dermal} = \frac{DA \times EV \times SA \times EF \times ED}{BW \times AT} \tag{13}$$

$$DA = K \times C \times t \times CF \tag{14}$$

$$SA = 239 \times H^{0.417} \times BW^{0.517} \tag{15}$$

where DA and SA are the exposure dosage of every single event in mg/cm^2 and the contacting area skin surface (cm^2), respectively. EV is the daily exposure frequency of dermal contact set at 1 for this study. ED is the exposure duration for carcinogenic risk, different from non-carcinogenic risk, and is set as 25,550 days for both adults and children. K is the coefficient of skin permeability (0.001 cm/h), t is the contact duration, which is set as 0.4 h/day for both adults and children [1,38]. CF is a conversion factor that equals 0.001, and H denotes the average height of the population estimated at 165.3 cm for males, 153.4 cm for females and 99.4 for children [1].

3. Results and Discussion

3.1. Physicochemical Parameters

Groundwater quality data were first checked for reliability and accuracy by calculating the correlation between EC and the sum of cations on one hand and with the sum of anions on the other hand. The results show a good correlation with $R^2 > 0.8$ (Figure 2a,b). The reliability of groundwater quality data was also checked by the ion charge balance between cations and anions as follows:

$$E(\%) = \frac{N_c - N_a}{N_c + N_a} \times 100 \tag{16}$$

where, N_c and N_a denote total concentrations of cations and anions of a sample in meq/L, respectively. The biggest value of E was 3.14%, which indicated that the samples were reliable, as the E value was between -5% and $+5\%$.

The physicochemical indices of groundwater samples were statistically analyzed, and the results are listed in Table 1. The pH values in this study ranged from 7.05 to 8.39, which were within the guidelines set by the WHO [42] for drinking water (6.5 to 8.5). Hem [56] concluded that the pH of groundwater was controlled by the equilibrium of CO_3^{2-}, CO_2 and HCO_3^-, and interpreted the chemical characteristics of natural water. The mean pH value of groundwater samples was 7.77, which was suitable for drinking purpose. Mechenich and Andrews [57] considered the range of pH values from 7.5–8.3 as an ideal values range for drinking water. Thus, it can be assumed that pH values for drinking water in Tongchuan City are good and ideal. However, 12 samples (25% of the total samples) showed slight alkalinity of the drinking water in the study area with pH ranging from 8 to 8.39. Alkalinity is not only associated with high pH values, but also with hardness and excessive TDS [33].

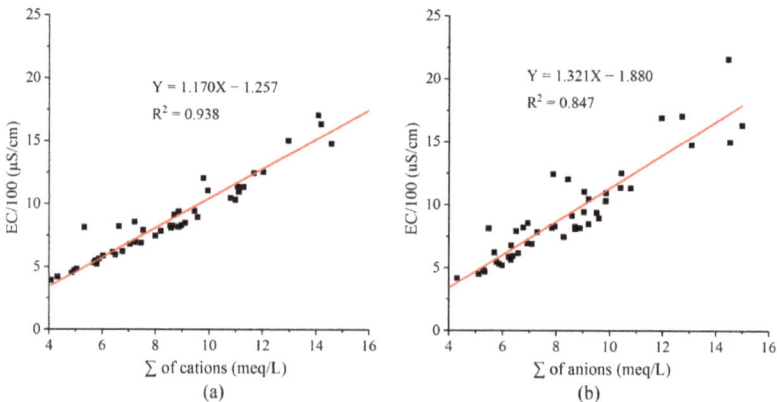

Figure 2. Ionic balance of groundwater data: (**a**) Σ of cations vs. EC/100; (**b**) Σ of anions vs. EC/100.

According to the average pH value, the groundwater in the study area can be used as drinking water. However, when comparing the detected TDS and TH values with the drinking water standards, there were two samples (4.2%) with TDS exceeding 1000 mg/L, and five samples (10.4%) with TH exceeding 500 mg/L. At the same time, referring to the drinking water quality guidelines recommended by the Ministry of Health of the People's Republic of China, there were eight samples (17%) whose TH exceeded 450 mg/L. This would be considered as hard water [1]. However, this classification is far different from the drinking water classification early made by Freeze and Cherry [58] (Table 2) based on TH. The groundwater classification on the basis of TDS and TH [14,31,58,59] in Tongchuan are as follows (Table 2): 35.4% and 64.6% of samples were hard water or very hard water; 47.9% were desirable and permissible for drinking; 95.8% were fresh water and 4.2% were brackish.

Table 2. TDS and TH-based classification of groundwater for drinking purpose in Tongchuan.

Parameters	Range	Water Type	% of Samples
TH	<75	Soft	0
	75–150	Moderately hard	0
	150–300	Hard	35.4
	>300	Very hard	64.6
TDS	<500	Desirable for drinking	47.9
	500–1000	Permissible for drinking	47.9
	<1000	Fresh water	95.8
	>1000	Brackish	4.2

In addition, the TH values of water are the measures of the dissolved Ca^{2+} and Mg^{2+} content, which are expressed in $CaCO_3$ mg/L and can be associated EC, which is normally twice the hardness for uncontaminated water [23,57]. Otherwise, if it is higher than that proportion, it provides information on the presence of components such as Na^+, Cl^- or SO_4^{2-} [57]. Through the analysis of the physical and chemical indicators of the samples in the study area, the average values of EC and TH were 869.75 μS/cm and 349.94 mg/L, respectively, and the conductivity was greater than two times of the TH, which indicated that slightly high concentrations of Na^+, Cl^-, and SO_4^{2-} were in some groundwater samples.

The order of major cations in the groundwater samples from the study area was Ca^{2+} > Na^+ > Mg^{2+} > K^+, with average values of 96.62, 51.81, 26.43, and 3.99 mg/L, respectively. The order of major anions of the samples was HCO_3^- > SO_4^{2-} > Cl^-, with average values of 389.29, 79.19, and 37.49 mg/L, respectively.

Indicated by the detected results of the samples, there was no HN_4^+ contamination in the groundwater of the study area because the maximum HN_4^+ concentration of the samples (0.13 mg/L) was in the range of natural levels of HN_4^+ in groundwater (below 0.2 mg/L), according to WHO [42]. The concentration of HN_4^+ in water is an indicator of possible bacterial, sewage, landfill, and animal waste pollution [30]. The concentration of Cl^- was not excessive in the analyzed samples from drinking water as it ranged from 2 to 144 mg/L. The WHO [42] has not set a health-based guideline value for Cl^-, but a concentration exceeding 250 mg/L can cause the water to be unsuitable for drinking as high Cl^- waters have a laxative effect for some people [33,55].

Although there is no health-based guideline value for Na^+ in potable water according to WHO [42], if its concentration exceeds 200 mg/L, it may taste bad, and excessive intake may cause hypertension [18]. Na^+ concentrations of four samples (8.3% of the total samples) slightly exceeded that threshold for the present study. A value of K^+ exceeding 12 mg/L in drinking water gives it a bitter taste [31]. In this study, only two samples (4.2%) exceeded this permissible limit. SO_4^{2-} was not excessive, except in one sample, where its concentration exceeded (572 mg/L) the SO_4^{2-} concentration limit proposed by WHO [30] for potable water, which is 500 mg/L.

To check the simultaneous occurrence of NO_3^- and NO_2^- in drinking water, the sum of the ratios of the concentration of each over its guideline value (GV) should not exceed 1 [42]:

$$\frac{C_{nitrate}}{GV_{nitrate}} + \frac{C_{nitrite}}{GV_{nitrite}} \leq 1 \tag{17}$$

where $C_{nitrate}$ is the concentration of NO_3^-, $C_{nitrite}$ is the concentration of NO_2^-, and $GV_{nitrate}$ and $GV_{nitrite}$ are the guideline values of NO_3^- and NO_2^-, respectively.

The application of this formula reveals that 16.6% of the groundwater samples were in the situation of simultaneous occurrence of NO_3^- and NO_2^- in drinking water. Furthermore, in the presence of microbial contamination, especially due to fecal contamination in drinking water, the health risk to infants is high [42].

In this study, 6.2% of the groundwater samples slightly exceeded the guideline value of permissible concentration in drinking water, which is 0.05 mg/L [43]. Fluoride is important for drinking water, with a concentration ranging from 0.7 to 1.2 mg/L, as it protects against dental cavities and strengthens the bones. When F^- concentration exceeds 1.5 mg/L, it causes teeth mottling, fluorosis or discoloration [33,42,60,61] as well as other health problems such as nervous system harm and urinary tract disease [62,63]. Although there were two samples with F^- concentration exceeding 1.5 mg/L, most of the samples (83.3%) were associated with low F^- concentrations below 0.7 mg/L. Therefore, to ensure the good health of the population in Tongchuan City, F^- should be added in drinking water to the majority of wells and be reduced in a few wells to avoid potential health hazards. In addition, 50 mg/L of the guideline value for NO_3^- was established by WHO [42] to protect the most sensitive populations. However, this population must be free of adverse health effects such as methemoglobinemia and thyroid effects at a concentration below 50 mg/L of NO_3^-. This health risk can seriously affect bottle-fed infants when mathemoglobinemia is complicated by the presence of microbial contamination and subsequent gastrointestinal infection that manifests as diarrhea.

Excessive boiling of water for microbiological safety purposes may increase the concentration of NO_3^-. Water for drinking should be heated until it reaches a rolling boil [42]. For NO_3^-, 45.8% of the samples exceeded the limits (20 mg/L) set by the Ministry of Health of the P.R. China [43].

3.2. Relationship between Depth of Wells and the Concentration of Physicochemical Parameters

Figure 3 shows the scatter plot of F^-, Na^+, and NO_2^- concentrations with groundwater level depth. It shows that the water samples were mostly concentrated in the shallow depth (less than 30 m). Fluoride is present in lower concentrations in shallow groundwater than in deep groundwater. This is because the dissolution of F-containing minerals such as fluorite is an important source of F^- in groundwater of the study area, and the amount of fluorite is higher in the deep aquifer. The alkaline pH can influence CaF_2 activity and favors the mobilization of F^- from soil and weathered rocks into groundwater. This assumption was also formulated by other researchers [64–66]. The enrichment of F^- can also be influenced by the ratio between HCO_3^-, Na^+ and Ca^{2+} in groundwater, as confirmed by Saxena and Ahmed [67], Rango et al. [68], and Kimambo et al. [64]. Na^+ concentration is lower in the shallow aquifer, which also supports the phenomenon of low F^- in shallow groundwater.

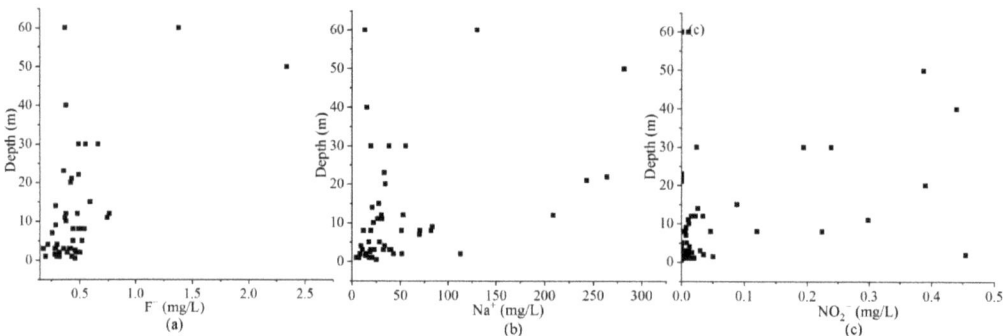

Figure 3. Relationship between fluoride and depth (**a**), sodium and depth (**b**), nitrite and depth (**c**).

Samples with low concentration of NO_2^- are usually observed in the shallow aquifer than in the deep aquifer. This may be due to the oxidation environment in the shallow aquifer that favors the transformation of NO_2^- to NO_3^-. Numerous studies have shown that human activities such as agriculture, industry, domestic sewage, landfills, and household waste influences shallow groundwater quality [1,32,69].

3.3. Hydrochemical Types of Groundwater

The Durov diagram depicted in Figure 4b reveals that most of the samples are concentrated in the field of HCO_3-Ca type and combined $HCO_3 \cdot SO_4$-Ca·Mg type. This situation may result from the dissolution of CO_3^- minerals and F^- [68]. As also discussed by Ravikumar et al. [70] and Lloyd and Heathcote [71], the HCO_3-Ca dominant frequently indicates that recharging waters in limestone and sandstone is associated with dolomite. To assess the water quality, a Piper diagram (Figure 4a) was used to characterize the hydrogeochemical facies of groundwater samples from the study area. The Piper plot shows that Ca^{2+}, Na^+, and Mg^{2+} are dominant cations in the region. Conversely, HCO_3^- and SO_4^{2-} dominate the facies, while Cl^- is quasi-inexistent. The general classification of all samples shows 81.25% Ca·Mg-HCO_3, 8% Ca·Mg-SO_4·Cl, 4.1% Na-Cl and 6.25% Na-HCO_3 water type (Figure 4a). The dominant Ca·Mg-HCO_3 type may indicate that the influence of dissolution on groundwater chemistry is more considerable, and it signifies the dominance of alkaline earths over alkalis; weak acids exceed strong acids. This observation was also found by other researchers, notably Xu et al. [72], Ravikumar et al. [70] and Singh et al. [16].

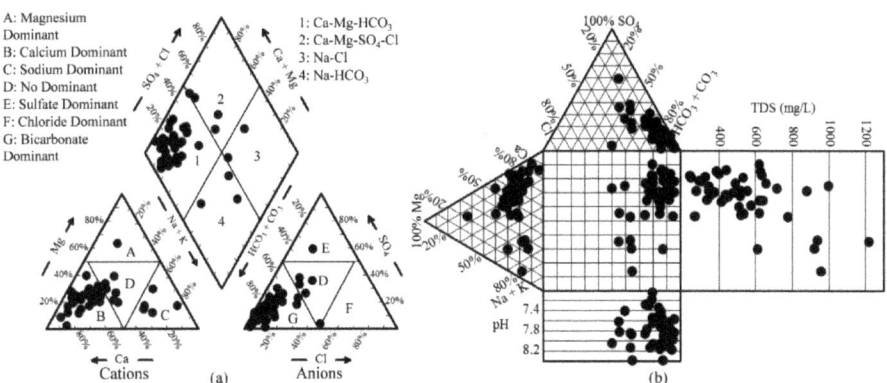

Figure 4. Piper (**a**) and Durov (**b**) diagrams showing the samples classifications.

3.4. Hydrochemical Correlation Analysis of Water Quality

To better understand the major hydrogeochemical processes that control the chemical characteristics, it is necessary to carry out a Pearson's correlation analysis that shows the relationship between each pair of physicochemical indices [39,73]. Table 3 gives the correlation values of physicochemical parameters of water samples.

Table 3. Pearson correlation matrix between physicochemical parameters of water samples.

	K^+	Na^+	Ca^{2+}	Mg^{2+}	NH_4^+	Cl^-	SO_4^{2-}	HCO_3^-	NO_3^-	NO_2^-	TDS	TH	pH	F^-	EC	Cr^{6+}
K^+	1	−0.030	0.164	0.056	0.099	0.036	0.061	0.150	0.163	−0.011	0.145	0.166	−0.110	−0.054	0.137	−0.06
Na^+		1	−0.186	0.313	−0.015	**0.727**	**0.745**	0.360	−0.097	0.108	**0.765**	−0.036	*0.287*	**0.602**	**0.742**	**0.375**
Ca^{2+}			1	0.103	0.346	0.255	0.225	0.334	**0.521**	0.017	**0.436**	*0.916*	*−0.721*	**−0.500**	**0.469**	*−0.354*
Mg^{2+}				1	**0.395**	**0.525**	*0.342*	0.271	**0.509**	0.335	**0.551**	**0.494**	0.143	0.188	**0.488**	0.23
NH_4^+					1	*0.311*	0.103	0.072	**0.431**	**0.482**	0.295	**0.462**	−0.115	−0.051	0.277	−0.06
Cl^-						1	**0.623**	0.243	**0.456**	0.304	*0.860*	**0.436**	0.083	0.343	**0.857**	0.14
SO_4^{2-}							1	0.148	−0.000	−0.025	**0.804**	0.335	−0.118	0.107	**0.804**	−0.02
HCO_3^-								1	0.012	0.054	**0.504**	**0.401**	−0.223	0.230	**0.469**	*0.340*
NO_3^-									1	0.364	**0.395**	**0.662**	−0.108	−0.045	**0.384**	−0.01
NO_2^-										1	0.207	0.151	0.259	0.347	0.141	*0.343*
TDS											1	**0.604**	−0.123	0.274	**0.980**	0.18
TH												1	**−0.571**	**−0.360**	**0.607**	−0.22
pH													1	**0.598**	−0.148	**0.379**
F^-														1	0.209	**0.703**
EC															1	0.10
Cr^{6+}																1

Bold number indicates that the correlation is significant at the 0.05 level (two-tailed). Italic number indicates that the correlation is significant at the 0.01 level (two-tailed).

As shown in Table 3, there is a strong correlation, which is explained by ions exchange between TDS and EC with $r = 0.980$ at the level of $p > 0.01$, Ca^{2+} content and TH with $r = 0.916$ at the level of $p > 0.01$, Cl^- and TDS with $r = 0.860$ at the level of $p > 0.01$, Cl^- and EC with $r = 0.857$ at the level of $p > 0.01$, and SO_4^{2-} correlates with TDS and EC with both $r = 0.804$ at the level of $p > 0.01$. In addition, a strong correlation exists between Na^+ and TDS, SO_4^{2-}, EC, and Cl^- with $r = 0.765, 0.745, 0.742$, and 0.727, respectively. Furthermore, there is a strong relationship between NO_3^- and TH with $r = 0.662$ at the level of $p > 0.01$, Cl^- and SO_4^{2-} with $r = 0.623$ at the level of $p > 0.01$, TH and EC with $r = 0.607$ at the level of $p > 0.01$, and Na^+ and F^- with $r = 0.602$ at the level of $p > 0.01$.

Although all the aforementioned correlations between parameters are positive, there is a strong negative correlation between Ca^{2+} and pH with $r = -0.721$ at the level of $p > 0.01$. Ca^{2+} and Mg^{2+} are significantly correlated to TH because they contribute to the water hardness.

A strong correlation between Cr^{6+} and F^- with $r = 0.703$ at level $p > 0.05$, which may be due to the oxidation mechanism of Cr^{3+} to Cr^{6+} in the presence of F^- in groundwater, was observed. Finally, a significant correlation between Cr^{6+} and both Na^+ and pH was also noticeable. All these parameters may have triggered the mobilization of Cr in the groundwater system [29].

3.5. Water Quality Index Assessment

Table 4 shows the weight assigned to each parameter, and the relative weights are calculated using Formula (3).

Table 4. Relative weight of physicochemical parameters. All units for all parameter indices are in mg/L, except pH (non-dimensional).

Parameters	Chinese Standards	Weight (w_i)	Relative Weight (W_i)
pH	6.5–8.5	4	0.0714
TH	450	5	0.0893
TDS	1000	5	0.0893
Na^+	200	2	0.0536
Cr^{6+}	0.05	5	0.0893
Cl^-	250	2	0.0357
SO_4^{2-}	250	4	0.0714
NO_3^-	20	5	0.0893
NH_4^+	0.5	5	0.0893
NO_2^-	1	4	0.0893
F^-	1	4	0.0714

Table 5 lists the water quality assessment results. As shown in Table 5, 37 of the samples (77.1%) are of excellent quality, 9 samples (18.7%) are of good quality, and 2 samples (4.2%) are of poor quality. The most significant parameters affecting the water quality in the study area are NO_3^-, F^-, and Cr^{6+}.

Water without excellent quality is dominated by wells with low depth represented by samples TW1-052, TW1-047, TW1-041, TW2-021, TW2-66 and TW2-67 with 2, 3, 2, 3, 10, and 8 m, respectively.

The contamination source of the wells represented by samples TW1-052, TW1-047, TW1-041 might be the ravines situated nearby. These ravines may bring contaminated water that leaks in the phreatic and shallow aquifer. The other concerned wells with low depth were possibly contaminated by human activities, as they are located in residential and agricultural areas.

As depicted in Figure 5, a major part of Tongchuan is dominated by excellent water and can be used for drinking purpose. However, in some towns such as Yuhua, Wangshiao, and Chenlu, for example, groundwater quality is not suitable for drinking. Therefore, water needs pretreatment before drinking, and taking effective measures to prevent groundwater pollution is imperative. Low deep wells should also be drilled deeply to avoid contamination by surface water leakage and pollution caused by human activities.

3.6. Health Risk Assessment

Table 6 presents the calculated health risk to adults and children when they are exposed to the contaminants in groundwater through drinking water intake. The total health risk due to contaminated drinking water intake ranges from 0.21 to 4.71, with a mean of 0.89 for adults. For children, the health risk is evaluated through the hazard quotient ranged from 0.35 to 7.85 with a mean of 1.52. Considering that HQ > 1 for non-carcinogenic risk indicates high potential health risk [1], water from wells represented by samples TW1-008, TW1-009, TW1-037, TW1-041, TW1-047, TW1-049 to TW1-054, TW1-059 to TW1-061, TW2-014 to TW2-067 was not safe, especially for children.

Table 5. Water quality index values and water types of the samples.

Samples	WQI	Water Quality	Samples	WQI	Water Quality	Samples	WQI	Water Quality
TW1-002	15.02	Excellent	TW1-038	22.99	Excellent	TW1-060	51.38	Good
TW1-003	14.75	Excellent	TW1-039	13.70	Excellent	TW1-061	26.24	Excellent
TW1-004	16.73	Excellent	TW1-041	51.76	Good	TW2-014	46.40	Excellent
TW1-005	20.98	Excellent	TW1-043	57.22	Good	TW2-018	64.38	Good
TW1-007	25.96	Excellent	TW1-046	24.53	Excellent	TW2-021	187.45	Poor
TW1-008	39.18	Excellent	TW1-047	76.75	Good	TW2-022	44.63	Excellent
TW1-009	41.89	Excellent	TW1-048	18.71	Excellent	TW2-037	17.05	Excellent
TW1-010	12.24	Excellent	TW1-049	51.04	Good	TW2-042	32.21	Excellent
TW1-012	30.86	Excellent	TW1-050	16.25	Excellent	TW2-043	13.98	Excellent
TW1-013	12.02	Excellent	TW1-051	14.03	Excellent	TW2-044	33.59	Excellent
TW1-014	18.41	Excellent	TW1-052	166.56	Poor	TW2-045	39.97	Excellent
TW1-023	13.20	Excellent	TW1-053	39.17	Excellent	TW2-057	23.71	Excellent
TW1-025	14.05	Excellent	TW1-054	59.63	Good	TW2-058	19.68	Excellent
TW1-032	16.13	Excellent	TW1-055	25.67	Excellent	TW2-066	54.70	Good
TW1-036	21.27	Excellent	TW1-058	23.88	Excellent	TW2-067	50.18	Good
TW1-037	38.25	Excellent	TW1-059	31.14	Excellent	TW2-069	15.90	Excellent

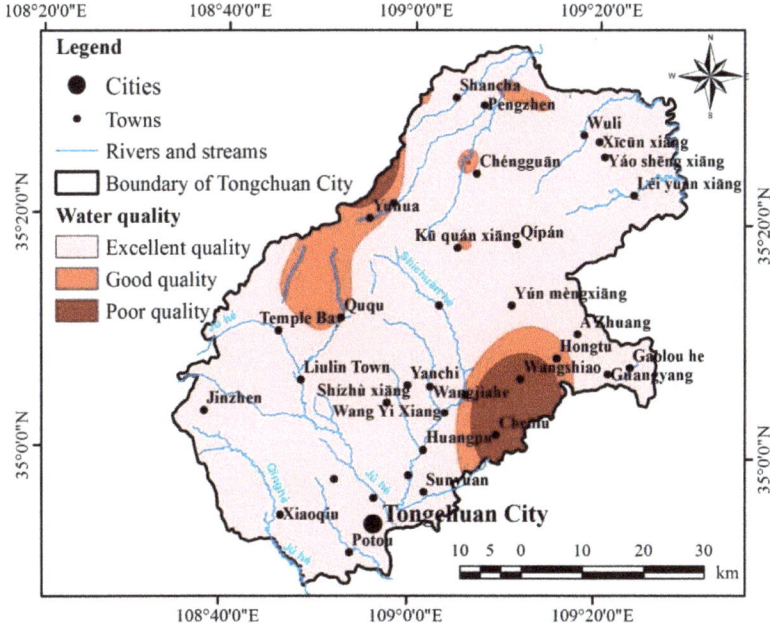

Figure 5. Water quality distribution in Tongchuan City.

As shown in Figure 6, NO_3^- contributes a considerable amount to non-carcinogenic risk for both adults and children and is followed by F^-, Cr^{6+}, and lastly, by NO_2^-.

The respective HQ mean values for adults are 0.54, 0.31, 0.02, 0.02, and 0.90, 0.52, 0.07, and 0.03 for children. HN_4^+ has zero contribution to health risk in this study area for both adults and children. High nitrate health risk is probably due to the anthropogenic activities, especially fertilizers in agriculture [21]. In addition, Wei et al. [34] reported that NO_3^- pollution was a major environmental geological problem in the groundwater for this region. Overall, 27.1% and 54.2% of the samples present a health risk through drinking water intake for adults and children, respectively.

Table 6. Calculated hazard quotient (HQ) of non-carcinogenic risk for adults and children.

Samples	Adults						Children					
	$HQ_{NH_4^+}$	$HQ_{NO_3^-}$	$HQ_{NO_2^-}$	HQ_{F^-}	$HQ_{Cr^{6+}}$	HQ_T	$HQ_{NH_4^+}$	$HQ_{NO_3^{2-}}$	$HQ_{NO_2^-}$	HQ_{F^-}	$HQ_{Cr^{6+}}$	HQ_T
TW1-002	0.00	0.06	0.00	0.15	0.00	0.21	0.00	0.10	0.01	0.24	0.00	0.35
TW1-003	0.00	0.06	0.00	0.34	0.00	0.41	0.00	0.10	0.00	0.57	0.00	0.68
TW1-004	0.00	0.00	0.12	0.25	0.11	0.48	0.00	0.00	0.19	0.42	0.21	0.82
TW1-005	0.00	0.00	0.01	0.34	0.00	0.36	0.00	0.00	0.02	0.57	0.00	0.59
TW1-008	0.00	0.37	0.00	0.31	0.00	0.68	0.00	0.61	0.01	0.52	0.00	1.13
TW1-009	0.00	0.26	0.01	0.50	0.00	0.76	0.00	0.43	0.01	0.84	0.00	1.28
TW1-010	0.00	0.00	0.00	0.27	0.00	0.27	0.00	0.00	0.00	0.45	0.00	0.45
TW1-012	0.00	0.11	0.00	0.28	0.00	0.39	0.00	0.18	0.00	0.47	0.00	0.65
TW1-013	0.00	0.07	0.00	0.26	0.00	0.33	0.00	0.12	0.00	0.43	0.00	0.55
TW1-014	0.00	0.18	0.00	0.13	0.00	0.32	0.00	0.31	0.00	0.22	0.00	0.53
TW1-023	0.00	0.00	0.01	0.28	0.00	0.29	0.00	0.00	0.01	0.47	0.00	0.48
TW1-025	0.00	0.12	0.00	0.29	0.00	0.41	0.00	0.19	0.00	0.48	0.00	0.68
TW1-032	0.00	0.25	0.00	0.20	0.00	0.45	0.00	0.42	0.01	0.33	0.00	0.76
TW1-036	0.00	0.15	0.00	0.24	0.00	0.39	0.00	0.25	0.00	0.40	0.00	0.65
TW1-037	0.00	0.61	0.01	0.18	0.00	0.81	0.00	1.01	0.02	0.31	0.00	1.34
TW1-038	0.00	0.29	0.00	0.21	0.00	0.50	0.00	0.48	0.00	0.35	0.00	0.84
TW1-039	0.00	0.09	0.00	0.30	0.00	0.40	0.00	0.15	0.01	0.51	0.00	0.67
TW1-041	0.00	1.08	0.01	0.21	0.00	1.30	0.00	1.80	0.02	0.35	0.00	2.16
TW1-043	0.00	0.33	0.00	0.32	0.00	0.66	0.00	0.55	0.00	0.54	0.00	1.09
TW1-046	0.00	0.07	0.00	0.19	0.00	0.26	0.00	0.12	0.00	0.32	0.00	0.44
TW1-047	0.00	2.15	0.01	0.24	0.00	2.39	0.00	3.58	0.01	0.40	0.00	3.99
TW1-048	0.00	0.00	0.00	0.33	0.00	0.33	0.00	0.00	0.00	0.55	0.00	0.55
TW1-049	0.00	1.16	0.00	0.25	0.00	1.42	0.00	1.94	0.01	0.42	0.00	2.36
TW1-050	0.00	0.12	0.00	0.28	0.00	0.40	0.00	0.21	0.00	0.46	0.00	0.67
TW1-051	0.00	0.00	0.00	0.30	0.00	0.30	0.00	0.00	0.00	0.51	0.00	0.51
TW1-052	0.01	3.12	0.12	0.21	0.00	3.46	0.01	5.20	0.20	0.35	0.00	5.76
TW1-053	0.00	0.75	0.00	0.25	0.00	1.01	0.00	1.25	0.01	0.42	0.00	1.68
TW1-054	0.00	1.12	0.00	0.17	0.00	1.30	0.00	1.87	0.00	0.29	0.00	2.16
TW1-055	0.00	0.36	0.00	0.20	0.00	0.55	0.00	0.59	0.00	0.33	0.00	0.92
TW1-058	0.00	0.25	0.00	0.12	0.00	0.37	0.00	0.41	0.01	0.20	0.00	0.61
TW1-059	0.00	0.51	0.00	0.18	0.00	0.69	0.00	0.85	0.00	0.31	0.00	1.16
TW1-060	0.00	1.32	0.01	0.19	0.00	1.52	0.00	2.21	0.01	0.32	0.00	2.54
TW1-061	0.00	0.38	0.00	0.24	0.00	0.63	0.00	0.63	0.00	0.41	0.00	1.05
TW2-014	0.00	0.78	0.00	0.91	0.27	1.96	0.00	1.30	0.00	1.52	1.04	3.86
TW2-018	0.00	0.28	0.10	1.54	0.23	2.16	0.00	0.47	0.17	2.58	0.90	4.11
TW2-021	0.00	4.32	0.06	0.32	0.00	4.71	0.00	7.21	0.11	0.54	0.00	7.85
TW2-022	0.00	0.98	0.00	0.24	0.00	1.23	0.00	1.64	0.00	0.41	0.00	2.05
TW2-037	0.00	0.20	0.01	0.44	0.00	0.64	0.00	0.33	0.01	0.73	0.00	1.07
TW2-042	0.01	0.40	0.05	0.36	0.00	0.83	0.01	0.67	0.09	0.61	0.00	1.37
TW2-043	0.00	0.08	0.00	0.29	0.00	0.37	0.00	0.14	0.00	0.48	0.00	0.62
TW2-044	0.00	0.43	0.10	0.28	0.02	0.83	0.00	0.71	0.17	0.46	0.09	1.43
TW2-045	0.00	0.69	0.08	0.49	0.00	1.26	0.00	1.15	0.13	0.81	0.00	2.10
TW2-057	0.00	0.34	0.02	0.39	0.00	0.75	0.00	0.57	0.04	0.65	0.00	1.26
TW2-058	0.00	0.19	0.00	0.36	0.02	0.57	0.00	0.32	0.00	0.59	0.09	1.01
TW2-066	0.00	1.13	0.01	0.32	0.00	1.45	0.00	1.88	0.01	0.53	0.00	2.42
TW2-067	0.00	0.62	0.06	0.32	0.25	1.25	0.00	1.03	0.10	0.54	0.97	2.64
TW2-069	0.00	0.00	0.00	0.29	0.00	0.29	0.00	0.00	0.00	0.48	0.00	0.48
Min	0.00	0.00	0.00	0.12	0.00	0.21	0.00	0.00	0.00	0.20	0.00	0.35
Max	0.01	4.32	0.12	1.54	0.27	4.71	0.01	7.21	0.20	2.58	1.04	7.85
Mean	0.00	0.54	0.02	0.31	0.02	0.89	0.00	0.90	0.03	0.52	0.07	1.52

In this study, Cr^{6+} was also considered as a carcinogenic risk pollutant. Considering the acceptable CR_{total} limit set as 1×10^{-6} by the Ministry of Environmental Protection of the P.R. China [30], the results shown in Table 7 revealed a critical carcinogenic risk by drinking and daily contact of water from six (12.5%) wells in the study area.

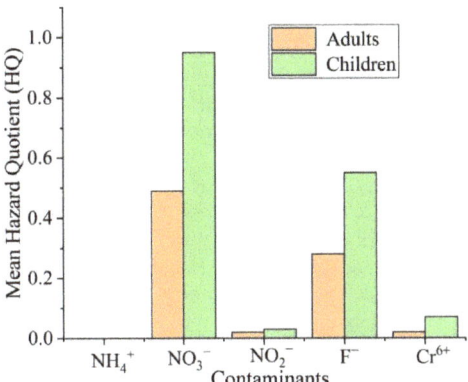

Figure 6. Representation of the mean HQ for non-carcinogenic (NH_4^+, NO_3^-, NO_2^-, F^-, Cr^{6+}) contaminants.

Table 7. Calculated carcinogenic risk due to Cr^{6+} in water intake and dermal contact.

Samples	Adults	Children	Samples	Adults	Children	Samples	Adults	Children
TW1-002	0	0	TW1-038	0	0	TW1-060	0	0
TW1-003	0	0	TW1-039	0	0	TW1-061	0	0
TW1-004	1.00×10^{-4}	2.00×10^{-4}	TW1-041	0	0	TW2-014	4.95×10^{-4}	8.16×10^{-4}
TW1-005	0	0	TW1-043	0	0	TW2-018	4.25×10^{-4}	7.01×10^{-4}
TW1-007	0	0	TW1-046	0	0	TW2-021	0	0
TW1-008	0	0	TW1-047	0	0	TW2-022	0	0
TW1-009	0	0	TW1-048	0	0	TW2-037	0	0
TW1-010	0	0	TW1-049	0	0	TW2-042	0	0
TW1-012	0	0	TW1-050	0	0	TW2-043	0	0
TW1-013	0	0	TW1-051	0	0	TW2-044	4.18×10^{-5}	6.89×10^{-5}
TW1-014	0	0	TW1-052	0	0	TW2-045	0	0
TW1-023	0	0	TW1-053	0	0	TW2-057	0	0
TW1-025	0	0	TW1-054	0	0	TW2-058	4.18×10^{-5}	6.89×10^{-5}
TW1-032	0	0	TW1-055	0	0	TW2-066	0	0
TW1-036	0	0	TW1-058	0	0	TW2-067	4.60×10^{-4}	7.58×10^{-4}
TW1-037	0	0	TW1-059	0	0	TW2-069	0	0

CR_{total} ranges from 4.18×10^{-5} to 4×10^{-4} for adults and from 6.89×10^{-5} to 8×10^{-4} for children. Similar results have also been found by He and Wu [74], Li et al. [75], Wu and Sun [38], Liu et al. [76], Ji et al. [77], and He et al. [78] in their study on groundwater quality and health risk assessment, which confirmed the health threats faced by the population, especially for children in the loess area of northwest China. According to WHO [42], the excessive Cr^{6+} concentration in drinking water can cause lung cancer via inhalation route. Groundwater from wells represented by samples TW1-004, TW2-014, TW2-018, TW2-044, TW2-058, and TW2-067 with CR_{total} values of more than 1×10^{-6} must be used with precaution for drinking purposes.

4. Conclusions

In the present study, water quality index (WQI), statistical analysis and graphical approaches were implemented to understand the status of groundwater quality in the Tongchuan area on the Loess Plateau, northwest China. In addition, GIS approaches helped to map the WQI results of this study. Daily average exposure dosage through oral pathway was taken into consideration to calculate health risks to the human body through drinking water intake. For dermal contact, the exposure dosage of every single event in mg/cm^2 and the skin surface (cm^2) were considered. The following conclusions can be achieved:

- In summary, the results of this study demonstrated that groundwater in the study area is suitable for drinking in general. WQI approach showed that 77.1% of the samples are of excellent quality, nine samples (18.7%) are of good quality, and two samples (4.2%) are of poor quality.
- NO_3^-, F^-, and Cr^{6+} are the most significant parameters affecting water quality in this study; 27.1% and 54.2% of the overall samples present a non-carcinogenic health risk through drinking water intake for adults and children, respectively. The CR_{total} of 12.5% of the samples ranges from 4.18×10^{-5} to 4×10^{-4} for adults and from 6.89×10^{-5} to 8×10^{-4} for children, which exceeded the acceptable limit (1×10^{-6}).
- NO_3^- considerably contributes to non-carcinogenic risk for both adults and children and is followed by F^-, Cr^{6+} and lastly by NO_2^-, with respective mean HQ of 0.49, 0.28, 0.02 and 0.02 for adults. For children, the mean HQ for NO_3^-, NO_2^-, F^- and Cr^{6+} are 0.95, 0.03, 0.55 and 0.07, respectively. HN_4^+ has zero contribution to health risk in this study area for both adults and children. The high concentration of NO_3^- in the study area is due to anthropogenic activities, especially fertilizers in agriculture as also discussed by previous researchers.
- WQI is not enough to conclude whether water is suitable or not for drinking. The assessment of carcinogenic and non-carcinogenic risk on the human body showed that groundwater in Tongchuan was not totally safe. Therefore, water pretreatment before drinking and taking effective measures to prevent groundwater pollution are imperative.

This study will be helpful to local decision makers for implementing measures, policy and strategies to protect groundwater resources and reduce the health risks of residents by groundwater consumption through oral and dermal pathways. It is also useful for international scholars who may find information for similar studies or its improvement.

Author Contributions: Research conceptualization, A.N. and P.L.; data curation, S.H., X.H., and A.N.; methodology, A.N.; writing—original draft, A.N.; writing—review and editing, P.L., S.H., X.H., S.M.K.A., and M.F., supervision, P.L. All authors have read and agreed to the published version of the manuscript.

Funding: This work was supported by the National Natural Science Foundation of China (42072286 and 41761144059), the Fundamental Research Funds for the Central Universities of CHD (300102299301 and 300102291507), the Fok Ying Tong Education Foundation (161098), and the Ten Thousand Talents Program (W03070125).

Institutional Review Board Statement: Not applicable.

Informed Consent Statement: Not applicable.

Data Availability Statement: All processed data generated or used during the study appear in the submitted article. Raw data may be provided on reasonable request from the corresponding author.

Acknowledgments: We acknowledge the members of the Key Laboratory of Subsurface Hydrology and Ecological Effects in Arid Region of the Ministry of Education, Chang'an University, Shaanxi, China for the groundwater samples processing. The useful and constructive comments from the editors and reviewers are sincerely acknowledged.

Conflicts of Interest: The authors declare no conflict of interest.

References

1. Li, P.; Meng, X.; Li, M.; Zhang, Y. Appraising groundwater quality and health risks from contamination in a semiarid region of Northwest China. *Expo. Health* **2016**, *8*, 361–379. [CrossRef]
2. Li, W.; Wu, J.; Zhou, C.; Nsabimana, A. Groundwater pollution source identification and apportionment using PMF and PCA-APCS-MLR receptor models in Tongchuan City, China. *Arch. Environ. Contam. Toxicol.* **2021**, *81*, 397–413. [CrossRef] [PubMed]

3. Wei, M.; Wu, J.; Li, W.; Zhang, Q.; Su, F.; Wang, Y. Groundwater geochemistry and its impacts on groundwater arsenic enrichment, variation, and health risks in Yongning County, Yinchuan Plain of northwest China. *Expo. Health* **2021**. [CrossRef]
4. Liu, J.; Gao, M.; Jin, D.; Wang, T.; Yang, J. Assessment of groundwater quality and human health risk in the aeolian-sand area of Yulin City, Northwest China. *Expo. Health* **2020**, *12*, 671–680. [CrossRef]
5. Wang, D.; Wu, J.; Wang, Y.; Ji, Y. Finding high-quality groundwater resources to reduce the hydatidosis incidence in the Shiqu county of Sichuan province, China: Analysis, assessment, and management. *Expo. Health* **2020**, *12*, 307–322. [CrossRef]
6. Zhang, Q.; Li, P.; Lyu, Q.; Ren, X.; He, S. Groundwater contamination risk assessment using a modified DRATICL model and pollution loading: A case study in the Guanzhong Basin of China. *Chemosphere* **2021**. [CrossRef]
7. Mfonka, Z.N.; Ngoupayou, J.R.N.; Kpoumie, A.; Ndjigui, P.D.; Zammouri, M.; Ngouh, A.N.; Mouncherou, O.F.; Mfochive, O.F.; Rakotondrabe, F. Hydrodynamic and groundwater vulnerability assessment of the shallow aquifer of the Foumban locality (Bamoun plateau, Western-Cameroon). *Arab J. Geosci.* **2019**, *12*, 165. [CrossRef]
8. Luque-Espinar, J.A.; Chica-Olmo, M. Impacts of anthropogenic activities on groundwater quality in a detritic aquifer in SE Spain. *Expo. Health* **2020**, *12*, 681–698. [CrossRef]
9. Mthembu, P.P.; Elumalai, V.; Brindha, K.; Li, P. Hydrogeochemical processes and trace metal contamination in groundwater: Impact on human health in the Maputaland coastal aquifer, South Africa. *Expo. Health* **2020**, *12*, 403–426. [CrossRef]
10. Wang, L.; Li, P.; Duan, R.; He, X. Occurrence, controlling factors and health risks of Cr^{6+} in groundwater in the Guanzhong Basin of China. *Expo. Health* **2021**. [CrossRef]
11. Li, Y.; Li, P.; Cui, X.; He, S. Groundwater quality, health risk and major influencing factors in the lower Beiluo River watershed of northwest China. *Hum. Ecol. Risk Assess.* **2021**, *27*, 1987–2013. [CrossRef]
12. Li, Y.; Li, P.; Liu, L. Source identification and potential ecological risk assessment of heavy metals in the topsoil of the Weining Plain (northwest China). *Expo. Health* **2021**. [CrossRef]
13. Moya, C.E.R.; Raiber, M.; Taulis, M.; Cox, M.E. Hydrochemical evolution and groundwater flow processes in the Galilee and Eromanga basins, Great Artesian Basin, Australia: A multivariate statistical approach. *Sci. Total Environ.* **2015**, *508*, 411–426. [CrossRef]
14. World Health Organization (W.H.O). *Guidelines for Drinking Water Quality*, 4th ed.; WHO: Geneva, Switzerland, 2011.
15. Li, P.; He, S.; Yang, N.; Xiang, G. Groundwater quality assessment for domestic and agricultural purposes in Yan'an City, northwest China: Implications to sustainable groundwater quality management on the Loess Plateau. *Environ. Earth Sci.* **2018**, *77*, 775. [CrossRef]
16. Singh, S.; Janaedhana, R.N.J.; Ramakrishna, C. Evaluation of groundwater quality and its suitability for domestic and irrigation use in parts of the Chandauli-Varanasi region, Uttar Pradesh, India. *J. Water Res. Prot.* **2015**, *7*, 572–582. [CrossRef]
17. Abiriga, D.; Vestgarden, L.; Klempe, H. Groundwater contamination from a municipal landfill: Effect of age, landfill closure, and season on groundwater chemistry. *Sci. Total Environ.* **2020**, *737*, 140307. [CrossRef]
18. Beyene, G.; Fufa, F.; Aberra, D. Evaluation of the suitability of groundwater for drinking and irrigation purposes in Jimma Zone of Oromia, Ethiopia. *Groundw. Sustain. Dev.* **2019**, *9*, 100216. [CrossRef]
19. Ketchemen-Tandia, B.; Boum-Nkot, S.; Ebondji, S.; Nlend, B.; Emvoutou, H.; Nzegue, O. Factors Influencing the shallow groundwater quality in four districts with different characteristics in urban area (Douala, Cameroon). *J. Geosci. Environ. Prot.* **2017**, *5*, 99–120. [CrossRef]
20. Smahi, D.; Hammoumi, O.; Fekri, A. Assessment of the impact of the landfill on groundwater quality: A case study of the Mediouna site, Casablanca, Morocco. *J. Water Res. Prot.* **2013**, *5*, 440–445. [CrossRef]
21. Li, P.; Tian, R.; Liu, R. Solute geochemistry and multivariate analysis of water quality in the Guohua Phosphorite Mine, Guizhou Province, China. *Expo. Health* **2019**, *11*, 81–94. [CrossRef]
22. Tiwari, A.K.; Orioli, S.; De Maio, M. Assessment of groundwater geochemistry and diffusion of hexavalent chromium contamination in an industrial town of Italy. *J. Contam. Hydrol.* **2019**, *225*, 103503. [CrossRef] [PubMed]
23. Wu, J.; Li, P.; Qian, H. Environmental chemistry of groundwater near an industrial area, Northwest China. *Asian J. Chem.* **2013**, *25*, 9795–9799. [CrossRef]
24. Zacchaeus, O.O.; Adeyemi, M.B.; Adedeji, A.A.; Adegoke, K.A.; Anumah, A.O.; Taiwo, A.M.; Ganiyu, S.A. Effects of industrialization on groundwater quality in Shagamu and Ota industrial areas of Ogun state, Nigeria. *Heliyon* **2020**, *6*, e04353. [CrossRef] [PubMed]
25. Papazotos, P.; Vasileiou, E.; Perraki, M. The synergistic role of agricultural activities in groundwater quality in ultramafic environments: The case of the Psachna basin, central Euboea, Greece. *Environ. Monit. Assess.* **2019**, *191*, 317. [CrossRef]
26. Vasileiou, E.; Papazotos, P.; Dimitrakopoulos, D.; Perraki, M. Hydrogeochemical processes and natural background levels of chromium in an ultramafic environment. The case study of Vermio mountain, Western Macedonia, Greece. *Water* **2021**, *13*, 2809. [CrossRef]
27. Chen, J.; Qian, H.; Wu, H.; Gao, Y.; Li, X. Assessment of arsenic and fluoride pollution in groundwater in Dawukou area, Northwest China, and the associated health risk for inhabitants. *Environ. Earth Sci.* **2017**, *73*, 314. [CrossRef]
28. Zissimos, A.M.; Christoforou, I.C.; Christofi, C.; Rigas, M.; Georgiadou, E.C.; Christou, A. Occurrence and distribution of hexavalent chromium in ground and surface waters in Cyprus. *Bull. Environ. Contam. Toxicol.* **2021**, *106*, 428–434. [CrossRef]

29. Papazotos, P.; Vasileiou, E.; Perraki, M. Elevated groundwater concentrations of arsenic and chromium in ultramafic environments controlled by seawater intrusion, the nitrogen cycle, and anthropogenic activities: The case of the Gerania Mountains, NE Peloponnese, Greece. *Appl. Geochem.* **2020**, *121*, 104697. [CrossRef]
30. Ministry of Environmental Protection of the P.R. China. *Technical Guidelines for Risk Assessment of Contaminated Sites, HJ 25.3-2014*; China Environmental Science Press: Beijing, China, 2014.
31. Adimalla, N.; Vasa, S.K.; Li, P. Evaluation of groundwater quality, Peddavagu in Central Telangana (PCT), South India: An insight of controlling factors of fluoride enrichment. *Modeling Earth Syst. Environ.* **2018**, *4*, 841–852. [CrossRef]
32. Choi, B.Y.; Yun, S.T.; Yu, S.Y.; Lee, P.K.; Park, S.S.; Chae, G.T. Hydrochemistry of urban groundwater in Seoul, South Korea: Effect of land use and pollutant recharge. *Environ. Geol.* **2005**, *48*, 979–990. [CrossRef]
33. Johnson, R. Drinking water quality: Testing and interpreting your results. In *NDSU*; USDA-NIFA: North Dakota State University: Fargo, ND, USA, 2019.
34. Wei, Y.; Fan, W.; Wang, W.; Deng, L. Identification of nitrate pollution sources of groundwater and analysis of potential pollution paths in loess regions: A case study in Tongchuan region, China. *Environ. Earth Sci.* **2017**, *76*, 423. [CrossRef]
35. Wang, Y.; Li, P. Appraisal of shallow groundwater quality with human health risk assessment in different seasons in rural areas of the Guanzhong Plain (China). *Environ. Res.* **2021**. [CrossRef]
36. Wu, C.; Fang, C.; Wu, X.; Zhu, G. Health-risk assessment of arsenic and groundwater quality classification using random forest in the Yanchi Region of Northwest China. *Expo. Health* **2020**, *12*, 761–774. [CrossRef]
37. Ni, F.; Liu, G.; Ye, J.; Ren, H.; Yang, S. ArcGIS-based rural drinking water quality health risk assessment. *J. Water Res. Prot.* **2009**, *1*, 351–361. [CrossRef]
38. Wu, J.; Sun, Z. Evaluation of shallow groundwater contamination and associated human health risk in an alluvial plain impacted by agricultural and industrial activities, mid-west China. *Expo. Health* **2016**, *8*, 311–329. [CrossRef]
39. Li, X.; Wu, H.; Qian, H.; Gao, Y. Groundwater chemistry regulated by hydrochemical processes and geological structures: A case study in Tongchuan, China. *Water* **2018**, *10*, 338. [CrossRef]
40. Xiao, J.; Jin, Z.D.; Zhang, F. Geochemical controls on fluoride concentrations in natural waters from the middle Loess Plateau, China. *J. Geochem. Explor.* **2015**, *159*, 252–261. [CrossRef]
41. World Health Organization (W.H.O). *Guidelines for Drinking Water Quality*, 3rd ed.; Incorporating the First and Second Addendum; WHO: Geneva, Switzerland, 2008.
42. World Health Organization (W.H.O). *Guidelines for Drinking-Water Quality*, 4th ed.; Incorporating the First Addendum; WHO: Geneva, Switzerland, 2017.
43. General Administration of Quality Supervision, Inspection and Quarantine of China, Standardization Administration of China. *Standards for Groundwater Quality (GB/T 14848–2017.)*; General Administration of Quality Supervision: Beijing, China, 2017.
44. Vasileiou, E.; Papazotos, P.; Dimitrakopoulos, D.; Perraki, M. Expounding the origin of chromium in groundwater of the Sarigkiol basin, Western Macedonia, Greece: A cohesive statistical approach and hydrochemical study. *Environ. Monit. Assess.* **2019**, *191*, 509. [CrossRef]
45. Piper, A.M. A graphic procedure in the geochemical interpretation of water analysis. *Trans. Am. Geophys. Union* **1954**, *25*, 914–928. [CrossRef]
46. Durov, S.A. Natural waters and graphic representation of their composition. *Dokl. Akad. Nauk. SSSR* **1948**, *59*, 87–90.
47. Boufekane, A.; Saighi, O. Assessing groundwater quality for irrigation using geostatistical method—Case of Wadi Nil Plain (North-East Algeria). *Groundw. Sustain. Dev.* **2019**, *8*, 179–186. [CrossRef]
48. Bordalo, A.A.; Teixeira, R.; Wiebe, W.J. A water quality index applied to an international shared river basin: The case of the Douro River. *Environ. Manag.* **2006**, *38*, 910–920. [CrossRef] [PubMed]
49. Li, P.; Qian, H.; Wu, J. Groundwater quality assessment based on improved water quality index in Pengyang County, Ningxia, Northwest China. *E-J. Chem.* **2010**, *7*, S209–S216. [CrossRef]
50. Şener, Ş.; Şener, E.; Davraz, A. Assessment of groundwater quality and health risk in drinking water basin using GIS. *J. Water Health* **2016**, *15*, 112–132. [CrossRef] [PubMed]
51. Varol, S.; Davraz, A. Evaluation of potential human health risk and investigation of drinking water quality in Isparta city center (Turkey). *J. Water Health* **2015**, *14*, 471–488. [CrossRef] [PubMed]
52. Singh, S.; Janardhana, R.N.J.; Ramakrishna, C. Assessment and monitoring of groundwater quality in semi-arid region. *Groundw. Sustain. Dev.* **2020**, *11*, 100381. [CrossRef]
53. Tirkey, P.; Bhattacharya, T.; Chakraborty, S. Water quality indices- important tools for water quality assessment: A review. *Int. J. Adv. Chem.* **2013**, *1*, 15–28. [CrossRef]
54. Hamed, S.; Omid, N.; Mahbobehv, G.; Abooalfazl, A.; Mansooreh, D.; Majid, R.; Mohammad, D.; Vahide, O.; Maryam, H. Groundwater quality evaluation and risk assessment of nitrate using monte carlo simulation and sensitivity analysis in rural areas of Divandarreh County, Kurdistan province, Iran. *Int. J. Environ. Anal. Chem.* **2020**. [CrossRef]
55. U.S.E.P.A. Supplemental Environmental Projects (SEP) Policy 2015 Update. EPA. 2015: Washington. Available online: https://www.epa.gov/sites/default/files/2015-04/documents/sepupdatedpolicy15.pdf (accessed on 13 October 2020).
56. Hem, J.D. Study and Interpretation of the Chemical Characteristics of Natural Water. US Geological Survey. 3 ed, Water-Supply, 1985. Paper 2254, 263. Available online: https://pubs.usgs.gov/wsp/wsp2254/html/pdf.html (accessed on 21 November 2020).
57. Mechenich, C.; Andrews, E. Evaluating the condition of your private water supply. *Interpreting* **1993**, *3558*, 1–8.

58. Freeze, R.A.; Cherry, J.A. *Grounndwater*; Prentice-Hall: Englewood Cliffs, NJ, USA, 1979; p. 604.
59. Davis, S.N.; De Wiest, R.J.M. *Hydrogeology*; John Wiley and Sons: Hoboken, NJ, USA, 1966; Volume 463, p. 824.
60. Adimalla, N. Occurrence, health risks, and geochemical mechanisms of fluoride and nitrate in groundwater of the rock-dominant semi-arid region, Telangana State, India. *Hum. Ecol. Risk Assess.* **2018**, *25*, 81–103. [CrossRef]
61. Mohan, D.S.R.; Singh, V.K.; Steele, P.; Pittman, C.U. Fluoride removal from water using bio-char, a green waste, low-cost adsorbent: Equilibrium uptake and sorption dynamics modeling. *Ind. Eng. Chem. Res.* **2012**, *51*, 900–914. [CrossRef]
62. Emenike, C.P.; Tenebe, I.T.; Jarvis, P. Fluoride contamination in groundwater sources in Southwestern Nigeria: Assessment using multivariate statistical approach and human health risk. *Ecotoxicol. Environ. Saf.* **2018**, *156*, 391–402. [CrossRef]
63. Kaoud, H.; Kalifa, B. Effect of fluoride, cadmium and arsenic intoxication on brain and learning–memory ability in rats. *Toxicol. Lett.* **2010**, *196*, S53. [CrossRef]
64. Kimambo, V.; Bhattacharya, P.; Mtalo, F.; Mtamba, J.; Ahmad, A. Fluoride occurrence in groundwater systems at global scale and status of defluoridation—State of the art. *Groundw. Sustain. Dev.* **2019**, *9*, 100223. [CrossRef]
65. Sivasankar, V.; Darchen, A.; Omine, K.; Sakthivel, R. Fluoride: A world ubiquitous compound, its chemistry, and ways of contamination. In *Surface Modified Carbons as Scavengers for Fluoride from Water*; Sivasankar, V., Ed.; Springer: Cham, Switzerland, 2016; pp. 5–32. [CrossRef]
66. Vithanage, M.; Bhattacharya, P. Fluoride in drinking water: Health effects and remediation. In *CO_2 Sequestration Biofuels and Depollution*; Lichtfouse, E., Schwarzbauer, J., Robert, D., Eds.; Springer: Cham, Switzerland, 2015; pp. 105–151. [CrossRef]
67. Saxena, V.; Ahmed, S. Inferring the chemical parameters for the dissolution of fluoride in groundwater. *Environ. Geol.* **2003**, *43*, 731–736. [CrossRef]
68. Rango, T.; Kravchenko, J.; Atlaw, B.; McCornick, P.G.; Jeuland, M.; Merola, B.; Vengosh, A. Groundwater quality and its health impact: An assessment of dental fluorosis in rural inhabitants of the Main Ethiopian Rift. *Environ. Int.* **2012**, *43*, 37–47. [CrossRef]
69. Marghade, D.; Malpe, D.B.; Zade, A.B. Major ion chemistry of shallow groundwater of a fast growing city of Central India. *Environ. Monit. Assess.* **2012**, *184*, 2405–2418. [CrossRef]
70. Ravikumar, P.; Somashekar, R.K.; Prakash, K.L. A comparative study on usage of Durov and Piper diagrams to interpret hydrochemical processes in groundwater from SRLIS river basin, Karnataka, India. *Elixir Earth Sci.* **2015**, *80*, 31073–31077.
71. Lloyd, J.A.; Heathcote, J.A. *Natural Inorganic Hydrochemistry in Relation to Groundwater: An Introduction*; O.U. Press: New York, NY, USA, 1985; p. 296.
72. Xu, P.; Feng, W.; Qian, H.; Zhang, Q. Hydrogeochemical characterization and irrigation quality assessment of shallow groundwater in the Central-Western Guanzhong Basin, China. *Int. J. Environ. Res. Public Health* **2019**, *16*, 1492. [CrossRef]
73. Singh, S.; Janardhana, R.N.J.; Ramakrishna, C. Hydrogeochemical assessment of surface and groundwater resources of Korba coalfield, Central India: Environmental implications. *Arab J. Geosci.* **2017**, *10*, 318. [CrossRef]
74. He, S.; Wu, J. Hydrogeochemical characteristics, groundwater quality, and health risks from hexavalent chromium and nitrate in groundwater of Huanhe formation in Wuqi County, Northwest China. *Expo. Health* **2019**, *11*, 125–137. [CrossRef]
75. Li, P.; He, X.; Guo, W. Spatial groundwater quality and potential health risks due to nitrate ingestion through drinking water: A case study in Yan'an City on the Loess Plateau of northwest China. *Hum. Ecol. Risk Assess.* **2019**, *25*, 11–31. [CrossRef]
76. Liu, L.; Wu, J.; He, S.; Wang, L. Occurrence and distribution of groundwater fluoride and manganese in the Weining Plain (China) and their probabilistic health risk quantification. *Expo. Health* **2021**. [CrossRef]
77. Ji, Y.; Wu, J.; Wang, Y.; Elumalai, V.; Subramani, T. Seasonal variation of drinking water quality and human health risk assessment in Hancheng City of Guanzhong Plain, China. *Expo. Health* **2020**, *12*, 469–485. [CrossRef]
78. He, X.; Wu, J.; He, S. Hydrochemical characteristics and quality evaluation of groundwater in terms of health risks in Luohe aquifer in Wuqi County of the Chinese Loess Plateau, northwest China. *Hum. Ecol. Risk Assess.* **2019**, *25*, 32–51. [CrossRef]

Article

Groundwater Quality and Associated Human Health Risk in a Typical Basin of the Eastern Chinese Loess Plateau

Jiao Li, Congjian Sun *, Wei Chen, Qifei Zhang, Sijie Zhou, Ruojing Lin and Yihan Wang

School of Geographical Science, Shanxi Normal University, 339 Taiyu Street, Taiyuan 030031, China; zimoqianji@126.com (J.L.); wan_xin_chen@126.com (W.C.); zhangqifei15@mails.ucas.ac.cn (Q.Z.); 15903424981sj@sina.com (S.Z.); lrj476985464@sina.com (R.L.); w1209243078@163.com (Y.W.)
* Correspondence: suncongjian@sina.com

Abstract: Groundwater is an important source for drinking, agricultural, and industrial purposes in the Linfen basin of the Eastern Chinese Loess Plateau (ECLP). To ensure the safety of drinking water, this study was carried out to assess the quality using the water quality index (WQI) and potential health risks of groundwater using the human health risk assessment model (HHRA). The WQI approach showed that 90% of the samples were suitable for drinking, and Pb, TH, F^-, SO_4^{2-}, and TDS were the most significant parameters affecting groundwater quality. The non-carcinogenic health risk results indicated that 20% and 80% of the samples surpassed the permissible limit for adult females and children. Additionally, all groundwater samples could present a carcinogenic health risk to males, females, and children. The pollution from F^-, Pb, and Cr^{6+} was the most serious for non-carcinogenic health risk. Cd contributed more than Cr^{6+} and As to carcinogenic health risks. Residents living in the central of the study area faced higher health risks than humans in other areas. The research results can provide a decision-making basis for the scientific management of the regional groundwater environment and the protection of drinking water safety and public health.

Keywords: water environment; human health risk; spatial distribution; Chinese Loess Plateau

Citation: Li, J.; Sun, C.; Chen, W.; Zhang, Q.; Zhou, S.; Lin, R.; Wang, Y. Groundwater Quality and Associated Human Health Risk in a Typical Basin of the Eastern Chinese Loess Plateau. *Water* 2022, 14, 1371. https://doi.org/10.3390/w14091371

Academic Editors: Jianhua Wu, Peiyue Li and Saurabh Shukla

Received: 25 March 2022
Accepted: 16 April 2022
Published: 22 April 2022

Publisher's Note: MDPI stays neutral with regard to jurisdictional claims in published maps and institutional affiliations.

Copyright: © 2022 by the authors. Licensee MDPI, Basel, Switzerland. This article is an open access article distributed under the terms and conditions of the Creative Commons Attribution (CC BY) license (https://creativecommons.org/licenses/by/4.0/).

1. Introduction

Groundwater is an indispensable part of human living space and the hydrological cycle, providing high-quality freshwater resources for human beings. It is important for domestic, industrial, and agricultural use globally [1–4]. For drinking purposes, approximately one-third of the world's population rely on groundwater as a water source [5–7]. Especially in arid and semi-arid areas where the precipitation is scarce and the surface water sources are limited, groundwater has become the main water source, or even the only one [2,8]. As the most important water source for human survival, groundwater quality is vital to human health. However, with the continuous population growth and rapid economic development, groundwater pollution has become an urgent problem endangering public health and has put pressure on groundwater resources worldwide [9,10]. For example, studies have shown that 2 types of birth defects and 15 types of cancer may be related to long-term exposure to NO_3^- contaminated groundwater [11–13]. Even at the same groundwater NO_3^- concentration, children and infants have greater health risks than adults, especially infants prone to a disease known as "blue baby syndrome", i.e., methemoglobinemia [14,15]. Fluoride is a major pollutant in groundwater on a global scale as about 260 million people suffer from endemic fluorosis and other diseases due to the intake of high fluoride in groundwater [16,17]. Potentially toxic elements (PTEs) in groundwater can cumulate in the human body throughout almost the human lifespan and cause many diseases, a matter of great concern for the past several years [18–21]. Anthropogenic sources of groundwater pollutants include fertilization, livestock waste, domestic sewage, landfill, metal industry, mining, and other industrial activities. Processes controlling concentrations of physicochemical parameters in groundwater are mainly the mineral dissolution, sorption and desorption processes, ion exchange, reduction and oxidation processes, and chemical weathering [22–26].

Groundwater environment assessment is the basis of sustainable utilization of regional groundwater resources and is of great significance to ecological environment protection. Various scientific approaches have been introduced to assess groundwater quality. Some of these methods include set pair analysis [27], hierarchical analysis [28], matter-element extension analysis [29], fuzzy comprehensive assessment method [2], and water quality index (WQI) [6,8]. The WQI is an efficient tool to access water quality and its suitability for drinking purposes. It was first developed by Horton [30] and since has been widely used in numerous water quality assessment works [31–35]. Varol and Davraz [6] used WQI and multivariate analysis to evaluate groundwater quality and its suitability for drinking and agricultural uses in the Tefenni plain, Turkey. Using an improved water quality index, Zhang et al. [36] considered that groundwater will be affected by the geological environment and human factors during the flow process in Guanzhong Basin, China. In recent years, groundwater quality assessment and spatial analysis based on combining Geographic Information System (GIS) with WQI methods have proven to be a powerful tool for spatial information management of groundwater resources [37–39].

Many scholars have also carried out human health risk assessments (HHRA) to directly and quantitatively reflect the negative health impacts of polluted water on human beings. This method has been widely used in the evaluation of different water bodies, such as rivers [40,41], lakes [42], and wetlands [43], which provide useful insight to ensure human health. For groundwater, Guo et al. [44] found that groundwater arsenic pollution caused by landfill leachate leakage poses unacceptable carcinogenic risks to people of all ages. Farmers continually applying fertilizers during the period between the rainy and dry season leads to the mobilization of NO_3^- and PTEs from cultivated soils to groundwater under favoring geochemical conditions in the dry season. Therefore, the non-carcinogenic risk in the dry season is higher than in the rainy season [1,45–48]. Kaur et al. [12] suggested that the hazard quotient values determined by deterministic and probabilistic approaches were nearly identical, and groundwater in most of the Panipat district in India is not suitable for direct drinking purposes.

The Chinese Loess Plateau (CLP) is a cradle of human civilization, where the groundwater plays an important role in the residents' lives and industrial and agricultural production. Due to the arid climate and increasing human activities, there is a serious shortage of water resources and a significant decline in water quality in the CLP [40]. Recently, health risks due to different water pollutants have been assessed on the CLP, such as fluoride [16], nitrogen [49], and arsenic [50]. However, these studies were mainly concentrated in the middle of the CLP. As for the Eastern Chinese Loess Plateau (Shanxi Province), the status of the groundwater environment and the threat of pollutants to human health are still unclear.

Therefore, this study was carried out to evaluate the quality and human health risks concerning groundwater in the Linfen Basin, a typical basin on the Eastern Chinese Loess Plateau. The objects of this study are (1) to analyze the hydrochemical characteristics of groundwater, (2) to evaluate groundwater quality using WQI, and (3) to assess the health risks of F^-, nitrogen, and PTEs (Fe, Mn, Hg, As, Cd, Cr^{6+} and Pb) to adults and children through drinking water intake and dermal contact. A spatial distribution map of groundwater quality and health risks in the study area was produced using Inverse distance weight (IDW) interpolation in GIS. This study can provide meaningful support for local governments in groundwater quality protection and groundwater resource management.

2. Materials and Methods

2.1. Study Area

Linfen Basin (35°23′–36°57′ N, 110°22′–112°34′ E) is situated in the southwest of Shanxi Province and includes Huozhou City, Hongtong County, Yaodu District, Quwo County, Xiangfen County, Yicheng County, and Houma City (Figure 1). It covers an area of ~4686 km². It is surrounded by the Hanhou Mountains to the north, the Emei platform to the south, the Taiyue and Zhongtiao Mountains to the east, and the Luoyun Mountains to the west. The area has been subjected to semi-arid and semi-humid monsoon climatic conditions, with mean annual precipitation of 420 to 550 mm, and mean annual temperatures of 10 °C [51]. The study area is not only an important irrigated agricultural area in the Loess Plateau but also the main supply center of energy sources in China. The area is rich in mineral resources, of which coal is the largest mineral resource. The main rivers in the study area are the Fenhe River, Xinshuihe River, Qinhe River, Huihe River, Ehe River, and Qingshuihe River. The total amount of regional water resources is 1.52 billion m³, of which the river runoff is 1.32 billion m³ (including 0.48 billion m³ of spring water), and the groundwater resource is 1.026 billion m³. The water resource in this area is scarce, with the per capita water resource occupancy being only 350 m³ [52].

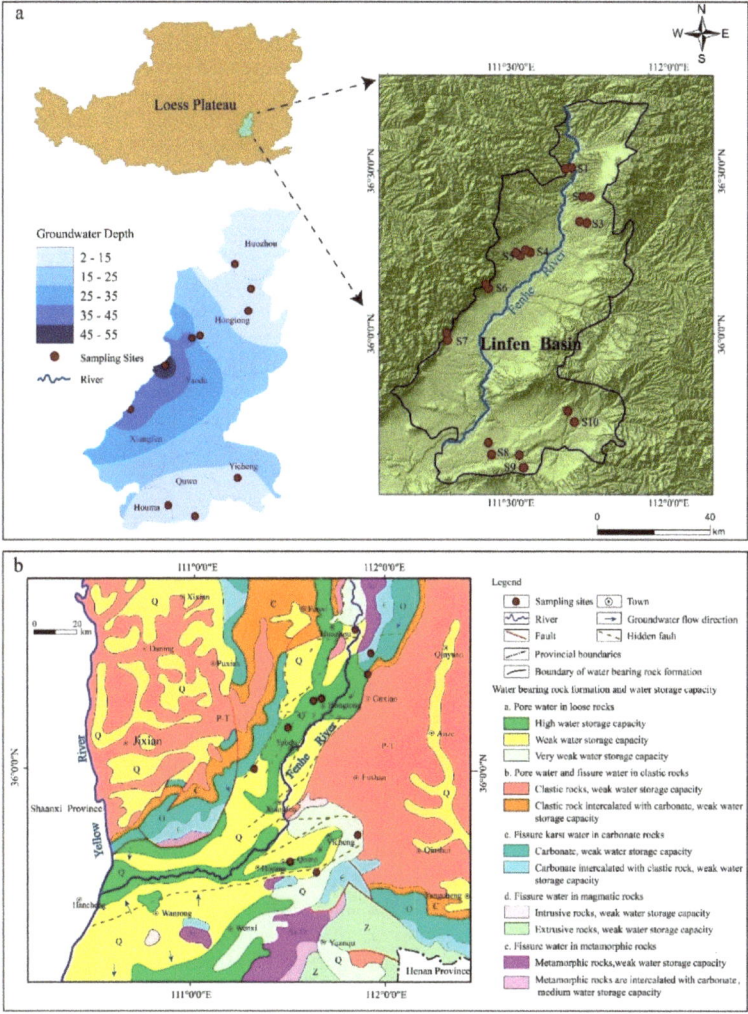

Figure 1. (a) Spatial distribution of groundwater sampling sites and groundwater depths and (b) Regional hydrogeological map of the study area.

The interior of the basin is dominated by Quaternary strata. The Lower Pleistocene is mainly yellowish-brown and grayish-yellow silty sand and sandy clay. This layer is widespread in the basin and is about 200 m thick in the middle of the basin. The Middle Pleistocene is a set of sand, sandy soil, and loam interbedded sediment, which has a thickness of ~150 m. The Upper Pleistocene in the piedmont inclined plain area is sand gravel mixed with sandy soil. Near the river valley, it is mostly sandy soil and loam deposited in river and lake facies, and the thickness of this layer is 30–50 m. The stratum lithology of Holocene is sandy soil, loam, sand, and gravel, which is mainly distributed in the Fenhe terrace. The exposed strata in the mountain area include gneiss, limestone, shale, sandstone, mudstone, sandy conglomerate, and loess [53]. The fault structure in the study area is complex, mostly being hidden faults, and the intersection of large faults is a favorable part of modern hot springs and mineralization [54].

According to the burial depth and hydraulic characteristics of the aquifers in the study area, the pore water of loose rocks in the study area can be divided into phreatic water, middle-layer confined water, and deep-layer confined water. Phreatic aquifers are mostly distributed in the middle of the basin, loess tableland, and piedmont inclined plain in a belt shape, and the aquifers are mainly medium and fine sand. Compared with the eastern piedmont and the central part of the basin, the middle-layer confined water aquifer in the western piedmont has a large thickness and coarse particles and has good water storage conditions. The distribution characteristics of deep-layer confined water are consistent with that of middle-layer water, and the aquifer is mainly sand and sand gravel. Groundwater recharge mainly includes lateral runoff, surface water seepage, and precipitation infiltration. The discharge of groundwater mainly depends on evaporation and artificial mining [54,55]. In the slow flow season, groundwater mainly belongs to the (SO_4^{2-}-Ca^{2+}) type, the (HCO_3^--Ca^{2+}-Na^+) type, and the (HCO_3^--Mg^{2+}-Na^+) type. In the quick flow season, groundwater is dominated by the (HCO_3^--Ca^{2+}-Na^+) type and the (HCO_3^--SO_4^{2-}-Ca^{2+}-Na^+) type [4]. In general, the groundwater depth in the study area shows a trend of high in the east and low in the west. The buried depth of groundwater in Yaodu and Xiangfen is generally deeper than that in other areas at 35–45 m (Figure 1).

2.2. Sampling and Analysis

Groundwater quality assessment and human health risk assessment based on 10 groundwater hydrological long-term monitoring wells set up by Shanxi Provincial Department of Water Resources in the study area. Groundwater samples were collected in 2017 and were used for the analysis of water quality parameters, including pH, total hardness (TH), total dissolved solids (TDS), sulfate (SO_4^{2-}), chloride (Cl^-), fluoride(F^-), cyanide, volatile phenols, chemical oxygen demand (COD_{Mn}), nitrate (NO_3-N), nitrite (NO_2-N), ammonia nitrogen (NH_4-N), and PTEs (Fe, Mn, Hg, As, Cd, Cr^{6+}, Pb) for each sample. Sample collection, preservation, transportation, and testing were carried out in strict accordance with the Technical Specifications for Environmental Monitoring of Groundwater [56]. Before sampling, wells were pumped for 10 min to remove stagnant water. All sampling containers were thoroughly cleaned with the groundwater to be sampled. To ensure the stability of the elements, the samples analyzed for TH, Fe, Mn, Cd, and Pb were mixed with HNO_3 solution, the samples for the analysis of NH_4-N were mixed with H_2SO_4 solution, and the samples for cyanide and Cr^{6+} analysis and for Hg and As analysis were mixed with NaOH and HCl, respectively. All samples were then sealed tightly and immediately sent to the laboratory of Linfen Hydrology and Water Resources Survey Branch for analysis (within 24 h). pH was measured directly in the field using a portable pH meter. TH was analyzed using the EDTA titration method. TDS was determined by the drying and weighing approach. SO_4^{2-}, Cl^-, F^-, NH_4^+, NO_3^-, and NO_2^- were tested using an ion chromatograph (ICS-600). Fe, Mn, Hg, As, Cd, Cr^{6+}, and Pb were measured using inductively coupled plasma-mass spectrometry (ICP-MS). Groundwater was filtered using a 0.45 μm filter before their analysis. During the analysis, distilled water and replicates were introduced to ensure the reliability of the results. The replicates had a relative error within ±5%, indicating acceptable analytical accuracy. IDW interpolation method has been widely used to study the spatial distribution of groundwater quality parameters. IDW uses the deterministic model method to calculate the unknown value according to the nearby points rather than the far-off ones. This interpolation method fits well for real-world parameters [37–39]. IDW interpolation results were verified by overlapping field survey data and laboratory analysis results. The pixel values of the IDW interpolation map match well with those of field verification data.

2.3. Water Quality Index

The WQI approach can not only comprehensively express the water quality information of groundwater but also quantitatively evaluate and compare the pollution degree of different water quality parameters [57]. This index is a mathematical instrument used to transform large quantities of water characterization data into a single number, representing the water quality level [32]. Firstly, each chemical parameter was assigned a weight (w_i) according to its impact on human health and groundwater quality. In this study, the highest weight of 5 was assigned to the parameters like TH, TDS, SO_4^{2-}, F^-, Fe, Mn, Cd, Cr^{6+}, and Pb due to their major importance in water quality assessment. These parameters are characterized by serious health effects and, when above critical concentration limits, may limit the usability of groundwater for domestic and drinking purposes [37,58]. Other parameters were assigned different weights ranging from 2 to 4. The relative weight is computed using the following formula:

$$W_i = \frac{w_i}{\sum_{i=1}^{n} w_i} \quad (1)$$

where W_i is the relative weight, w_i is the weight of each parameter, and n is the number of parameters.

Then, the quality rating for each parameter is assigned by dividing its concentration in each water sample by its limit defined by the Chinese national standards [59] and multiplying the result by 100:

$$q_i = \frac{C_i}{S_i} \times 100 \quad (2)$$

where q_i is the quality rating, and C_i is the concentration of each parameter in each water sample. S_i is the drinking water standard for each parameter set by the Chinese national standard [59].

To calculate the WQI, the SI_i has to be determined firstly:

$$SI_i = W_i \times q_i \quad (3)$$

$$WQI = \sum_{i=1}^{n} SI_i \quad (4)$$

where SI_i is the subindex of the ith parameter. The WQI values are classified into five categories: excellent water (<50), good water (50–100), poor water (100–200), very poor water (200–300), and unsuitable water (>300) [57].

2.4. Human Health Risk Assessment

The human health risk assessment is the basis for controlling groundwater pollution and ensuring a safe drinking water supply [49,60–62]. Groundwater can affect human health through various exposure pathways, the most common of which are drinking water intake and dermal contact [63]. The model recommended by the Ministry of Ecology and Environment of the P.R. China [64] based on the United States Environmental Protection Agency models [65] was adopted in this study. There are many agricultural and industrial production activities in the study area. Therefore, representative pollutants F^-, nitrogen (NO_3-N, NO_2-N, NH_4-N), and PTEs (Fe, Mn, Hg, As, Cd, Cr^{6+}, and Pb) are selected as the parameters of risk assessment. Due to differences in the physiology of males, females, and children, this study separately evaluated the health risks of oral intake and dermal intake.

The non-carcinogenic risk through drinking water intake is calculated as follows [60,63]:

$$\text{Intake}_{oral} = \frac{C \times IR \times EF \times ED}{BW \times AT} \quad (5)$$

$$HQ_{oral} = \frac{\text{Intake}_{oral}}{RfD_{oral}} \quad (6)$$

The non-carcinogenic risk through dermal contact is expressed as follows [42]:

$$\text{Intake}_{dermal} = \frac{DA \times EV \times SA \times EF \times ED}{BW \times AT} \quad (7)$$

$$DA = K \times C \times t \times CF \quad (8)$$

$$SA = 239 \times H^{0.417} \times BW^{0.517} \quad (9)$$

$$HQ_{dermal} = \frac{\text{Intake}_{dermal}}{RfD_{dermal}} \quad (10)$$

$$RfD_{dermal} = RfD_{oral} \times ABS_{gi} \quad (11)$$

where Intake_{oral}, Intake_{dermal}, HQ_{oral}, HQ_{dermal}, RfD_{oral}, and RfD_{dermal} represent the chronic daily dose via ingestion and dermal contact (mg/(kg day)), the hazard quotient through oral and dermal exposure pathways, reference dose for oral and dermal contact pathways (mg/(kg day)), respectively. C, DA, SA, and ABS_{gi} are the pollutant concentration of groundwater (mg/L), exposure dose (mg/cm^2), skin surface area (cm^2), and gastrointestinal absorption factor, respectively. The definitions and values of other parameters are shown in Tables 1 and 2.

Table 1. Definition and values of key parameters for human health risk assessment.

Parameters	Units	Values		
		Males	Females	Children
Ingestion rate (IR)	L/day	1 [a]	1 [a]	0.7 [a]
Exposure frequency (EF)	day/a	350 [a]	350 [a]	350 [a]
Exposure duration (ED)	a	24 [a]	24 [a]	6 [a]
Body weight (BW)	kg	69.6 [b]	59 [b]	19.2 [b]
Average time (AT)	day	8400 [a]	8400 [a]	2190 [a]
Skin permeability coefficient (K)	cm/h	0.002 for Cr^{6+} and 0.001 for other parameters [c]		
Contact duration (t)	h/day	0.4 [d]		
Conversion factor (CF)	-	0.001 [c]		
Average height (H)	cm	169.7 [b]	158 [b]	113.15 [b]
Daily exposure frequency (EV)	-	1 [a]		

[a] refer to [64]; [b] refer to [66]; [c] refer to [67]; [d] refer to [63].

Table 2. The values of RfD, ABS_{gi}, and SF for different ions.

Parameters	Non-Carcinogenic		Carcinogenic		ABSgi
	RfDoral	RfDdermal	SForal	SFdermal	
Cr^{6+}	0.003	0.000075	0.42	16.8	0.025
As	0.0003	0.0003	1.5	1.5	1
Cd	0.001	0.00005	6.1	122	0.05
F$^-$	0.04	0.04			1
NO$_3$-N	1.6	1.6			1
NO$_2$-N	0.1	0.1			1
NH$_4$-N	0.97	0.97			1
Fe	0.3	0.3			1
Mn	0.14	0.14			1
Hg	0.0003	0.000021			0.07
Pb	0.0014	0.0014			1

The total non-carcinogenic risks are calculated as follows [1,29,45]:

$$HQ_i = HQ_{oral} + HQ_{dermal} \tag{12}$$

$$HI_{total} = \sum_{i=1}^{n} HQ_i \tag{13}$$

where HQ is the non-carcinogenic hazard quotient and HI is the hazard index. i represents the risk assessment parameters. When the value of HQ and HI is less than 1, it is safe for human health. HI > 1 indicates unacceptable risk, and residents are exposed to non-carcinogenic risks [60,64].

In addition to non-carcinogenic risk, As, Cd, and Cr^{6+} can also create carcinogenic risks for humans [7,41]. The carcinogenic risk through drinking water intake and dermal contact is calculated as follows:

$$CR_{oral} = Intake_{oral} \times SF_{oral} \tag{14}$$

$$CR_{dermal} = Intake_{dermal} \times SF_{dermal} \tag{15}$$

$$SF_{dermal} = \frac{SF_{oral}}{ABS_{gi}} \tag{16}$$

$$CR_{total} = CR_{oral} + CR_{dermal} \tag{17}$$

where CR denotes the carcinogenic risk. SF is the slope factor for the carcinogenic contaminants $(mg/(kg\ day))^{-1}$. The SF_{oral} values for As, Cd, and Cr^{6+} are shown in Table 2. The average time (AT) for carcinogenic risk is set at 27,740 days for both adults and children, as the harm to human health caused by cadmium, chromium, and arsenic will last a lifetime [64]. The acceptable limit for CR is 1×10^{-6}.

3. Results and Discussion

3.1. Hydrochemical characteristics of Groundwater

The statistical results of water quality for groundwater samples are given in Table 3. pH is one of the most important parameters for evaluating the suitability of drinking water [60]. The Chinese national standard proposes that the pH value of groundwater suitable for drinking is 6.5–8.5 [59]. As Table 3 shows, pH values of the groundwater range from 7.27 to 7.85, with a mean value of 7.59. Therefore, the groundwater in the study area is weakly alkaline water that can be used for drinking.

TH represents dissolved Ca^{2+} and Mg^{2+} in groundwater. High TH in groundwater may affect the taste of drinking water and reduce the efficacy of detergents [34]. In addition, regarding human health, the long-term drinking of extremely hard water may increase the incidence of urolithiasis, anencephaly, prenatal mortality, and some cancer-related cardiovascular diseases [68]. In this study, TH varies between 167 and 869 mg/L with a mean of 426 mg/L. According to the national Chinese drinking water standards, samples S1, S9, and S10 are extremely hard water, with TH exceeding the acceptable limit of 450 mg/L for drinking. These samples are predominantly distributed in the southern part of the study area (Figure 2a). TH enrichment in groundwater may be due to the dissolution of soluble salts and minerals, as well as to human intervention [2].

TDS is one of the major water quality parameters, mainly representing the various minerals present in the water [6]. TDS varies in a wide range of 280–1312 mg/L, with a mean value of 689 mg/L (Table 3). Based on TDS content, Liu et al. [69] categorized waters as freshwater (TDS < 1000 mg/L) and brackish water (TDS > 1000 mg/L). Only sample S10 in Yicheng is brackish water (Figure 2b). Generally speaking, higher TDS usually indicates stronger water-rock interaction and may also be affected by domestic wastewater, irrigation return flow, and fertilization [1,70]. High TDS in groundwater is generally harmless in healthy people and may cause constipation or have a laxative effect, but it may have a greater impact on people with kidney and heart disease [6,33,71].

Cl^- and SO_4^{2-} in groundwater are mainly related to the regional lithological conditions and are also affected by anthropogenic sources [68]. The concentration of Cl^- is between 7.93 and 88.1 mg/L and is lower than the Chinese national standard of 250 mg/L. The concentration of SO_4^{2-} in the study area ranged from 68 to 536 mg/L, with a mean of 182.16 mg/L. Samples S5 and S10 exceeded the acceptable limit of SO_4^{2-} for drinking. High SO_4^{2-} concentration is observed in the Yaodu and Yicheng parts of the central and south of the study area (Figure 2c). The Ordovician karst aquifers widely distributed in the study area are affected by gypsum dissolution, and the hydrochemical type of groundwater is $SO_4^{2-} \cdot HCO_3^- \text{-Ca} \cdot Mg$. In addition, the oxidation of sulfur in coal-bearing strata ($S + O_2 + 2H_2O \rightarrow SO_4^{2-} + 4H^+$) will also cause increased sulfate concentration in groundwater [72]. Therefore, the high mean value of SO_4^{2-} in this study is probably due to the high natural background value rather than pollution.

F^- in drinking water is essential for human health at low concentrations, such as protecting teeth from caries [2]. However, excessive fluoride intake can cause dental fluorosis, skeletal fluorosis, and thyroid disease in adults [17,73]. The Chinese national standard stipulates that F^- concentration in drinking water should be less than 1.0 mg/L. In this study, F^- is in the range of 0.25–1.71 mg/L, with an average value of 0.75 mg/L. Two groundwater samples in Yaodu did not meet the requirement of the national standard (Figure 2d). The high concentration of fluoride in groundwater may be mainly related to the lithology of the region, especially the dissolution of fluoride-bearing minerals [16,74].

Both cyanide and volatile phenol are toxic organics. The concentration of cyanide in all groundwater samples is less than 0.0004 mg/L. For volatile phenols, except for sample S10 in Yicheng, whose value is 0.002 mg/L, the other samples are 0.0003 mg/L. COD_{Mn} is an indicator that can indirectly reflect the organic pollution of groundwater [44,74]. The COD_{Mn} values for the samples are observed to be from 0.1 to 0.9 mg/L, with an average of 0.25 mg/L. Sample S10 in Yicheng has the highest volatile phenol and COD_{Mn} values. As shown in Table 3, the concentrations of cyanide, volatile phenol, and COD_{Mn} are all within the drinking water standard limit stipulated by the national standard, indicating that the groundwater is less affected by organic pollution.

Table 3. Statistical analysis results for hydrochemical parameters of groundwater.

Parameters	Min	Max	Mean	Median	SD	C.V (%)	Chinese Standards	P [a] (%)
pH	7.27	7.85	7.59	7.63	0.175	2.306	6.5–8.5	0
TH	167	869	426	359	202.768	47.554	450	30
TDS	280	1312	689	637	282.271	40.968	1000	10
SO_4^{2-}	68	536	182	136	135.689	74.489	250	20
Cl^-	7.93	88.10	47.6	49.3	26.188	55.055	250	0
F^-	0.25	1.71	0.75	0.69	0.382	51.226	1	20
cyanide	0.0004	0.0004	0.0004	0.0004	0.000	0	0.05	0
volatile phenols	0.0003	0.002	0.00047	0.0003	0.001	108.511	0.002	0
COD_{Mn}	0.1	0.9	0.3	0.2	0.229	91.652	3	0
NO_3-N	0.002	11.300	4.760	2.255	4.264	89.567	20	0
NO_2-N	0.004	0.070	0.015	0.004	0.023	150.277	1	0
NH_4-N	0.025	0.160	0.065	0.034	0.048	73.457	0.5	0
Fe	0.03	1.41	0.18	0.03	0.411	232.533	0.3	10
Mn	0.010	0.139	0.023	0.010	0.039	163.918	0.1	10
Hg	0.00001	0.00006	0.000017	0.00001	0.00002	91.319	0.001	0
As	0.0002	0.0002	0.0002	0.0002	0.000	0.000	0.01	0
Cd	0.002	0.002	0.002	0.002	0.000	0.000	0.005	0
Cr^{6+}	0.004	0.034	0.009	0.006	0.009	93.858	0.05	0
Pb	0.011	0.011	0.011	0.011	0.000	0.000	0.01	100

[a] percentage of the sample exceeding the permissible limits. Units for all parameters are in mg/L, except for pH (non-dimensional).

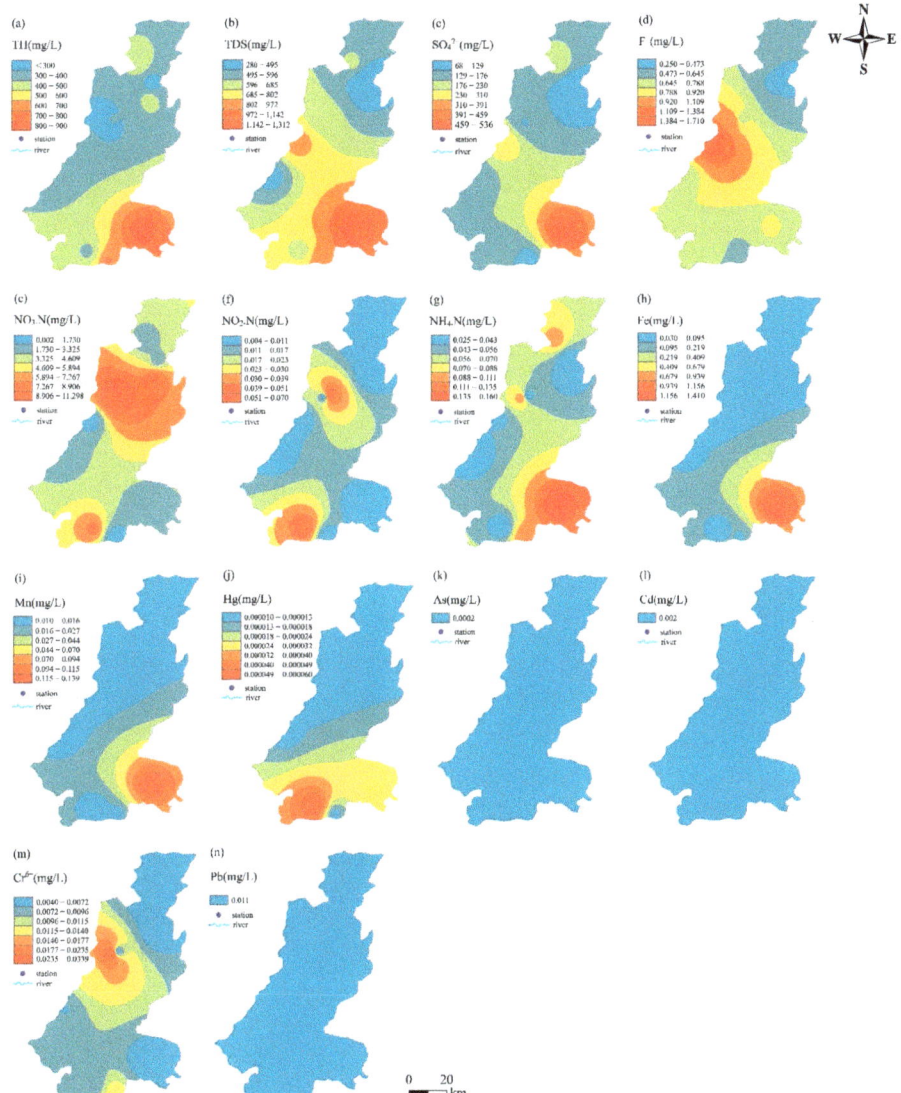

Figure 2. Spatial distributions of mass concentrations of groundwater hydrochemical parameters: (**a**) TH, (**b**) TDS, (**c**) SO_4^{2-}, (**d**) F^-, (**e**) NO_3-N, (**f**) NO_2-N, (**g**) NH_4-N, (**h**) Fe, (**i**) Mn, (**j**) Hg, (**k**) As, (**l**) Cd, (**m**) Cr^{6+}, and (**n**) Pb.

In recent years, nitrogen pollution (NO_3-N, NO_2-N, and NH_4-N) has become a hot issue for many researchers due to its adverse effects on groundwater quality and human health [2,12,14,49,74–76]. The extensive use of nitrogenous fertilizers in agricultural activities is one of the most common sources of nitrogen pollution in groundwater [1,63]. Measured values of NO_3-N, NO_2-N, and NH_4-N are in the range of 0.002–11.3, 0.004–0.7 and 0.025–0.16 mg/L, respectively. Higher NO_3-N and NO_2-N concentrations are observed in the central and southwest parts of the area, while a high value of NH_4-N is mainly distributed around Yicheng (Figure 2e–g). According to the Chinese standards, groundwater is unacceptable for drinking when the NO_3-N, NO_2-N, and NH_4-N concentration in groundwater is higher than 20, 1, and 0.5 mg/L, respectively. Therefore, the groundwater in the study area is less contaminated with nitrogen and is suitable for drinking.

PTEs content in groundwater is usually low. However, even in very low concentrations, -they can create biological toxicity and pose serious threats to aquatic ecosystems and human health [20,21,41]. As shown in Table 3, the Fe, Mn, Hg, and Cr^{6+} concentrations range from 0.03 to1.41, 0.01–0.139, 0.00001–0.00006, and 0.004–0.034 mg/L, respectively. The concentrations of As, Cd, and Pb are 0.0002, 0.002, and 0.011 mg/L, respectively. The mean concentration of metals is in the following order: Fe > Mn > Pb > Cr^{6+} > Cd > As > Hg. All metals, except for Fe, Mn and Pb, are within the permissible levels for drinking water. Samples with high concentrations of Fe and Mn are mainly found in the southeastern parts of the basin (Figure 2h,i). Fe and Mn have similar geochemical behavior. Their dissolution and migration to groundwater are affected by reduction conditions, residence time, well depth, and salinity [77]. The similarity in the spatial distribution of Cr^{6+} and NO_3^--N concentrations may be related to the synergistic role of nitrogen (N)-bearing fertilizers to elevated Cr^{6+} concentration in groundwater. This may be due to the production of H^+ and soil acidification during the nitrification process of NH_4^+ oxidation to NO_3^-, favoring the increased dissolution of Cr^{3+} which is subsequently oxidized into Cr^{6+} by natural and/or anthropogenic factors [46,47].

3.2. Groundwater Quality Assessment

In this study, pH, TDS, TH, SO_4^{2-}, Cl^-, F^-, volatile phenols, NO_3-N, NO_2-N, NH_4-N, Fe, Mn, Hg, Cd, Cr^{6+} and Pb are selected as the parameters to evaluate the overall groundwater quality, using the WQI introduced previously. The values of cyanide, arsenic, and chemical oxygen demand in groundwater are very low, so they have little impact on water quality and can be ignored in water quality assessment. The weights and relative weights assigned to each parameter are shown in Table 4.

Table 4. Relative weight of hydrochemical parameters.

Parameters	Chinese Standards	Weight(w_i)	Relative Weight (W_i)
pH	6.5–8.5	4	0.0588
TDS	1000	5	0.0735
TH	450	5	0.0735
SO_4^{2-}	250	5	0.0735
Cl^-	250	2	0.0294
F^-	1	5	0.0735
Volatile phenols	0.002	2	0.0294
NO_3-N	20	4	0.0588
NO_2-N	1	4	0.0588
NH_4-N	0.5	4	0.0588
Fe	0.3	5	0.0735
Mn	0.1	5	0.0735
Hg	0.001	3	0.0441
Cd	0.005	5	0.0735
Cr^{6+}	0.05	5	0.0735
Pb	0.01	5	0.0735
		$\sum w_i = 68$	$\sum W_i = 1$

Units for all parameters are in mg/L, except pH (non-dimensional).

The calculated WQI values and water types are presented in Table 5. The results of WQI range from 23.63 to 105.96. Out of 10 groundwater samples, sample S5 is categorized as good water. Sample S10 is classified as poor water. The other 8 samples are excellent water. For the study area, the most significant parameters affecting groundwater quality are Pb, TH, F^-, SO_4^{2-}, and TDS.

Table 5. Water quality index values and water types in the study area.

Sample	WQI	Water Quality	Sample	WQI	Water Quality
S1	37.91	Excellent water	S6	40.77	Excellent water
S2	23.63	Excellent water	S7	32.88	Excellent water
S3	32.20	Excellent water	S8	38.36	Excellent water
S4	38.83	Excellent water	S9	42.18	Excellent water
S5	50.17	Good water	S10	105.96	Poor water

From the spatial distribution of groundwater quality index results, it can be seen that poor quality water area is mainly located near Yicheng in the southeastern area of the study (Figure 3). The main pollutants in the groundwater in this area are TH, TDS, SO_4^{2-}, Fe, Mn, and Pb, all of which exceed the upper limit for drinking purposes. The poor groundwater quality in Yicheng may be related to the buried depth of groundwater. Generally, when the groundwater is buried deeper, it takes longer for the surface pollutants to reach the aquifer. Thus, the possibility of the pollutants being adsorbed and diluted during the infiltration process becomes greater, and the degree of pollution in the groundwater system will decrease. The buried groundwater depth in the Yicheng area is shallow at 2–15 m. In addition, the lithology of the buried deep aquifer is mainly coarse sand and medium-coarse sand. The better permeability of the aquifer makes it easier for surface pollutants to seep into the groundwater, resulting in groundwater pollution. Fe and Mn in groundwater come from coal and metal deposits, especially iron ore. High TDS leads to increased ionic strength and decreased activity coefficient, which will dissolve more Fe and Mn in groundwater. In addition, the organic matter released from surface pollutants into groundwater can quickly deplete the dissolved oxygen in groundwater, resulting in a reductive hydrochemical environment more conducive to the dissolution of Fe and Mn [77].

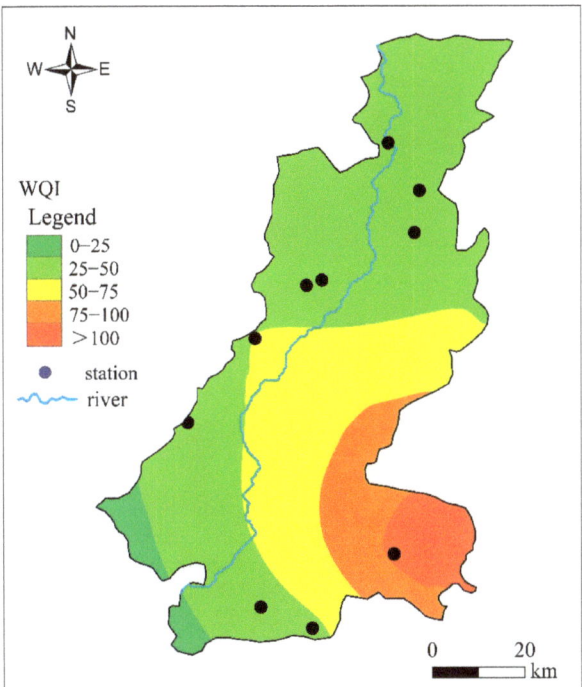

Figure 3. Spatial distribution of groundwater quality based on water quality index values.

The assessment results indicate that the groundwater in the study area is dominated by excellent water that can be used for drinking purposes. For the Yicheng area with poor quality groundwater unsuitable for drinking, groundwater pollution remediation and safe water supply measures should be implemented as soon as possible.

3.3. Human Health Risk Assessment

The health risks of groundwater in the study area were assessed based on the model introduced previously. The calculated health risks for adults and children through drinking water and dermal contact are shown in Table 6. For adult males, the HQ_{oral} values range from 0.285 to 0.827, with a mean of 0.521. The HQ_{oral} values for adult females and children range from 0.336 to 0.976 and 0.693–2.012, with an average of 0.615 and 1.269, respectively. The HQ_{dermal} values are smaller than the HQ_{oral}, ranging from 0.017 to 0.104 for males, 0.018 to 0.109 for females, and 0.026 to 0.156 for children, with means of 0.034, 0.036, and 0.051, respectively. This suggests that non-carcinogenic risk is mainly caused by oral exposure. The HI_{total} values for males and females range from 0.302 to 0.902 and 0.354–1.051, with means of 0.555 and 0.651, respectively. For children, the HI_{total} values are 0.719–2.100, with an average value of 1.320. For males, females, and children, HI_{total} values of 0%, 20%, and 80% of the samples exceed 1, indicating that males in the study area do not have associated non-carcinogenic health risks. In contrast, females and children face higher non-carcinogenic risks. Females and children have smaller body weights and therefore have higher average daily exposure dose of contaminants than males [50,60].

Table 6. The non-carcinogenic and carcinogenic risk results from drinking water and dermal contact.

Sample	The Non-Carcinogenic Risk								
	HQ_{oral}			HQ_{dermal}			HI_{total}		
	Males	Females	Children	Males	Females	Children	Males	Females	Children
S1	0.393	0.463	0.956	0.018	0.019	0.027	0.411	0.482	0.983
S2	0.285	0.336	0.693	0.017	0.018	0.026	0.302	0.354	0.719
S3	0.436	0.514	1.060	0.018	0.019	0.028	0.454	0.533	1.088
S4	0.529	0.625	1.288	0.019	0.020	0.029	0.548	0.644	1.317
S5	0.827	0.976	2.012	0.043	0.045	0.065	0.870	1.021	2.077
S6	0.799	0.942	1.943	0.104	0.109	0.156	0.902	1.051	2.100
S7	0.455	0.536	1.106	0.027	0.028	0.040	0.481	0.564	1.146
S8	0.528	0.623	1.285	0.030	0.032	0.046	0.558	0.655	1.330
S9	0.401	0.473	0.975	0.043	0.045	0.065	0.443	0.518	1.040
S10	0.562	0.663	1.368	0.019	0.020	0.029	0.582	0.684	1.397
Mean	0.521	0.615	1.269	0.034	0.036	0.051	0.555	0.651	1.320

Sample	The Carcinogenic Risk								
	CR_{oral}			CR_{dermal}			CR_{total}		
	Males	Females	Children	Males	Females	Children	Males	Females	Children
S1	6.169×10^{-5}	7.278×10^{-5}	3.914×10^{-5}	1.201×10^{-5}	1.263×10^{-5}	4.725×10^{-6}	7.371×10^{-5}	8.541×10^{-5}	4.386×10^{-5}
S2	6.169×10^{-5}	7.278×10^{-5}	3.914×10^{-5}	1.201×10^{-5}	1.263×10^{-5}	4.725×10^{-6}	7.371×10^{-5}	8.541×10^{-5}	4.386×10^{-5}
S3	6.169×10^{-5}	7.278×10^{-5}	3.914×10^{-5}	1.201×10^{-5}	1.263×10^{-5}	4.725×10^{-6}	7.371×10^{-5}	8.541×10^{-5}	4.386×10^{-5}
S4	6.169×10^{-5}	7.278×10^{-5}	3.914×10^{-5}	1.201×10^{-5}	1.263×10^{-5}	4.725×10^{-6}	7.371×10^{-5}	8.541×10^{-5}	4.386×10^{-5}
S5	7.631×10^{-5}	9.000×10^{-5}	4.841×10^{-5}	2.054×10^{-5}	2.160×10^{-5}	8.079×10^{-6}	9.686×10^{-5}	1.116×10^{-4}	5.648×10^{-5}
S6	1.165×10^{-4}	1.370×10^{-4}	7.391×10^{-5}	4.400×10^{-5}	4.625×10^{-5}	1.730×10^{-5}	1.605×10^{-4}	1.837×10^{-4}	9.121×10^{-5}
S7	6.718×10^{-5}	7.924×10^{-5}	4.261×10^{-5}	1.521×10^{-5}	1.599×10^{-5}	5.983×10^{-6}	8.239×10^{-5}	9.524×10^{-5}	4.860×10^{-5}
S8	6.900×10^{-5}	8.140×10^{-5}	4.377×10^{-5}	1.628×10^{-5}	1.711×10^{-5}	6.402×10^{-6}	8.528×10^{-5}	9.851×10^{-5}	5.018×10^{-5}
S9	7.814×10^{-5}	9.218×10^{-5}	4.957×10^{-5}	2.161×10^{-5}	2.272×10^{-5}	8.498×10^{-6}	9.975×10^{-5}	1.149×10^{-4}	5.807×10^{-5}
S10	6.169×10^{-5}	7.278×10^{-5}	3.914×10^{-5}	1.201×10^{-5}	1.263×10^{-5}	4.725×10^{-6}	7.371×10^{-5}	8.541×10^{-5}	4.386×10^{-5}
Mean	7.156×10^{-5}	8.442×10^{-5}	4.540×10^{-5}	1.777×10^{-5}	1.868×10^{-5}	6.989×10^{-6}	8.933×10^{-5}	1.031×10^{-4}	5.239×10^{-5}

Contaminants in groundwater contribute differently to health risks. Concerning each water quality parameter, the non-carcinogenic HQ values of F^-, NO_3-N, NO_2-N, NH_4-N, Fe, Mn, Hg, Pb, Cr^{6+}, As, and Cd are in the ranges of 0.090–1.501, 1.809×10^{-5}–0.248, 5.789×10^{-4}–0.025, 3.730×10^{-4}–5.793×10^{-3}, 1.447×10^{-3}–0.165, 1.034×10^{-3}–3.487×10^{-2}, 5.288×10^{-4}–7.444×10^{-3}, 0.114–0.276, 0.030–0.540, 9.648×10^{-3}–0.023 and 0.033–0.076, respectively. This result suggests that besides F^-, the non-carcinogenic risks of other contaminants are acceptable to both adults and children. As shown in Figure 4, the contribution of pollutants in groundwater to the HI_{total} value is observed in the following order: F^- > Pb > Cr^{6+} > NO_3-N > Cd > As > Fe > Mn > NO_2-N > NH_4-N > Hg. F^- contributes the most to non-carcinogenic risk (46.86%), followed by Pb (22.78%) and Cr^{6+} (11.22%). Contribution of other pollutants to the non-carcinogenic risk is less than 10%, indicating that F^-, Pb, and Cr^{6+} may be drivers of adverse effects on human health.

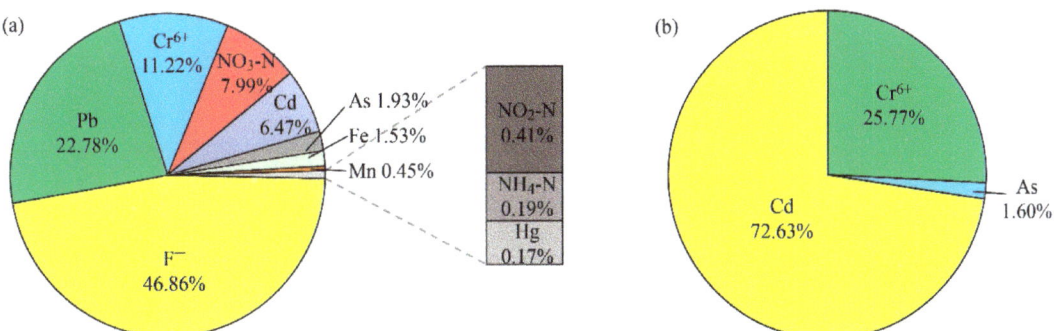

Figure 4. Contributive ratios of contaminants in groundwater to health risks (**a**) non-carcinogenic risk; and (**b**) carcinogenic risk.

The spatial distribution of HI_{total} values for males, females, and children is consistent with fluoride concentration (Figure 5). Higher HI_{total} and F^- concentration mainly appear in the Yaodu, midwest of the study area. Groundwater with high F^- is found in semi-arid and arid areas of northern China, such as the middle Loess Plateau [16], Ningxia plain [61], Guanzhong Plain [1,36], Hetao Plain [78], and Tianjin [17]. The respective HQ mean values of F^- for males, females, and children are 0.503, 0.593, and 1.220 in the Yaodu, indicating that children are exposed to health risks from fluoride. Fluoride-bearing minerals are enriched in magmatic rocks and aluminosilicates exposed to the surface of the area, such as fluorite (CaF_2), villiaumite (NaF), and biotite [79,80]. There are also many active fault zones in and around Yaodu, and fluorine-containing volatile gas or hydrothermal fluid migrates upward along the faults and penetrates groundwater, increasing the fluorine content [79]. In addition, areas with high F^- in groundwater have a higher population density, and industries such as coal mining, metallurgy, and coking are concentrated. Discharge of domestic sewage and industrial wastewater is the other reason for the increase in F^- concentration in groundwater. Although the Cr^{6+} concentration of all groundwater samples is within the desirable limit for drinking, it contributes more than 10% to the health risk, similar to Pb.

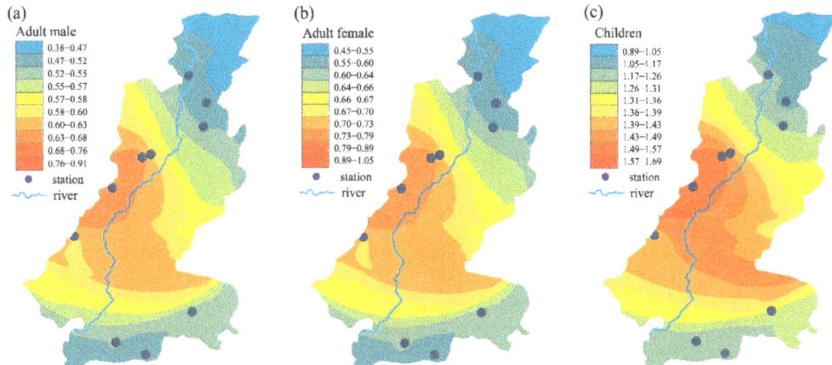

Figure 5. Spatial distribution of non-carcinogenic health risks for males (**a**), females (**b**), and children (**c**).

The carcinogenic risks due to exposure to As, Cd, and Cr^{6+} through drinking water and dermal contact are shown in Table 6. The ranges of the CR_{oral} for males, females, and children are 6.169×10^{-5}–1.165×10^{-4}, 7.278×10^{-5}–1.370×10^{-4}, and 3.914×10^{-5}–7.391×10^{-5}, with means of 7.156×10^{-5}, 8.442×10^{-5}, and 4.540×10^{-5}, respectively. The results of the CR_{dermal} are slightly smaller than CR_{oral}, ranging from 1.201×10^{-5}–4.400×10^{-5} for males, 1.263×10^{-5}–4.625×10^{-5} for females, and 4.725×10^{-6}–1.730×10^{-5} for children, with means of 1.777×10^{-5}, 1.868×10^{-5}, and 6.989×10^{-6}, respectively. As a result, the CR_{total} values for males and females are 7.371×10^{-5}–1.605×10^{-4} and 8.541×10^{-5}–1.837×10^{-4}, with means of 8.933×10^{-5} and 1.031×10^{-4}. Concerning children, the CR_{total} values range from 4.386×10^{-5}–9.121×10^{-5} with an average value of 5.239×10^{-5}. The carcinogenic risk values of all samples exceed the acceptable limit (1×10^{-6}) recommended by the Ministry of Ecology and Environment of the P. R. China [64] for both adults and children. Additionally, the carcinogenic risk for adults is higher than for children, especially females. Similar results have also been found by Li et al. [60] and Zhang et al. [50] in Weining Plain and Guanzhong Plain, respectively.

For each contaminant, only the carcinogenic risk of As to children is below the acceptable limit, with an average of 8.317×10^{-7}. As per the average values of the CR_{total}, Cd contributes 72.63% to the total CR, Cr^{6+} and As account for 25.77% and 1.60% of the CR_{total}, respectively. From the spatial distribution map of carcinogenic risk, it can be seen that the south-central part of the study area has a higher CR_{total} value for both adults and children, especially in Xiangfen and the west Yaodu areas. The coal-bearing formations are distributed all over the Linfen Basin, except Huoshan Mountain in the east of Yaodu and Ta'ershan-Erfengshan Mountain in the south of the study area. The areas with lower CR_{total} values in Figure 6 correspond to regions lacking coal-bearing formations. This result suggests that the carcinogenic risk is closely related to the regional geological environment. The natural leaching process and human mining activities will cause many hazardous substances to enter the groundwater. In agricultural activities, the application of N-bearing fertilizers and phosphorous (P)-bearing fertilizers will increase PTEs concentrations such as Cd, Cr, As, and Pb in groundwater under the appropriate favoring geochemical conditions [46,47]. Long-term drinking of such groundwater by residents will increase the risk of visceral cancers such as lung, liver, skin, and kidney [18].

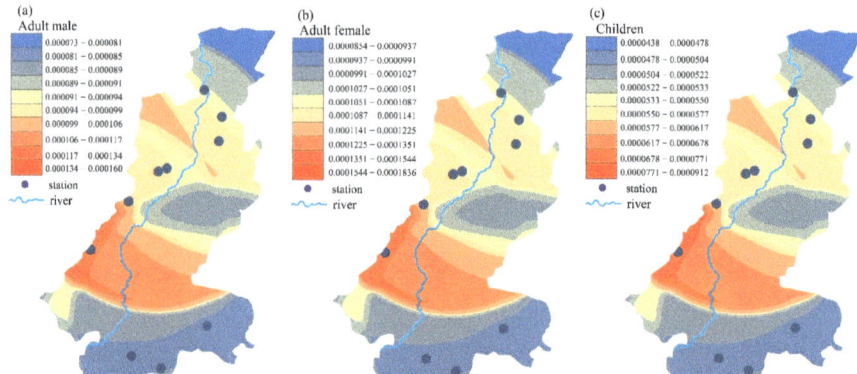

Figure 6. Spatial distribution of carcinogenic health risks for males (**a**), females (**b**), and children (**c**).

Residents living in the central part of the study area face high health risks ($HI_{total} > 1$ and $CR_{total} > 1 \times 10^{-6}$) due to the groundwater being affected by the geological environment and human activities. Therefore, government officials should pay more attention to PTEs pollution in groundwater caused by mining and the production of mineral resources. Furthermore, supplying residents with high-quality drinking water with safe concentrations of F^- should be the goal of sustainable groundwater management in the Yaodu area. It is urgent to take various measures to treat the polluted groundwater before direct consumption by residents to ensure water safety and people's health.

Additionally, compared with the results of groundwater quality assessment, we found that although most of the water quality of the study area is in good condition, both adults and children face great health risks, especially carcinogenic. Therefore, the overall groundwater assessment should be accompanied by a health risk assessment to better evaluate the suitability of groundwater for drinking.

4. Conclusions

In this study, groundwater samples from Linfen Basin were collected and analyzed for physicochemical parameters. The water quality index was used to evaluate the groundwater quality, while the health risk was assessed for adults and children concerning different exposure pathways. The main conclusions of the study are as follows:

1. The groundwater in the study area is weakly alkaline, with TH and TDS ranging between 167–869 and 280–1312 mg/L. Compared with the Chinese national standards, 30%, 10%, 20%, 20%, 10%, 10%, and 100% of the total samples exceeded the standard limits of drinking water in terms of TH, TDS, SO_4^{2-}, F^-, Fe, Mn, and Pb. Higher TH, TDS, SO_4^{2-}, Fe, and Mn are mainly distributed in the southeastern part of the study area, while a high concentration of F^- was observed in the central area of the study.

2. Most groundwater has good water quality and can be used as drinking water. Pb, TH, F^-, SO_4^{2-}, and TDS are the most significant parameters affecting groundwater quality. The poor quality of groundwater near Yicheng might be due to the shallow buried depth of groundwater and the good permeability of the aquifer.

3. Contaminated groundwater in the study area can pose human health risks to residents through multiple exposure pathways, including drinking water intake and dermal contact. The total non-carcinogenic health risks for males, females, and children range from 0.302 to 0.902, 0.354–1.051, and 0.719–2.100, respectively. Males do not have associated non-carcinogenic health risks, while females and children face higher non-carcinogenic risks than males. The ranges of the total carcinogenic health risks for males, females, and children are 7.371×10^{-5}–1.605×10^{-4}, 8.541×10^{-5}–1.837×10^{-4}, and 4.386×10^{-5}–9.121×10^{-5}, respectively. The carcinogenic risk exceeds the acceptable limit recommended by the Ministry of Ecology and Environment of the P. R. China for both adults and children. The great risks ($HI_{total} > 1$ and $CR_{total} > 1 \times 10^{-6}$) for adults and children all occur in the central study area.

Author Contributions: Conceptualization, J.L. and C.S.; investigation, W.C. and Q.Z.; data curation, J.L. and R.L.; writing—original draft preparation, J.L.; writing—review and editing, J.L., C.S., W.C., and Q.Z.; visualization, S.Z. and Y.W.; supervision, C.S. All authors have read and agreed to the published version of the manuscript.

Funding: This research was funded by the Basic Research Program of Shanxi Province (Free Exploration), grant number 20210302123265 and 20210302123261, and the National Natural Science Foundation of China (41807445).

Institutional Review Board Statement: Not applicable.

Informed Consent Statement: Not applicable.

Data Availability Statement: All processed data used in the study have been shown in the article. Raw data may be available on request from the corresponding author.

Acknowledgments: We thank the editors and reviewers for their valuable comments and suggestions to improve the quality of the paper.

Conflicts of Interest: The authors declare no conflict of interest.

References

1. Wang, Y.; Li, P. Appraisal of shallow groundwater quality with human health risk assessment in different seasons in rural areas of the Guanzhong Plain (China). *Environ. Res.* **2022**, *207*, 112210. [CrossRef]
2. Wegahita, N.K.; Ma, L.; Liu, J.; Huang, T.; Luo, Q.; Qian, J. Spatial Assessment of Groundwater Quality and Health Risk of Nitrogen Pollution for Shallow Groundwater Aquifer around Fuyang City, China. *Water* **2020**, *12*, 3341. [CrossRef]
3. Karunanidhi, D.; Aravinthasamy, P.; Deepali, M.; Subramani, T.; Bellows, B.C.; Li, P. Groundwater quality evolution based on geochemical modeling and aptness testing for ingestion using entropy water quality and total hazard indexes in an urban-industrial area (Tiruppur) of Southern India. *Environ. Sci. Pollut. Res.* **2021**, *28*, 18523–18538. [CrossRef] [PubMed]
4. Sun, C.; Chen, W.; Shen, Y. The seasonal and spatial distribution of hydrochemical characteristics of groundwater and its controlling factors in the eastern Loess Plateau. *Earth Sci. Inform.* **2021**, *14*, 2293–2308. [CrossRef]
5. Xing, L.; Guo, H.; Zhan, Y. Groundwater hydrochemical characteristics and processes along flow paths in the North China Plain. *J. Asian Earth Sci.* **2013**, *70–71*, 250–264. [CrossRef]
6. Varol, S.; Davraz, A. Evaluation of the groundwater quality with WQI (Water Quality Index) and multivariate analysis: A case study of the Tefenni plain (Burdur/Turkey). *Environ. Earth Sci.* **2015**, *73*, 1725–1744. [CrossRef]
7. Qiao, J.; Zhu, Y.; Jia, X.; Shao, M.; Niu, X.; Liu, J. Distributions of arsenic and other heavy metals, and health risk assessments for groundwater in the Guanzhong Plain region of China. *Environ. Res.* **2020**, *181*, 108957. [CrossRef] [PubMed]
8. Nsabimana, A.; Li, P.; He, S.; He, X.; Alam, S.M.K.; Fida, M. Health Risk of the Shallow Groundwater and Its Suitability for Drinking Purpose in Tongchuan, China. *Water* **2021**, *13*, 3256. [CrossRef]
9. Shukla, S.; Saxena, A. Appraisal of Groundwater Quality with Human Health Risk Assessment in Parts of Indo-Gangetic Alluvial Plain, North India. *Arch. Environ. Contam. Toxicol.* **2021**, *80*, 55–73. [CrossRef]
10. Su, F.; Wu, J.; He, S. Set pair analysis-Markov chain model for groundwater quality assessment and prediction: A case study of Xi'an city, China. *Hum. Ecol. Risk Assess.* **2019**, *25*, 158–175. [CrossRef]
11. Knobeloch, L.; Salna, B.; Hogan, A.; Postle, J.; Anderson, H. Blue babies and nitrate-contaminated well water. *Environ. Health Perspect.* **2000**, *108*, 675–678. [CrossRef]
12. Kaur, L.; Rishi, M.S.; Siddiqui, A.U. Deterministic and probabilistic health risk assessment techniques to evaluate non-carcinogenic human health risk (NHHR) due to fluoride and nitrate in groundwater of Panipat, Haryana, India. *Environ. Pollut.* **2020**, *259*, 113711. [CrossRef] [PubMed]
13. Gulis, G.; Czompolyova, M.; Cerhan, J.R. An ecologic study of nitrate in municipal drinking water and cancer incidence in Trnava District, Slovakia. *Environ. Res.* **2002**, *88*, 182–187. [CrossRef]
14. Gao, S.; Li, C.; Jia, C.; Zhang, H.; Guan, Q.; Wu, X.; Wang, J.; Lv, M. Health risk assessment of groundwater nitrate contamination: A case study of a typical karst hydrogeological unit in East China. *Environ. Sci. Pollut. Res.* **2020**, *27*, 9274–9287. [CrossRef] [PubMed]
15. Skold, A.; Cosco, D.L.; Klein, R. Methemoglobinemia: Pathogenesis, diagnosis, and management. *South. Med. J.* **2011**, *104*, 757–761. [CrossRef] [PubMed]
16. Xiao, J.; Jin, Z.; Zhang, F. Geochemical controls on fluoride concentrations in natural waters from the middle Loess Plateau, China. *J. Geochem. Explor.* **2015**, *159*, 252–261. [CrossRef]
17. Zhang, L.; Zhao, L.; Zeng, Q.; Fu, G.; Feng, B.; Lin, X.; Liu, Z.; Wang, Y.; Hou, C. Spatial distribution of fluoride in drinking water and health risk assessment of children in typical fluorosis areas in north China. *Chemosphere* **2020**, *239*, 124811. [CrossRef]
18. Qasemi, M.; Shams, M.; Sajjadi, S.A.; Farhang, M.; Erfanpoor, S.; Yousefi, M.; Zarei, A.; Afsharnia, M. Cadmium in Groundwater Consumed in the Rural Areas of Gonabad and Bajestan, Iran: Occurrence and Health Risk Assessment. *Biol. Trace Elem. Res.* **2019**, *192*, 106–115. [CrossRef]
19. Li, R.; Kuo, Y.M.; Liu, W.W.; Jang, C.S.; Zhao, E.; Yao, L. Potential health risk assessment through ingestion and dermal contact arsenic-contaminated groundwater in Jianghan Plain, China. *Environ. Geochem. Health* **2018**, *40*, 1585–1599. [CrossRef]
20. Pourret, O.; Hursthouse, A. It's time to replace the term "heavy metals" with "potentially toxic elements" when reporting environmental research. *Int. J. Environ. Res. Public Health* **2019**, *16*, 4446. [CrossRef]
21. Papazotos, P. Potentially toxic elements in groundwater: A hotspot research topic in environmental science and pollution research. *Environ. Sci. Pollut. Res.* **2021**, *28*, 47825–47837. [CrossRef] [PubMed]
22. Appelo, C.A.J.; Postma, D. *Geochemistry, Groundwater and Pollution*, 2nd ed.; CRC Press: Boca Raton, FL, USA, 2004; pp. 1–540.
23. Langmuir, D. *Aqueous Environmental Geochemistry*; Prentice Hall: Upper Saddle River, NJ, USA, 1997; pp. 50–429.
24. Apollaro, C.; Marini, L.; De Rosa, R. Use of reaction path modeling to predict the chemistry of stream water and groundwater: A case study from the Fiume Grande valley (Calabria, Italy). *Environ. Geol.* **2007**, *51*, 1133–1145. [CrossRef]
25. Fuoco, I.; De Rosa, R.; Barca, D.; Figoli, A.; Gabriele, B.; Apollaro, C. Arsenic polluted waters: Application of geochemical modelling as a tool to understand the release and fate of the pollutant in crystalline aquifers. *J. Environ. Manag.* **2022**, *301*, 113796. [CrossRef] [PubMed]
26. Fuoco, I.; Figoli, A.; Criscuoli, A.; Brozzo, G.; De Rosa, R.; Gabriele, B.; Apollaro, C. Geochemical modeling of chromium release in natural waters and treatment by RO/NF membrane processes. *Chemosphere* **2020**, *254*, 126696. [CrossRef] [PubMed]
27. Tian, R.; Wu, J. Groundwater quality appraisal by improved set pair analysis with game theory weightage and health risk estimation of contaminants for Xuecha drinking water source in a loess area in Northwest China. *Hum. Ecol. Risk Assess.* **2019**, *25*, 132–157. [CrossRef]

28. Egbueri, J.C. Groundwater quality assessment using pollution index of groundwater (PIG), ecological risk index (ERI) and hierarchical cluster analysis (HCA): A case study. *Groundw. Sustain. Dev.* **2020**, *10*, 100292. [CrossRef]
29. Zhou, Y.; Li, P.; Chen, M.; Dong, Z.; Lu, C. Groundwater quality for potable and irrigation uses and associated health risk in southern part of Gu'an County, North China Plain. *Environ. Geochem. Health* **2021**, *43*, 813–835. [CrossRef]
30. Horton, R.K. An index number system for rating water quality. *J. Water Pollut. Control. Fed.* **1965**, *37*, 300–306.
31. Bordalo, A.A.; Teixeira, R.; Wiebe, W.J. A Water Quality Index applied to an international shared river basin: The case of the Douro River. *Environ. Manag.* **2006**, *38*, 910–920. [CrossRef]
32. Sánchez, E.; Colmenarejo, M.F.; Vicente, J.; Rubio, A.; García, M.G.; Travieso, L.; Borja, R. Use of the water quality index and dissolved oxygen deficit as simple indicators of watersheds pollution. *Ecol. Indic.* **2007**, *7*, 315–328. [CrossRef]
33. Li, P.; Hui, Q.; Wu, J. Groundwater quality assessment based on improved water quality index in Pengyang County, Ningxia, Northwest China. *J. Chem.* **2010**, *7*, S209–S216. [CrossRef]
34. Wu, J.; Zhang, Y.; Zhou, H. Groundwater chemistry and groundwater quality index incorporating health risk weighting in Dingbian County, Ordos basin of northwest China. *Geochemistry* **2020**, *80*, 125607. [CrossRef]
35. Asadi, E.; Isazadeh, M.; Samadianfard, S.; Ramli, M.F.; Mosavi, A.; Nabipour, N.; Shamshirband, S.; Hajnal, E.; Chau, K.W. Groundwater Quality Assessment for Sustainable Drinking and Irrigation. *Sustainability* **2019**, *12*, 177. [CrossRef]
36. Zhang, Q.; Xu, P.; Qian, H. Groundwater Quality Assessment Using Improved Water Quality Index (WQI) and Human Health Risk (HHR) Evaluation in a Semi-arid Region of Northwest China. *Expo. Health* **2020**, *12*, 487–500. [CrossRef]
37. Sener, S.; Sener, E.; Davraz, A. Evaluation of water quality using water quality index (WQI) method and GIS in Aksu River (SW-Turkey). *Sci. Total Environ.* **2017**, *584–585*, 131–144. [CrossRef]
38. Kawo, N.S.; Karuppannan, S. Groundwater quality assessment using water quality index and GIS technique in Modjo River Basin, central Ethiopia. *J. Afr. Earth Sci.* **2018**, *147*, 300–311. [CrossRef]
39. Ram, A.; Tiwari, S.K.; Pandey, H.K.; Chaurasia, A.K.; Singh, S.; Singh, Y.V. Groundwater quality assessment using water quality index (WQI) under GIS framework. *Appl. Water Sci.* **2021**, *11*, 46. [CrossRef]
40. Xiao, J.; Wang, L.; Deng, L.; Jin, Z. Characteristics, sources, water quality and health risk assessment of trace elements in river water and well water in the Chinese Loess Plateau. *Sci. Total Environ.* **2019**, *650*, 2004–2012. [CrossRef]
41. Yang, X.; Duan, J.; Wang, L.; Li, W.; Guan, J.; Beecham, S.; Mulcahy, D. Heavy metal pollution and health risk assessment in the Wei River in China. *Environ. Monit. Assess.* **2015**, *187*, 111. [CrossRef]
42. Li, P.; Feng, W.; Xue, C.; Tian, R.; Wang, S. Spatiotemporal Variability of Contaminants in Lake Water and Their Risks to Human Health: A Case Study of the Shahu Lake Tourist Area, Northwest China. *Expo. Health* **2017**, *9*, 213–225. [CrossRef]
43. Qin, L.T.; Pang, X.R.; Zeng, H.H.; Liang, Y.P.; Mo, L.Y.; Wang, D.Q.; Dai, J.F. Ecological and human health risk of sulfonamides in surface water and groundwater of Huixian karst wetland in Guilin, China. *Sci. Total Environ.* **2020**, *708*, 134552. [CrossRef] [PubMed]
44. Guo, Y.; Li, P.; He, X.; Wang, L. Groundwater Quality in and Around a Landfill in Northwest China: Characteristic Pollutant Identification, Health Risk Assessment, and Controlling Factor Analysis. *Expo. Health* **2022**. [CrossRef]
45. Ji, Y.; Wu, J.; Wang, Y.; Elumalai, V.; Subramani, T. Seasonal Variation of Drinking Water Quality and Human Health Risk Assessment in Hancheng City of Guanzhong Plain, China. *Expo. Health* **2020**, *12*, 469–485. [CrossRef]
46. Papazotos, P.; Vasileiou, E.; Perraki, M. The synergistic role of agricultural activities in groundwater quality in ultramafic environments: The case of the Psachna basin, central Euboea, Greece. *Environ. Monit. Assess.* **2019**, *191*, 1–32. [CrossRef] [PubMed]
47. Papazotos, P.; Vasileiou, E.; Perraki, M. Elevated groundwater concentrations of arsenic and chromium in ultramafic environments controlled by seawater intrusion, the nitrogen cycle, and anthropogenic activities: The case of the Gerania Mountains, NE Peloponnese, Greece. *Appl. Geochem.* **2020**, *121*, 104697. [CrossRef]
48. Kubier, A.; Hamer, K.; Pichler, T. Cadmium background levels in groundwater in an area dominated by agriculture. *Integr. Environ. Assess. Manag.* **2020**, *16*, 103–113. [CrossRef]
49. Zhang, Y.; Wu, J.; Xu, B. Human health risk assessment of groundwater nitrogen pollution in Jinghui canal irrigation area of the loess region, northwest China. *Environ. Earth Sci.* **2018**, *77*, 273. [CrossRef]
50. Zhang, Y.; Xu, B.; Guo, Z.; Han, J.; Li, H.; Jin, L.; Chen, F.; Xiong, Y. Human health risk assessment of groundwater arsenic contamination in Jinghui irrigation district, China. *J. Environ. Manag.* **2019**, *237*, 163–169. [CrossRef]
51. Tian, Q.; Yin, J.; Hao, X. MIS3 climate change assessed according to loess deposition in the Linfen Basin, China. *Arid. Zone Res.* **2022**, *39*, 10–20. (In Chinese)
52. Lin, R.; Sun, C.; Chen, W.; Qiao, P.; Wang, Y. Groundwater quality characteristics in downstream valley of Fenhe river. *J. Arid. Land Resour. Environ.* **2021**, *35*, 135–142. (In Chinese)
53. Zhang, J.J. Structural characteristics of Linfen Basin. *Huabei Land Resour.* **2017**, *3*, 27–31. (In Chinese)
54. Bai, L.P.; Wang, Y.Y.; Wang, J.S. The numerical model-based groundwater level early-warning system: A case study of Linfen basin. *Geol. China* **2009**, *36*, 246–253. (In Chinese)
55. Wang, Y.; Bai, L.; Wang, J. Influences of climatic anomaly on the groundwater system in Linfen Basin. *Resour. Sci.* **2009**, *31*, 1168–1174. (In Chinese)
56. Ministry of Ecology and Environment of the People's Republic of China. *Technical Specifications for Environmental Monitoring of Groundwater (HJ/T 164-2004)*; China Environmental Science Press: Beijing, China, 2004; pp. 1–40. (In Chinese)

57. Sahu, P.; Sikdar, P.K. Hydrochemical framework of the aquifer in and around East Kolkata Wetlands, West Bengal, India. *Environ. Geol.* **2008**, *55*, 823–835. [CrossRef]
58. Yidana, S.M.; Yidana, A. Assessing water quality using water quality index and multivariate analysis. *Environ. Earth. Sci.* **2010**, *59*, 1461–1473. [CrossRef]
59. Ministry of Health of the P.R. China; Standardization Administration of the P.R. China. *Standards for Groundwater Guality (GB14848-2017)*; China Standard Press: Beijing, China, 2017; pp. 1–14. (In Chinese)
60. Li, P.; Li, X.; Meng, X.; Li, M.; Zhang, Y. Appraising Groundwater Quality and Health Risks from Contamination in a Semiarid Region of Northwest China. *Expo. Health* **2016**, *8*, 361–379. [CrossRef]
61. Chen, J.; Wu, H.; Qian, H.; Gao, Y. Assessing Nitrate and Fluoride Contaminants in Drinking Water and Their Health Risk of Rural Residents Living in a Semiarid Region of Northwest China. *Expo. Health* **2017**, *9*, 183–195. [CrossRef]
62. Zhu, L.; Yang, M.; Chen, X.; Liu, J. Health Risk Assessment and Risk Control: Drinking Groundwater in Yinchuan Plain, China. *Expo. Health* **2019**, *11*, 59–72. [CrossRef]
63. Wu, J.; Sun, Z. Evaluation of Shallow Groundwater Contamination and Associated Human Health Risk in an Alluvial Plain Impacted by Agricultural and Industrial Activities, Mid-west China. *Expo. Health* **2016**, *8*, 311–329. [CrossRef]
64. Ministry of Ecology and Environment of the P.R. China. *Technical Guidelines for Risk Assessment of Soil Contamination of Land for Construction (HJ 25.3-2019)*; China Environmental Science Press: Beijing, China, 2019; pp. 1–48. (In Chinese)
65. USEPA. *Risk Assessment Guidance for Superfund, Volume I: Human Health Evaluation Manual (Part A)*; EPA/540/1-89/002; USEPA: Washington, DC, USA, 1989.
66. National Health Commission of the People's Republic of China. *Report on Nutrition and Chronic Disease Status of Chinese Residents*; The State Council Information Office of the People's Republic of China: Beijing, China, 2020. (In Chinese)
67. USEPA. *Risk Assessment Guidance for Superfund, Volume I: Human Health Evaluation Manual (Part E, Supplemental Guidance for Dermal Risk Assessment) Final*; EPA/540/R-99/005; USEPA: Washington, DC, USA, 2004.
68. Mohamed, A.K.; Dan, L.; Kai, S.; Mohamed, M.A.A.; Aldaw, E.; Elubid, B.A. Hydrochemical Analysis and Fuzzy Logic Method for Evaluation of Groundwater Quality in the North Chengdu Plain, China. *Int. J. Environ. Res. Public Health* **2019**, *16*, 302. [CrossRef]
69. Liu, Y.L.; Luo, K.L.; Lin, X.X.; Gao, X.; Ni, R.X.; Wang, S.B.; Tian, X.L. Regional distribution of longevity population and chemical characteristics of natural water in Xinjiang, China. *Sci. Total Environ.* **2014**, *473–474*, 54–62. [CrossRef]
70. Karakuş, C.B. Evaluation of groundwater quality in Sivas province (Turkey) using water quality index and GIS-based analytic hierarchy process. *Int. J. Environ. Health Res.* **2019**, *29*, 500–519. [CrossRef] [PubMed]
71. Ramakrishnaiah, C.R.; Sadashivaiah, C.; Ranganna, G. Assessment of water quality index for the groundwater in Tumkur Taluk, Karnataka State, India. *J. Chem.* **2009**, *6*, 523–530. [CrossRef]
72. Xie, H.; Liang, Y.; Li, J.; Zou, S.; Shen, H.; Zhao, C.; Wang, Z. Distribution Characteristics and Health Risk Assessment of Metal Elements in Groundwater of Longzici Spring Area. *Environ. Sci.* **2021**, *42*, 4257–4266. (In Chinese)
73. Korner, P.; Georgis, L.; Wiedemeier, D.B.; Attin, T.; Wegehaupt, F.J. Potential of different fluoride gels to prevent erosive tooth wear caused by gastroesophageal reflux. *BMC Oral Health* **2021**, *21*, 183. [CrossRef] [PubMed]
74. Su, H.; Kang, W.; Xu, Y.; Wang, J. Assessing Groundwater Quality and Health Risks of Nitrogen Pollution in the Shenfu Mining Area of Shaanxi Province, Northwest China. *Expo. Health* **2018**, *10*, 77–97. [CrossRef]
75. Hou, C.; Chu, M.L.; Botero-Acosta, A.; Guzman, J.A. Modeling field scale nitrogen non-point source pollution (NPS) fate and transport: Influences from land management practices and climate. *Sci. Total Environ.* **2021**, *759*, 143502. [CrossRef] [PubMed]
76. Chen, J.; Wu, H.; Qian, H. Groundwater Nitrate Contamination and Associated Health Risk for the Rural Communities in an Agricultural Area of Ningxia, Northwest China. *Expo. Health* **2016**, *8*, 349–359. [CrossRef]
77. Zhang, Z.; Xiao, C.; Adeyeye, O.; Yang, W.; Liang, X. Source and Mobilization Mechanism of Iron, Manganese and Arsenic in Groundwater of Shuangliao City, Northeast China. *Water* **2020**, *12*, 534. [CrossRef]
78. He, X.; Ma, T.; Wang, Y.; Shan, H.; Deng, Y. Hydrogeochemistry of high fluoride groundwater in shallow aquifers, Hangjinhouqi, Hetao Plain. *J. Geochem. Explor.* **2013**, *135*, 63–70. [CrossRef]
79. Shen, L.; Wang, W. Distribution of fluorin ion in Yao du District of Lin fen City and causes of formation. *Water Resour. Prot.* **2005**, *21*, 76–78. (In Chinese)
80. Adimalla, N.; Wu, J. Groundwater quality and associated health risks in a semi-arid region of south India: Implication to sustainable groundwater management. *Hum. Ecol. Risk Assess.* **2019**, *25*, 191–216. [CrossRef]

Article

Assessment of the Hydrochemical Characteristics and Formation Mechanisms of Groundwater in A Typical Alluvial-Proluvial Plain in China: An Example from Western Yongqing County

Xueshan Bai [1], Xizhao Tian [1], Junfeng Li [2], Xinzhou Wang [1,*], Yi Li [1] and Yahong Zhou [3]

[1] Hebei Key Laboratory of Environment Monitoring and Protection of Geological Resources, Hebei Geological Environment Monitoring Institute, Shijiazhuang 050021, China; hjjcybaixs@163.com (X.B.); 13513399710@163.com (X.T.); 13933129093@163.com (Y.L.)
[2] Hebei Institute of Hydrogeology and Engineering Geology, Shijiazhuang 050000, China; lijie7098@163.com
[3] School of Water Resources and Environment, Hebei GEO University, Shijiazhuang 050031, China; zhyh327@163.com
* Correspondence: xzwang050031@163.com

Citation: Bai, X.; Tian, X.; Li, J.; Wang, X.; Li, Y.; Zhou, Y. Assessment of the Hydrochemical Characteristics and Formation Mechanisms of Groundwater in A Typical Alluvial-Proluvial Plain in China: An Example from Western Yongqing County. *Water* 2022, 14, 2395. https://doi.org/10.3390/w14152395

Academic Editor: Thomas M. Missimer

Received: 24 June 2022
Accepted: 28 July 2022
Published: 2 August 2022

Publisher's Note: MDPI stays neutral with regard to jurisdictional claims in published maps and institutional affiliations.

Copyright: © 2022 by the authors. Licensee MDPI, Basel, Switzerland. This article is an open access article distributed under the terms and conditions of the Creative Commons Attribution (CC BY) license (https://creativecommons.org/licenses/by/4.0/).

Abstract: The geographic location of Yongqing County is optimal, covering the center of the Beijing, Tianjin, and Baoding triangle. However, the economic and social development of Yongqing County in recent years has resulted in negative impacts on groundwater. Therefore, investigating the current status of groundwater chemistry in Yongqing County is of great significance to provide a useful basis for future studies on groundwater quality assessment. The aim of this study is to assess the hydrochemical characteristics and formation mechanisms of the unconfined aquifers of Yongqing County using descriptive statistical and multivariate statistical methods. In addition, ionic ratios, Piper diagram, Gibbs diagrams, and PHREEQC software were used in this study to determine the main factors influencing the hydrochemical characteristics of the unconfined aquifers. The results suggested slightly alkaline groundwater of the unconfined aquifers in the western part of Yongqing County, belonging to the fresh-brackish groundwater type. In addition, the hydrochemistry facies types in the study area are complex, consisting of four facies types, namely $HCO_3^- - Mg \cdot Ca$, $HCO_3^- - Na$, $HCO_3^- - Na \cdot Ca$, and $HCO_3^- - Na \cdot Mg$. On the other hand, the main factors influencing the hydrochemical characteristics of groundwater are mineral dissolution followed by some anthropogenic pollution. Rock dominance was the main influencing factor, demonstrated by the dissolution of silicate and carbonate rock minerals. In addition, the alternating adsorption of cations occurring in the aquifer plays a non-negligible effect on the hydrochemical characteristics of the unconfined aquifers in the study area. In fact, the validation results using PHREEQC inverse hydrogeochemical simulations demonstrated consistent conclusions with those mentioned above. According to the findings obtained, the dissolution of carbonate and silicate minerals as well as Na^+, K^+, and Ca^{2+} ion exchange in the aquifer are the main factors influencing the hydrochemical characteristics of the unconfined aquifers of Yongqing County. The recommendations suggest put forward in this research are helpful to understand the formation mechanism of hydrochemistry in typical alluvial proluvial plain and provide insights for decision makers to protect the groundwater resources.

Keywords: groundwater quality; water chemistry characteristics; multivariate statistical analysis; PHREEQC; typical mountain front tilted plain

1. Introduction

Water and environmental issues have attracted increasing attention over the past few decades due to the development of the economy and society. In fact, although the total amount of water resources in China is high, the per capita water use is less than a quarter of the world's per capita water use [1]. Therefore, China has been listed by the United Nations among the 13 water-poor countries. In the current situation of water scarcity in China, groundwater resource is becoming particularly important as they are considered the main source of water supply for most cities. To assess the current status of the regional groundwater environment, it is crucial to conduct studies on the hydrochemical characteristics and formation mechanisms of groundwater. These studies can, indeed, provide a general understanding of the hydrochemical characteristics of aquifers and the main factors influencing groundwater for future research.

At present, most studies on groundwater have been focused on hydrochemical characteristics and formation mechanisms of groundwater, hydrogeochemical simulation [2], human health risk assessment associated with organic and mineral groundwater pollution, groundwater pollution remediation technologies, and source of groundwater recharge and pollution using isotope technology. For example, Zhao et al. [3] assessed the spatiotemporal distribution of hydrochemical characteristics and formation mechanisms as well as the renewal ability of a confined aquifer in Hangzhou Bay New Area and showed that the aquifer was formed in the Late Pleistocene, with the absence of any hydraulic connection between this aquifer and other aquifers, which suggests a low regeneration capacity. Sang et al. [4] used the PHREEQC software to assess the hydrogeochemical characteristics of confined and unconfined aquifers in the delta area of the Nakdong River basin, Busan, Korea, which is located in the southeast of Beijing, China, and revealed unsaturated salt and supersaturated dolomite and calcite indices, which indicates that dissolution of carbonate rocks and ion exchange of major ions are the main hydrogeochemical processes in groundwater. Liu et al. assessed the groundwater quality and human health risk of groundwater samples collected from Yulin City and revealed good groundwater quality in the study area, with the presence of nitrate (NO_3^-) contamination in groundwater in agricultural areas [5], which suggests that reasonable groundwater management strategy should be established. Propp et al. [6] assessed groundwater quality in 20 historic landfills in Ontario, Canada, and indicated that most groundwater was strongly influenced by waste leachate, as landfills are long-term sources of several types of contaminants in groundwater. On the other hand, Pham et al. developed a new technique for the remediation of persistent contaminants, such as dense nonaqueous phase liquids in groundwater [7]. Moreover, innovative methods involving the sustained release of a selected reagent, namely persulfate, through pellets made from inorganic materials (e.g., zeolite, diatomite, and silica flour) are reported through a proof-of-concept study. This study demonstrated the potential feasibility of sustained persulfate release from inert matrices for groundwater treatment. Phan et al. used isotopic tracing techniques, namely δ^2H, $\delta^{18}O$, $\delta^{13}C$, δ^3H, and $\delta^{14}C$ activities, to assess the groundwater recharge, runoff, and discharge conditions of groundwater in the High Plains region of northeastern New Mexico, USA, which suggests that groundwater in the study area may be a mixture of Holocene groundwater and modern water co-existence [8]. Surface features (e.g., alluvial channels) promote the groundwater recharge, resulting in higher recharge rates in the region than the regional average rates.

This study aims to assess the regional hydrochemical characteristics of groundwater in a typical pre-hill alluvial plain area [9]. In addition, an inverse simulation technique was performed using the PHREEQC software to achieve qualitative and quantitative analyses of regional groundwater chemistry in the study area [10]. In this study, the chemical characteristics and formation mechanism of groundwater in the phreatic aquifer in a typical pre-hill tilted plain area were studied [11,12]. The study area is located in the eastern plain area of the Taihang Mountains, which is a typical pre-mountain sloping plain. The study area is a part of a typical alluvial-proluvial plain, characterized by a single groundwater facies type and good groundwater quality. However, the increasingly frequent human activities in recent years have affected quantitatively and qualitatively the groundwater in the study area. This study provides a solid theoretical basis for ensuring the safety of drinking water in Yongqing County and achieving the sustainable use of water resources.

2. Study Area Overview
2.1. Physical Geography Overview

Yongqing County is a county-level city in Langfang (116°22′–116°43′ E and 39°07′–39°28′ N), covering the middle of Hebei Province, the hinterland of the North China Plain, and the middle section of the Beijing-Tianjin Golden Corridor, with a total surface area of 776 km². The centers of the triangle of Beijing, Tianjin, and Baoding are located in the hinterland of Beijing and the economic circle around the Bohai Sea (Figure 1). The study area is located 60 km north of the capital Beijing, 60 km east of Tianjin City, 80 km from the capital airport, and 100 km from Tianjin Xingang. The study area is mainly located in the western part of Yongqing County and the eastern plain area of the Taihang Mountains, covering four townships, namely Yongqing Township, Longhuzhuang Township, Houyi Township, and Liujie Township, covering a total area of about 374 km². The northwestern and southeastern parts of the study area are characterized by high and low topography, respectively. In addition, the study is not flat and has some irregularities due to the interactive deposition of rivers, with the presence of several lowlands and depressions distributed at river crossings. The rivers are characterized by low erodibility since they are plain-type. According to the type of geomorphogenesis and surface morphology, the study area is located in a secondary geomorphic unit of alluvial the microtilt plain area.

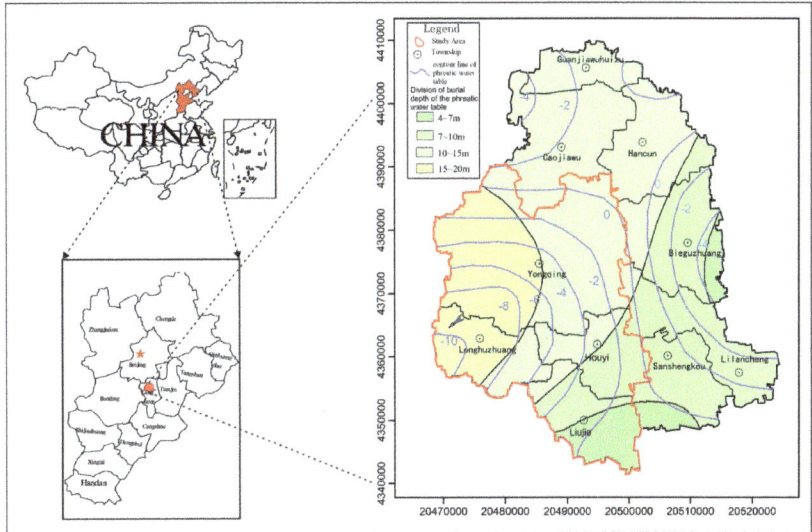

Figure 1. Groundwater burial depth classification and iso-water level map in the study area.

2.2. Socio-Economic Conditions

Yongqing is located in the new urban area of Langfang and a new space in Beijing, where the economy and society are developing rapidly. According to the statistics, the county's gross regional product was estimated at 22.01 billion yuan in 2020, while the general public budget revenue was estimated at 1.699 billion yuan. In addition, the per capita disposable income of urban and rural residents was 40,411 and 18,576 yuan, respectively. According to the data of the Seventh National Census Bulletin, released in November 2020, the resident population of Yongqing County is about 384,767 people. On the other hand, besides the largest gas-fired industrial zone in the northern part of Yongqing County, eight leading industries are present in Yongqing County, including chemical, pharmaceutical, cable, musical instrument, and building materials industries. Agriculture activities consist mainly of planting and animal husbandry. Yongqing has become a national pollution-free vegetable base county.

2.3. Hydrogeological Overview

The exposed strata in the study area consist of Cenozoic Quaternary loose deposits, with a thickness range of 350–500 m. The genetic types are complex, with a dominance of alluvial and lacustrine floodplains and their transitional types. However, the underlying strata of the Quaternary System consist of the Neoproterozoic and Paleocene strata.

The study area is located in the Yongding River area, including the alluvial-proluvial plain area and paleochannel zone. The groundwater system in the study area is classified into four aquifer groups, of which the first and second groups are shallow groundwater aquifers, consisting primarily of fine-to-medium sands and silty sand with gravel. The bottom boundary depth and aquifer thickness ranges of these aquifer groups are 160–180 m and 30–50 m, respectively. The third aquifer group is the deep groundwater aquifer, with bottom boundary depth and aquifer thickness ranges of 350–385 m and 60–100 m, respectively. The lithology of this aquifer changes from gravelly sand and medium sand to fine sand from north–south. The fourth aquifer group is the deep groundwater aquifer group, with bottom boundary depth and aquifer thickness ranges of 420–520 m and 20–40 m, respectively. There is no isolation factor among the aquifers, but the water flow is not quite the same in each layer because the aquifers are not homogeneous. The lithology of this aquifer consists mainly of fine-to-medium sands. On the other hand, the main recharge source of the shallow aquifers is atmospheric precipitation, followed by infiltration of irrigation and surface water and lateral runoff. Whereas mining activities are the main discharge source of the shallow aquifers, followed by runoff discharge downstream. The general shallow groundwater runoff trend is Northwest to Southeast, with a hydraulic gradient range of 0.8–1.4%. However, due to the existence of a local groundwater funnel, the direction of groundwater flow in some areas has changed (Figure 1).

3. Materials and Methods

3.1. Sample Collection and Analysis

In this study, a total of 14 groundwater samples were collected from shallow aquifer I+II during the monsoon period (June–August, 2011), with sampling buried depths of groundwater levels ranging from 50 to 150 m. The groundwater samples were collected appropriately according to the Code of Practice for Groundwater Environmental Monitoring (HJ 164-2020). The collected samples were first stored in a refrigerated box and then sent to the laboratory for chemical analysis. Samples requiring additives, added before sampling is completed. Finally, the status of the samples is checked regularly. The spatial distribution map of groundwater sampling points area was generated using the MAPGIS 6.7 software (Figure 2). It can be seen from Figure 2 that the sampling points were evenly distributed, covering the aquifer area in the study area.

The analytical data include the results of analyses of more than 20 parameters, in which the groundwater depth, longitude, latitude, smell and taste, turbidity, and naked-eye visible matter were measured and recorded in-situ. Other hydrochemical parameters, namely pH, hardness, total dissolved solids (TDS), mineralization, silicon dioxide (SiO_2), potassium (K^+), sodium (Na^+), calcium (Ca^{2+}), magnesium (Mg^{2+}), sulfate (SO_4^{2-}), chloride (Cl^-), bicarbonate (HCO_3^-), iron (Fe^{2+}/Fe^{3+}), fluorine (F^-), nitrite (NO_2^-), nitrate (NO_3^-), arsenic (As), and manganese (Mn). However, Fe^{3+} was not considered in this study due to its low concentration in groundwater, which was below the detection limit.

All the analytical methods used in the analyses were carried out according to the standard methods reported by Nsabimana et al. [13]. On the other hand, in order to check the reliability of the water quality analysis results, the ionic balance was used calculated according to the following formula:

$$E(\%) = \frac{\sum N_c - \sum N_a}{\sum N_c + \sum N_a} \times 100 \quad (1)$$

where E is the relative error; N_c is the concentration of the cation in the groundwater sample (meq/L); N_a is the concentration of the anion (meq/L).

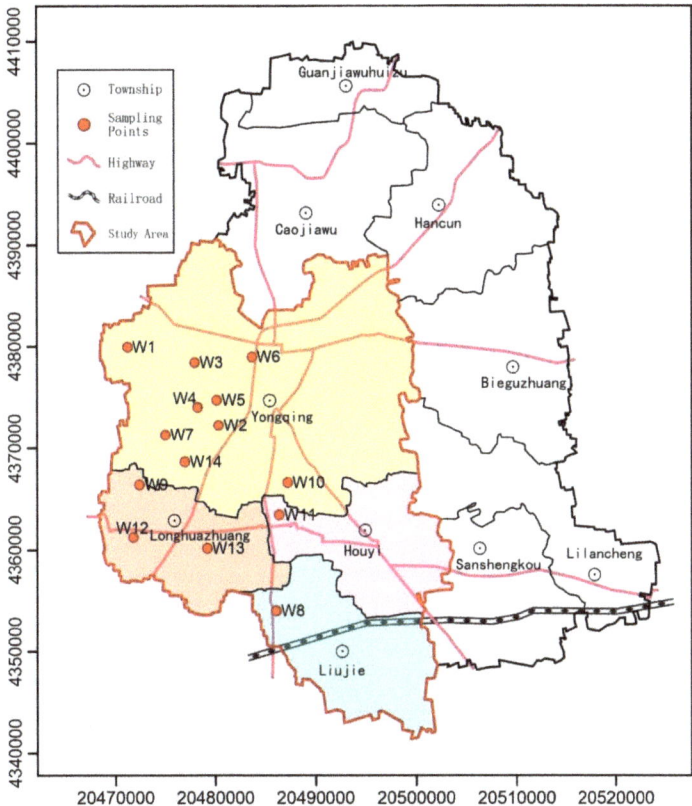

Figure 2. Spatial distribution of sampling points in the study area.

According to the results obtained, all the analytical data of groundwater samples showed E values less than ±5%, which suggests appropriate analytical methods.

3.2. Data Analysis Methods

In addition to, descriptive statistics, the principal component analysis test was performed in this study using the SPSS software to determine the main influencing factors affecting the hydrochemical characteristics of groundwater in Yongqing County [14,15]. In addition, ionic ratios were used to analyze the alternate adsorption of cations and investigate the main sources of hydrochemical elements in groundwater. On the other hand, the PHREEQC software was used for hydrogeochemical inversion simulations.

4. Results and Discussion

4.1. Descriptive Statistical Analysis

In order to determine the hydrochemical characteristics of groundwater, the SPSS statistical analysis software was used to perform descriptive statistical analysis (minimum, median, maximum, mean, and standard deviation) on the results data of 18 hydrochemical parameters of groundwater samples (Table 1).

According to the results obtained, pH values ranged from 7.5 to 8.3, with a mean value of 7.8, which indicates that the pH of groundwater in the western part of Yongqing County is slightly alkaline. In addition, the mineralization ranged from 345.5 to 1800.9 mg/L, with a mean value of 896.8 mg/L. In fact, some groundwater samples showed high mineralization values (above 1 g/L), indicating fresh-brackish groundwater. The TDS concentration range was 243.7–1333.6 mg/L, with a mean value of 638.4 mg/L, showing significant spatial variation in the TDS concentrations in the study area with the presence of higher concentrations than that of the National Standard for Drinking Water (GB 5749-2006). On the other hand, Na^+ was the most abundant cation in groundwater, with a mean concentration value of 116.764 mg/L, followed, respectively, by Ca^{2+}, Mg^{2+}, K^+, and Fe^{2+}. Whereas HCO_3^- was the highest anion in groundwater, with a mean concentration value of 516.77 mg/L, followed, respectively, by SO_4^{2-}, Cl^-, NO_3^-, F^-, and NO_2^-.

Table 1. Descriptive statistics of chemical parameters of groundwater.

Item	Min(mg/L)	Med(mg/L)	Max(mg/L)	Mean(mg/L)	SD
TDS	243.683	669.979	1333.622	638.442	331.674
HCO_3^-	203.589	545.767	934.549	516.770	255.924
TH	28.912	358.710	840.592	328.933	224.884
Na^+	44.100	106.100	220.700	116.764	66.109
SO_4^{2-}	26.500	57.300	244.200	87.821	65.843
Ca^{2+}	7.585	52.095	113.771	53.407	32.334
Cl^-	9.227	46.490	170.345	50.546	42.496
Mg^{2+}	2.421	54.967	143.836	47.487	38.019
SiO_2	13.290	18.502	23.930	18.671	2.458
pH	7.500	7.770	8.310	7.806	0.239
NO_3^-	1.420	3.780	9.590	3.687	2.077
F^-	0.175	0.641	3.083	0.896	0.780
K^+	0.400	0.600	2.600	0.754	0.573
Fe^{2+}	0.025	0.175	1.110	0.283	0.340
Mn	0.005	0.171	0.477	0.164	0.161
NO_2^-	0.002	0.006	0.064	0.014	0.017
As	0.001	0.002	0.004	0.002	0.001

By comparing the average concentration values of the hydrochemical parameters of groundwater with the Drinking Water Standards' (GB 5749-2006), it was found some ions exceeded the drinking limit values, which suggests a deterioration of the groundwater quality in the study area [16,17]. The main manifestations were SO_4^{2-}, NO_3^-, Mn, F^-, Na^+, and two combined indicators TH and TDS.

4.2. Basic Characteristics of Groundwater Chemistry

The groundwater facies types in the study area were determined by calculating first the relative content of anion and cation in the collected groundwater, and then drawn the Piper diagram (Figure 3). Actually, the groundwater facies type can be determined by plotting groundwater sample points in all three zones (2 triangles and 1 diamond) of the Piper diagram [18].

Through the comparison between the calculation and Piper diagram [19], it was found that among the 14 groundwater sample points, 2, 2, 3, and 7 sampling points fall in HCO_3^-–Mg·Ca, HCO_3^-–Na, HCO_3^-–Na·Ca, and HCO_3^-–Na·Mg type zones, respectively, which suggests complex hydrochemical characteristics of groundwater in the study area [20–22]. The spatial distribution of groundwater facies types in the study area is shown in Figure 4.

It can be seen from Figure 4 that the HCO_3^--Na·Mg facies type of groundwater was observed in a major part of the study area, covering mainly the east-central part of Yongqing Township, Houyi Township, Liujie Township, and the eastern part of Longhuzhuang Township. Whereas HCO_3^-–Na, HCO_3^-–Mg·Ca, and HCO_3^-–Na·Ca facies types were mainly distributed in the southwestern part of the urban area of Yongqing Town, the western part of Yongqing town, and the western part of Longhuzhuang Township, respectively. In addition, the groundwater facies types in the study area are more complex, which suggests influences of human activities on the hydrochemical characteristics of groundwater in the shallow aquifer in the study area [23].

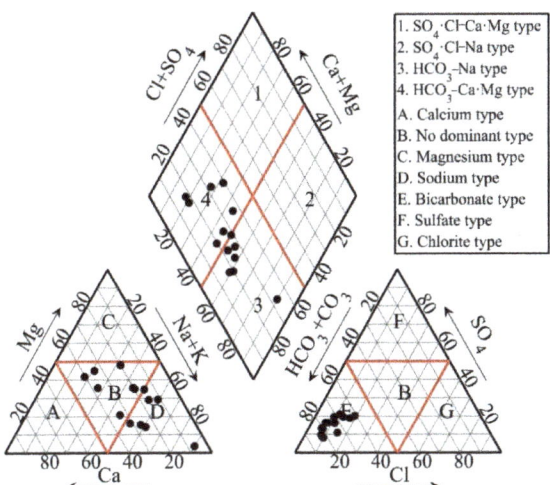

Figure 3. Piper diagram of phreatic groundwater in the study area.

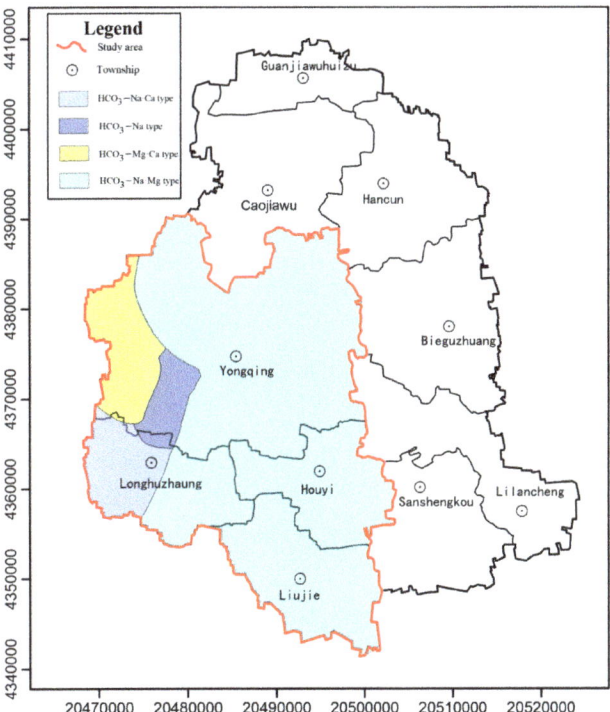

Figure 4. Zoning map of groundwater facies types in the study area.

Overall, the relative contents of HCO_3^-, Ca^{2+}, and Mg^{2+} were higher in fresh and brackish water with low TDS. Water facies types and ion abundances in groundwater were consistent with these hydrochemical results. Yongqing County is located in the alluvial fan plain in front of the Taihang Mountains. The general flow trend of the shallow groundwater is from northwest to southeast runoff. However, the local groundwater flow direction in some areas in Yongqing County is northeast-southwest due to the impact of the local groundwater landing funnel [24]. Groundwater facies types are mainly influenced by rainfall infiltration and lateral runoff, resulting in low TDS and mineralization of groundwater. The groundwater facies type was mainly represented by the bicarbonate facies type, which suggests the dissolution of calcite, dolomite, and other carbonate minerals from the Taihang Mountains.

4.3. Analysis of the Evolution Mechanism

In this study, multivariate statistics and water chemistry analysis methods were carried out to investigate the main factors influencing the hydrochemical characteristics of groundwater Yongqing County [25,26].

4.3.1. PCA (Principal Component Analysis)

Factor analysis is a statistical method used to effectively reduce the number of variables (dimensionality reduction) to minimize the loss of information in the original dataset, thus achieving a comprehensive analysis of data [27]. This statistical method has been widely used in various fields of research, including hydrogeological research. In this study, the factor analysis test was used to investigate the factors influencing the hydrochemical characteristics of groundwater.

It can be seen from the total variance explained that the first 4 principal components explained 87% of the total variance. In addition, the eigen values of the first 4 principal components were all greater than 1. Therefore, the first five principal components were selected. The main influencing factors were assessed to use the factors loading of the principal components (Table 2 and Figure 5). Factor loadings indicate the correlation between variables and principal components (PCs).

Table 2. Total variance explained by the factor analysis.

PCs	Initial Eigenvalue		
	Total	Variance Percentage	Cumulative Percentage
1	9.315	51.751	51.751
2	3.054	16.966	68.716
3	1.721	9.562	78.279
4	1.622	9.009	87.288
5	0.831	4.618	91.906
6	0.586	3.254	95.160
7	0.424	2.354	97.514
8	0.216	1.203	98.717
9	0.121	0.675	99.392
10	0.070	0.390	99.782
11	0.024	0.135	99.917
12	0.014	0.080	99.997
13	0.001	0.003	100.000
14	0.000	0.000	100.000
15	0.000	0.000	100.000
16	0.000	0.000	100.000
17	0.000	0.000	100.000
18	0.000	0.000	100.000

According to Tables 2 and 3, it can be seen that the variance contribution of Factor 1 (F1) accounted for 51.751% of the total variance, with strong positive loadings with HCO_3^-, Mg^{2+}, TH, TDS, and mineralization (M), thus the findings indicate that these five parameters are the most important, representing the dissolution of calcite, dolomite carbonate and salt minerals, and silicate minerals. The results showed that F1 represents the main factors influencing the hydrochemical characteristics of shallow groundwater in the study area. The cumulative variance contribution of factor 2 (F2) was 68.716%. This factor revealed strong positive loadings with Na^+ and F^-, which indicates that the high importance of these two parameters in F2. Therefore, it was suggested that F2 was related to the anthropogenic factors represented by F^- concentrations in groundwater. On the other hand, Factor 3 (F3) and Factor 4 (F4) explained smaller proportions of variance of 9.562 and 9.009%, respectively. The result showed strong positive loadings of NO_2^- and As on F3, which indicates the high importance of these parameters in F3. Whereas NO_3^- and Fe^{2+} revealed strong positive and negative loadings on F4, respectively, which indicates the high importance of these two parameters in F4. The results suggested that F3 and F4 are related to industrial and agricultural pollution. The provincial-level industrial Park in Yongqing County, approved by the People's Government of Hebei Province in 2003, may affect the groundwater quality due to industrial wastewater discharges. Moreover, the agricultural area accounts for about 75% of the total area of the study area, which suggests a significant negative impact of agricultural activities on the groundwater quality in Yongqing County.

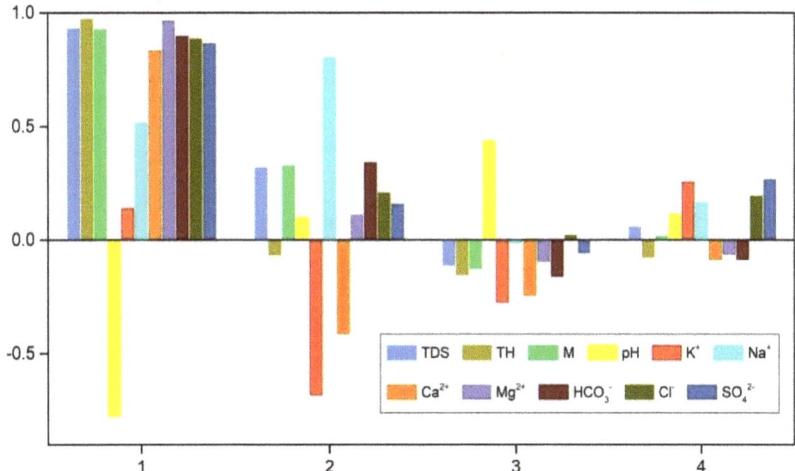

Figure 5. Factor loading values between variable and principal components.

Table 3. Factor loading values between variables and principal components.

Chemical Parameters	PCs			
	F1	F2	F3	F4
TDS	0.934	0.323	−0.116	0.062
TH	0.975	−0.071	−0.158	−0.081
M	0.930	0.332	−0.130	0.020
pH	−0.779	0.101	0.439	0.116
K^+	0.142	−0.686	−0.277	0.258
Na^+	0.520	0.806	−0.017	0.169
Ca^{2+}	0.836	−0.418	−0.248	−0.091
Mg^{2+}	0.970	0.114	−0.099	−0.069
HCO_3^-	0.900	0.345	−0.164	−0.090
Cl^-	0.889	0.211	0.024	0.196
SO_4^{2-}	0.868	0.161	−0.062	0.271
F^-	0.326	0.822	−0.232	0.154
NO_3^-	0.184	0.193	0.019	0.743
NO_2^-	−0.212	−0.112	0.883	0.173
Mn	0.883	−0.067	0.041	−0.292
Fe^{2+}	0.199	0.265	0.019	−0.786
As	−0.039	0.097	0.946	−0.125
SiO_2	0.259	−0.478	−0.473	−0.575

Note: Extraction method: principal component analysis; Rotation method: Caesar normalized maximum variance method a; a. The rotation was converged after 6 iterations.

4.3.2. Gibbs Diagram Analysis

Gibbs diagrams (Figure 6) have been widely used to reveal the ionic characteristics and determine the sources of the hydrochemical characteristics of river water [28,29]. In addition, they have been commonly used to analyze the hydrochemical characteristics of groundwater. Gibbs diagrams can be used to assess the relationship between TDS and $Na^+/(Na^++Ca^{2+})$ and between TDS and $Cl^-/(Cl^-+HCO_3^-)$ and identify the main sources of ions as well as the main factors influencing the hydrochemical characteristics of groundwater.

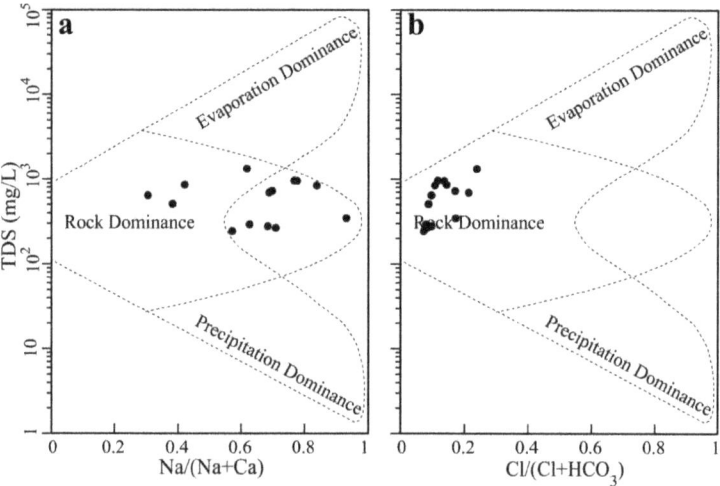

Figure 6. Gibbs diagrams, (**a**) TDS versus $Na/(Na+Ca)$, (**b**) TDS versus $Cl/(Cl+HCO_3)$.

It can be seen from Figure 6 that all water sampling points fall in the rock dominance control zone, which indicates that that rock dominance was the main influencing factor controlling the hydrochemical characteristics of groundwater in the study area. As shown in Figure 6a, Na^+ has a wide distribution and exhibit high proportions in some samples, indicating that sodium ions may be generated by a variety of sources. However, the proportion of Cl^- is small and the distribution is concentrated, indicating that all Cl^- is produced from the similar source or through the same geochemical process (Figure 6b). It should be noted that the Gibbs diagrams revealed only the natural factors influencing the groundwater chemical characteristics, while the human factors were not considered.

4.3.3. Inter-Ion Distance Diagram Analysis

According to the results of factor analysis and Gibbs diagram, it is evident that rock dominance and mineral dissolution were the main factors controlling the hydrochemical characteristics of groundwater. Therefore, to further investigate the mineral species derived from rock weathering and dissolution and to verify the results obtained using the principal component analysis, the HCO_3^-/Na^+, Ca^{2+}/Na^+, Mg^{2+}/Na^+, and Ca^{2+}/Na^+ ratios were used to distinguish between the influences of different rock and mineral weathering on groundwater components.

The results showed that most of the water sampling points were plotted between silicate weathering and carbonate dissolution zones, which suggests significant influences of the weathering of silicate and carbonate minerals on the hydrochemical characteristics of groundwater (Figures 7 and 8).

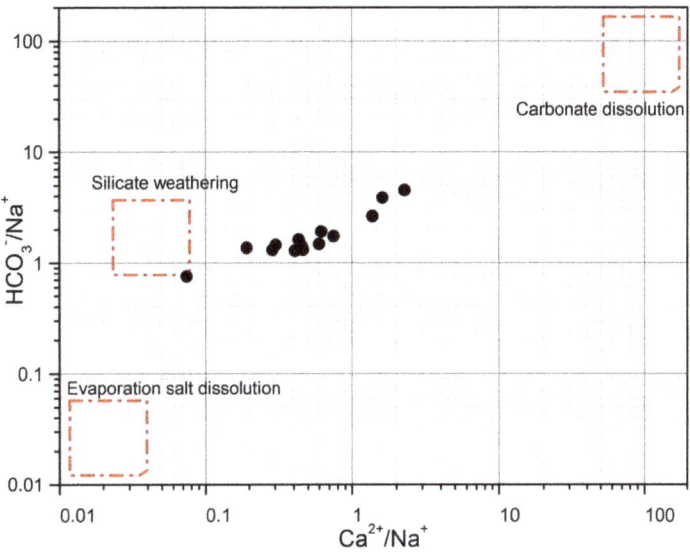

Figure 7. Relationship between Ca^{2+}/Na^+ and HCO_3^-/Na^+ of the phreatic aquifer in Yongqing County.

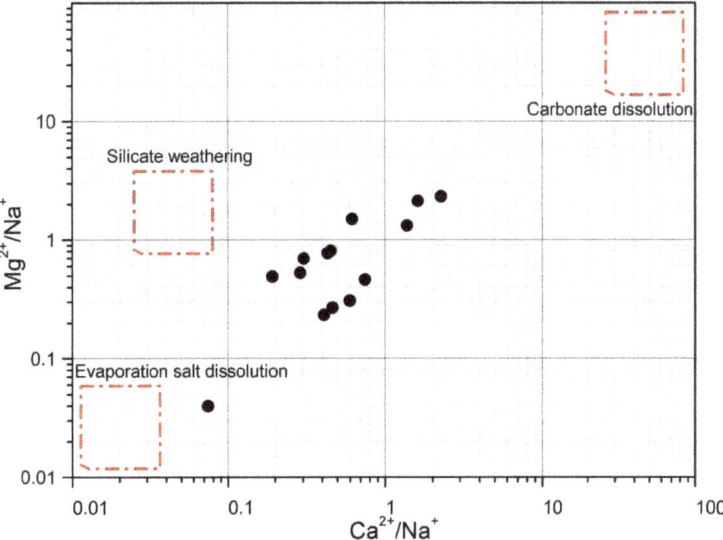

Figure 8. Relationship between Ca^{2+}/Na^+ and Mg^{2+}/Na^+ of phreatic aquifer in Yongqing County.

In order to further check whether the ion exchange process occurred in groundwater, the relationship diagram between $Ca^{2+}+Mg^{2+}-HCO_3^--SO_4^{2-}$ and $Na^++K^+-Cl^-$ were used for discrimination [30]. Water sampling points in the first quadrant of graphs (positive X and Y-coordinate values) suggest that rock salt dissolution is not the source of Na^+ and K^+, while the dissolution of peritectic minerals is not the source of Ca^{2+} and Mg^{2+} in groundwater, which is explained by the presence of a higher amount of Na^++K^+ than Cl^- and lower amount of $Ca^{2+}+Mg^{2+}$ than $HCO_3^--SO_4^{2-}$ (Figure 9). In addition, equal ions concentrations suggest that a cation exchange process occurred in groundwater.

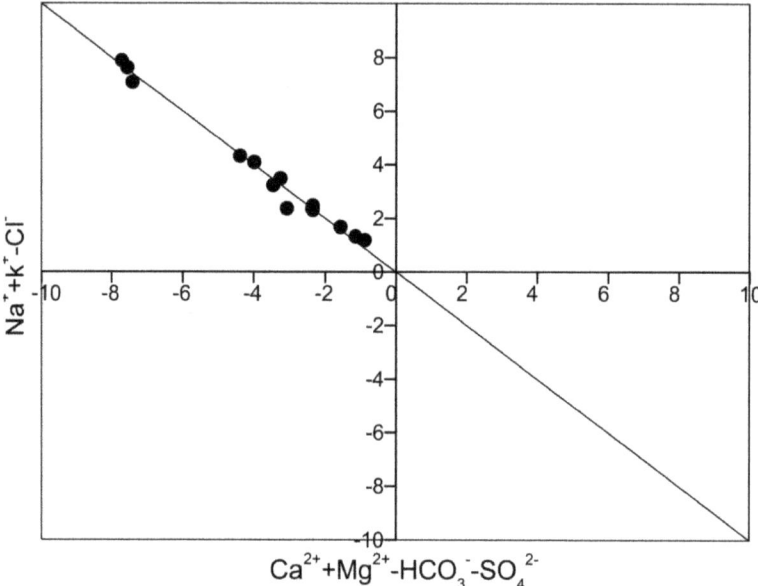

Figure 9. Relationship between $Ca^{2+}+Mg^{2+}-HCO_3^--SO_4^{2-}$ and $Na^++K^+-Cl^-$ ions of phreatic aquifer in Yongqing County.

According to the results obtained, all the water sampling points were plotted on the 1:1 line, indicating that the cation exchange process occurred in the unconfined aquifers of the study area.

4.4. HydrogeochemicalInverse Simulation

Yongqing County is located in the pre-mountain alluvial and flood plain of the Taihang Mountains. The unconfined aquifers in the study area are mainly recharged from atmospheric precipitation and lateral runoff recharge. In general, the unconfined aquifers of pre-mountain alluvial and floodplain have been controlled by lateral runoff recharge and are characterized by good groundwater quality and a single groundwater facies type, which is inconsistent with the results of this study. In fact, the results revealed complex groundwater facies types of the unconfined aquifers in the study area, the results suggest that numerous factors influencing the hydrochemical characteristics of groundwater quality. Therefore, the PHREEQC software was used in this study to comprehensively assess the hydrogeochemical process, investigate the major complex facies types, and analyze the water-rock interaction in the unconfined aquifers in the study area, providing quantitative analysis results.

4.4.1. Simulation Path Selection

The simulation path was selected in this study based on the geological and hydrogeological settings of the study area to better represent the evolution characteristics of groundwater in the entire study area [31,32]. Since the groundwater flow direction is northeast-southwest due to the influence of the local groundwater funnel, a simulation path was selected along this direction of groundwater flow, taking into the sampling points that are located in this area (Figure 10).

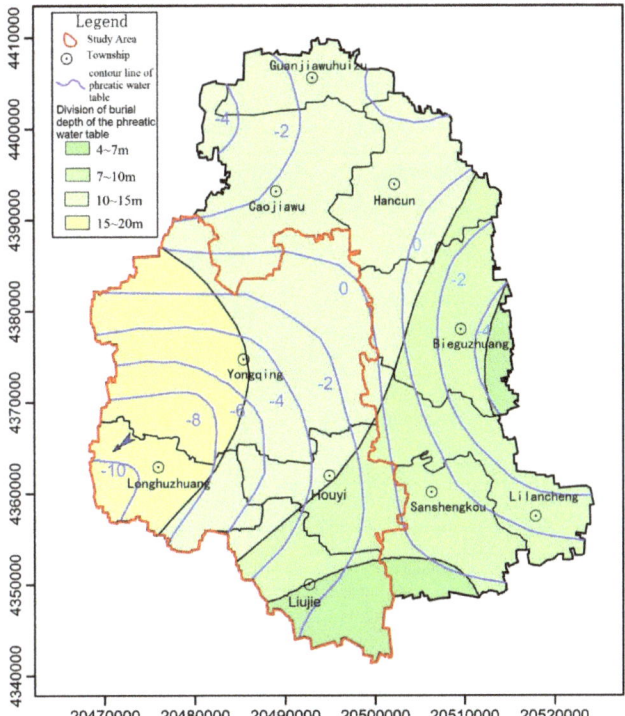

Figure 10. Hydrogeochemical inverse simulation path in the study area.

W6 and W12 were selected as the upstream and downstream, respectively. The hydrochemical characteristic results observed at these two sampling points were used as input data in the PHREEQC simulation software to perform inverse hydrogeochemical simulation. In addition, the uncertainty limit was set at 0.05.

4.4.2. Selection of Possible Mineral Phases

The hydrogeochemical process of the unconfined aquifers along the groundwater flow direction is closely related to the properties of surrounding rocks. Apparently, the mineral of surrounding rocks can provide insight into the hydrochemical components of groundwater. Therefore, the selection of mineral phases plays a key role in the accuracy of the simulation results.

The study area is located in the North China Plain, where the overlying strata are mainly Quaternary deposits. The unconfined aquifers are of Quaternary loose rock type. CO_2, as a possible mineral phase, can be used in the model due to the presence of gas-exchange surface area. In addition, the proportional relationship between ions suggested the occurrence of an exchange cation process in the aquifer, which implies that the reason for considering the ion exchange as a possible mineral phase in the simulation. On the hand, the results of the principal component analysis show that carbonate and silicate dissolutions were the main factors influencing the hydrochemical characteristics of groundwater. Therefore, carbonate minerals (e.g., calcite and dolomite) and aluminosilicate minerals (e.g., potassium feldspar and sodium feldspar) were considered as possible mineral phases in the simulation.

4.4.3. Analysis of Simulation Results

The simulation results of the hydrogeochemical processes of groundwater in the study area are shown in Tables 4 and 5.

Table 4. Upstream and downstream water saturation indices along the reaction path.

Mineral	Upstream	Downstream
Anhydrite	−2.08	−2.38
Calcite	0.65	0.77
Chalcedony	0.01	0.12
$CO_2(g)$	−1.80	−2.15
Dolomite	1.69	1.78
Fluorite	−0.34	−1.56
Gypsum	−1.78	−2.08
Halite	−6.42	−7.48
Hematite	16.24	19.50
Manganite	−5.65	−3.92
Quartz	0.44	0.55
Sylvite	−8.72	−8.94

Table 5. Results of the inverse simulation along the reaction path.

Phase	Formula	Mole Transfers (mol/L)
Gypsum	$CaSO_4:2H_2O$	-9.29×10^{-4}
Calcite	$CaCO_3$	2.20×10^{-3}
Calcium exchange	CaX_2	3.49×10^{-3}
Sodium exchange	NaX	-6.34×10^{-3}
Magnesium exchange	MgX_2	-3.19×10^{-4}
Carbon Dioxide(g)	$CO_2(g)$	-2.73×10^{-3}
Albite	$Na_2O \cdot Al_2O_3 \cdot 6SiO_2$	2.32×10^{-5}
K-feldspar	$KAlSi_3O_8$	1.93×10^{-5}
K-mica	$KAl_3Si_3O_{10}(OH)_2$	-1.42×10^{-5}
Fluorite	CaF_2	-6.49×10^{-5}
Manganite	$MnO(OH)$	3.75×10^{-6}
Hematite	Fe_2O_3	1.73×10^{-5}

Note: Positive and negative values indicate dissolution and precipitation, respectively.

The saturation indices of minerals in the upstream and downstream water are reported in Table 4. Positive and negative values indicate oversaturated and unsaturated groundwater samples, respectively. The results showed that calcite, chalcedony, dolomite, hematite, and quartz were all oversaturated. This finding can be explained by the fact that calcite and dolomite are the main minerals of the unconfined aquifers in the study area in the eastern plains in the Taihang Mountain front the remaining minerals revealed negative values, thus indicating unsaturation.

Table 5 shows the molar transfer of possible mineral phases along the reaction path. The results showed precipitation of gypsum, carbon dioxide, K-mica, and fluorite in the unconfined aquifers of the study area. Whereas Calcite, sodium exchange, K-feldspar, manganite, and hematite were in dissolved forms in the unconfined aquifers, with dissolved amounts of 2.20×10^{-5}, 1.93×10^{-5}, 3.75×10^{-6}, and 1.73×10^{-5} mol/L, respectively. On the other hand, the cation exchange in the unconfined aquifer was characterized by the exchange of Ca^{2+}, Na^+, and Mg^{2+}, which suggests adsorption of Na^+ and Mg^{2+} in groundwater, while Ca^{2+} in the exchange medium was in a dissolved form in groundwater. The amount values of Na^+, Mg^{2+}, and Ca^{2+} ion exchange were 6.34×10^{-3}, 3.19×10^{-4}, and 3.49×10^{-3} mol/L, respectively.

The results of the inverse hydrogeochemical simulations performed using the PHREEQC software are consistent with those obtained using the principal components analysis and water chemistry analysis. The main factors influencing the hydrochemical characteristics of groundwater are the dissolution of carbonate minerals (calcite and dolomite) and silicate minerals (sodium feldspar and potassium feldspar).

5. Conclusions

In this study, 14 groundwater samples were collected from the unconfined aquifers of Western Yongqing County to assess the hydrochemical characteristics of groundwater using descriptive, multivariate statistics, and water chemistry analysis methods. The following main conclusions were drawn from this study:

(1) The unconfined aquifers in the western part of Yongqing County revealed weakly alkaline groundwater. In addition, the results suggested fresh and brackish groundwater in the study area. The abundance of cations and anions followed the orders of $Na^+ > Ca^{2+} > Mg^{2+} > K^+ > Fe^{2+}$ and $HCO_3^- > SO_4^{2-} > Cl^- > NO_3^- > F^- > NO_2^-$, respectively.

(2) The relative content of anions and cations in groundwater and Piper's trilinear diagram demonstrated complex hydrochemical facies types of groundwater in the study area. The groundwater facies types were HCO_3^-–Mg·Ca, HCO_3^-–Na, HCO_3^-–Na·Ca, and HCO_3^-–Na·Mg.

(3) The main factors influencing the hydrochemical characteristics of groundwater were determined to use the principal component analysis, Gibbs diagrams, and ionic ratios. The results revealed that mineral dissolution, as well as some anthropogenic factors, are the main factors influencing the groundwater chemistry in the study area. In addition, Gibbs diagrams and ionic ratios revealed that silicates and carbonates were the main minerals influencing the hydrochemical characteristics of unconfined aquifers in the study area, followed by the alternating cation adsorption.

(4) Simulation of water-rock interactions in the unconfined aquifers was performed using the PHREEQC software. The simulation results are consistent with those obtained using the principal component analysis, Gibbs diagram, and ionic ratios. According to the obtained results, the dissolution of carbonate minerals (calcite and dolomite) and silicate minerals (sodium feldspar and potassium feldspar) were the main factors influencing the hydrochemical characteristics of the unconfined aquifers in the study area.

In conclusion, although there are anthropogenic activities in some areas of Yongqing County, the groundwater quality of the unconfined aquifers in the western part of Yongqing County is relatively good. Therefore, relevant departments need to strengthen the monitoring and management of groundwater quality to ensure scientific management and sustainable utilization of groundwater in Yongqing County.

Author Contributions: Conceptualization, X.B.; methodology, Y.L.; software, J.L.; validation, Y.Z.; formal analysis, X.T.; investigation, X.W.; resources, X.B.; data curation, X.T.; writing—original draft preparation, Y.Z.; writing—review and editing, J.L. All authors have read and agreed to the published version of the manuscript.

Funding: This research is funded by Open project of Hebei key laboratory of geological resources and environmental monitoring and protection (JCYKT202101); Natural Science Foundation of Hebei Province of China (D2022403016); Hebei University Science and technology research project (ZD2022119); Science and technology innovation team project of Hebei GEO University (KJCXTD-2021-14); Introduction of foreign intelligence project in Hebei province in 2021 (2021ZLYJ-1); Hebei water conservancy science and technology plan project (2021-45).

Informed Consent Statement: Not applicable.

Data Availability Statement: The data are all monitored by the Hebei Geological Environment Monitoring Institute, which are true and reliable.

Acknowledgments: The authors acknowledge the support provided by Hebei GEO University; Hebei Key Laboratory of Environment Monitoring and Protection of Geological Resources, Hebei Geological Environment Monitoring Institute, Shijiazhuang; Hebei Institute of hydrogeology and engineering geology.

Conflicts of Interest: The authors declare no conflict of interest.

References

1. Duan, R.; Li, P.; Wang, L.; He, X.; Zhang, L. Hydrochemical characteristics, hydrochemical processes and recharge sources of the geothermal systems in Lanzhou City, northwestern China. *Urban Clim.* **2022**, *43*, 101152. [CrossRef]
2. Qian, H.; Li, P.; Ding, J.; Yang, C.; Zhang, X. Formation of the River Water Chemistry in the Middle Section of Dousitu River, China. *E-J. Chem.* **2011**, *8*, 727–738.
3. Zhao, Y.; Wang, C.; Xiang, W.; Zhang, S. Evaluation of the hydrochemical evolution characteristics and renewable capacity of deep fresh groundwater in the Hangzhou Bay New Zone, China. *Environ. Earth Sci.* **2019**, *78*, 644. [CrossRef]
4. Sang, Y.; Rajesh, R.; Venkatramanan, S.; Selvam, S.; Paramasivam, C.; Yun, Y.; Hussam, E. Processes and characteristics of hydrogeochemical variations between unconfined and confined aquifer systems: A case study of the Nakdong River Basin in Busan City, Korea. *Environ. Sci. Pollut. Res.* **2020**, *27*, 10087–10102.
5. Liu, J.; Gao, M.; Jin, D.; Wang, T.; Yang, J. Assessment of Groundwater Quality and Human Health Risk in the Aeolian-Sand Area of Yulin City, Northwest China. *Expo. Health* **2020**, *12*, 671–680. [CrossRef]
6. Propp, V.; Silva, A.; Spencer, C.; Brown, S.; Catingan, S.; Smith, J.; Roy, J. Organic contaminants of emerging concern in leachate of historic municipal landfills. *Environ. Pollut.* **2021**, *276*, 116474. [CrossRef]
7. Pham, P.; Federico-Perez, R.; Fine, K.; Matzek, L.; Xue, Z. Sustained release of persulfate from inert inorganic materials for groundwater remediation. *Chemosphere* **2020**, *259*, 127508. [CrossRef]
8. Phan, V.; Zeigler, K.; Vinson, D. High Plains groundwater isotopic composition in northeastern New Mexico (USA): Relationship to recharge and hydrogeologic setting. *Hydrogeol. J.* **2021**, *29*, 1445–1461. [CrossRef]
9. Ren, C.; Zhang, Q. Groundwater Chemical Characteristics and Controlling Factors in a Region of Northern China with Intensive Human Activity. *Int. J. Environ. Res. Public Health* **2020**, *17*, 9126. [CrossRef]
10. Islam, M.N.; Jo, Y.T.; Park, J.H. Leaching and redistribution of Cu and Pb due to simulated road runoff assessed by column leaching test, chemical analysis, and PHREEQC modeling. *Environ. Earth Sci.* **2016**, *75*, 1041. [CrossRef]
11. Qian, H.; Li, P. Hydrochemical characteristics of groundwater in Yinchuan plain and their control factors. *Asian J. Chem.* **2011**, *23*, 2927–2938.
12. Li, J.; Zhang, Z.; Feng, Y.; Zhang, X. Use of genetic-algorithm-optimized back propagation neural network and ordinary kriging for predicting the spatial distribution of groundwater quality parameter. *Proc. SPIE Int. Soc. Opt. Eng.* **2013**, *3*, 8768.
13. Nsabimana, A.; Li, P.; He, S.; He, X.; Alam, S.M.; Fida, M. Health Risk of the Shallow Groundwater and Its Suitability for Drinking Purpose in Tongchuan, China. *Water* **2021**, *13*, 3256. [CrossRef]
14. Li, P.; Qian, H.; Wu, J. Hydrochemical Characteristics and Evolution Laws of Drinking Groundwater in Pengyang County, Ningxia, Northwest China. *E-J. Chem.* **2011**, *8*, 565–575.
15. Li, P.; Qian, H.; Wu, J. Hydrochemical formation mechanisms and quality assessment of groundwater with improved TOPSIS method in Pengyang County Northwest China. *E-J. Chem.* **2011**, *8*, 1164–1173.
16. Wu, J.; Li, P.; Qian, H. Hydrochemical characterization of drinking groundwater with special reference to fluoride in an arid area of China and the control of aquifer leakage on its concentrations. *Environ. Earth Sci.* **2015**, *73*, 8575–8588. [CrossRef]
17. Li, P.; Wu, J.; Qian, H. Hydrochemical appraisal of groundwater quality for drinking and irrigation purposes and the major influencing factors: A case study in and around Hua County, China. *Arab. J. Geosci.* **2016**, *9*, 15. [CrossRef]
18. Zhou, Y.; Hao, L.; Li, J. Chemical characteristics and formation mechanism of groundwater in the water source of Soro tree in Shandong Province. *South North Water Divers. Water Sci. Technol.* **2014**, *12*, 5.
19. Wang, Y.; Cheng, X.; Zhang, M.; Qi, X. Hydrochemical characteristics and formation mechanisms of Malian River in Yellow River basin during dry season. *Environ. Chem.* **2018**, *37*, 164–172.
20. Du, D.; Liu, H.; Zhang, J.; Miao, J.; Cao, X. Groundwater Chemical Characteristics and Salinization Mechanism in the Coastal Plain of the South Bank of Laizhou Bay. *IOP Conf. Ser. Earth Environ. Sci.* **2021**, *697*, 012026. [CrossRef]
21. Li, Z.; Zhang, C.; Zhou, Y. Spatio-temporal evolution characteristics and influencing factors of carbon emission reduction potential in china. *Environ. Sci. Pollut. Res.* **2021**, *28*, 59925–59944. [CrossRef] [PubMed]
22. Li, X.; Wang, R.; Li, J. Study on hydrochemical characteristics and formation mechanism of shallow groundwater in eastern Songnen Plain. *J. Groundw. Sci. Eng.* **2018**, *6*, 161–170.
23. An, L.; Zhao, Q.; Ye, S.; Liu, G.; Ding, X. Chemical characteristics and formation role of shallow groundwater in the Yellow River Delta. *Environ. Sci.* **2012**, *33*, 370–378.
24. Yang, T.; Wang, S. Analysis on the Chemical Characteristics of Shallow Groundwater and Causes of Formation in the Area around Poyang Lake. *Meteorol. Environ. Res.* **2011**, *2*, 77–80.
25. Wen, Y.; Qiu, J.; Cheng, S.; Xu, C.; Gao, X. Hydrochemical evolution mechanisms of shallow groundwater and its quality assessment in the estuarine coastal zone: A case study of qidong, china. *Int. J. Environ. Res. Public Health* **2020**, *17*, 3382. [CrossRef] [PubMed]
26. Li, P.; He, S.; He, X.; Tian, R. Seasonal hydrochemical characterization and groundwater quality delineation based on matter element extension analysis in a paper wastewater irrigation area, northwest China. *Expo. Health* **2018**, *10*, 241–258. [CrossRef]
27. Xu, F.; Li, P.; Du, Q.; Yang, Y.; Yue, B. Seasonal hydrochemical characteristics, geochemical evolution, and pollution sources of Lake Sha in an arid and semiarid region of northwest China. *Expo. Health* **2022**, 1–14. [CrossRef]
28. Roy, D.; Najafian, K.; Schleyer, P. Chemical evolution: The mechanism of the formation of adenine under prebiotic conditions. *Proc. Natl. Acad. Sci. USA* **2007**, *104*, 17272–17277. [CrossRef] [PubMed]

29. Li, D.; Gan, S.; Li, J.; Dong, Z.; Long, Q.; Qiu, S.; Zhou, Y.; Lu, C.; Li, C. Hydrochemical Characteristics and Formation Mechanism of Strontium-Rich Groundwater in Shijiazhuang, North China Plain. *J. Chem.* **2021**, *2021*, 5547924. [CrossRef]
30. Shi, W.; Li, L.; Zhang, L.; Tian, X. Analysis of groundwater chemistry characteristics in Qi Lihai emergency water source of Tianjin. *J. Water Resour. Water Eng.* **2016**, *27*, 98–102+107.
31. Li, W.; Wang, M.; Liu, L.; Yan, Y. Assessment of Long-Term Evolution of Groundwater Hydrochemical Characteristics Using Multiple Approaches: A Case Study in Cangzhou, Northern China. *Water* **2015**, *7*, 1109–1128. [CrossRef]
32. Mthembu, P.; Elumalai, V.; Brindha, K.; Li, P. Hydrogeochemical processes and trace metal contamination in groundwater: Impact on human health in the Maputaland coastal aquifer, South Africa. *Expo. Health* **2020**, *12*, 403–426. [CrossRef]

Article

Delineation of Hydrochemical Characteristics and Tracing Nitrate Contamination of Groundwater Based on Hydrochemical Methods and Isotope Techniques in the Northern Huangqihai Basin, China

Jing Jin [1,2,*], Zihe Wang [2], Yiping Zhao [2], Huijun Ding [3] and Jing Zhang [2]

1. Yinshanbeilu Grassland Eco-Hydrology National Field Observation and Research Station, Beijing 100038, China
2. Institute of Water Resources for Pastoral Area, MWR, Hohhot 010020, China
3. Geological Environment Monitoring Institute of Inner Mongolia, Hohhot 010020, China
* Correspondence: jinjingmwr@163.com; Tel.: +86-1535-4829-832

Abstract: Hydrochemical research and identification of nitrate contamination are of great significant for the endorheic basin, and the Northern Huangqihai Basin (a typical endorheic basin) was comprehensively researched. The results showed that the main hydrochemical facies were HCO_3–Mg·Ca and HCO_3–Ca·Mg. Spatial variation coefficients of most indices were greater than 60%, which was probably caused by human activities. The hydrochemical evolution was mainly affected by rock weathering and also by cation exchange. The D–^{18}O relationship of groundwater was $\delta D = 5.93\delta^{18}O − 19.18$, and the *d–excess* range was −1.60~+6.01‰, indicating that groundwater was mainly derived from precipitation and that contaminants were very likely to enter groundwater along with precipitation infiltration. The $NO_3(N)$ contents in groundwater exceeded the standard. Hydrochemical analyses indicated that precipitation, industrial activities and synthetic NO_3 were unlikely to be the main sources of nitrate contamination in the study area. No obvious denitrification occurred in the transformation process of nitrate. The $\delta^{15}N(NO_3)$ values ranged from +0.29‰ to +14.39‰, and the $\delta^{18}O(NO_3)$ values ranged from −6.47‰ to +1.24‰. Based on the $\delta^{15}N(NO_3)$ – $\delta^{18}O(NO_3)$ dual isotope technique and hydrochemical methods, manure, sewage and NH_4 fertilizers were identified to be the main sources of nitrate contamination. This study highlights the effectiveness of the integration of hydrochemical and isotopic data for nitrate source identification, and is significant for fully understanding groundwater hydrochemistry in endorheic basins and scientifically managing and protecting groundwater.

Keywords: endorheic basin; hydrochemistry; isotope technique; integration; nitrate

1. Introduction

Water quantity affects the extent of water exploitation, while water quality determines the value of water use [1]. In the natural environment, the specific hydrochemical characteristics of groundwater are formed over time in response to the comprehensive influence of climate, topography, aquifer lithology and other factors. The groundwater in some areas is low salinity freshwater and is rich in trace elements that are good for human health (such as Sr, Li and H_2SiO_3). This kind of water has very high value for use. However, in some areas, the groundwater is naturally inferior, characterized by high levels of salinity, fluorine and arsenic [2,3], which may aggravate water shortages due to poor water quality, especially for the endorheic basins that water resources are rare. Due to the intensification of human activities, the hydraulic head field and hydrochemical evolution process have been disturbed to a certain extent [4,5]. Variation in the vadose zone thickness changes the oxidation–reduction environment of the dissolved minerals during the leaching process. Coupled with the input of artificial contaminants, the hydrochemical characteristics may change. Some activities may improve groundwater quality, while others cause groundwater pollution, such as nitrate contamination [6–9]. A high content of nitrate in water leads to eutrophication of water bodies and degradation of ecosystems. Drinking groundwater with a high nitrate content for a long time can cause serious diseases, such as methemoglobinemia, blue babies and gastric cancer [9]. At present, nitrogen contamination in water has become an international problem and needs to be solved in order to ensure the safety of drinking water and sustain the ecological health.

Approximately 20% of the Earth's land is covered by endorheic basins, but the basins account for only 2.3% of the total worldwide annual river runoff, and the hydrochemical research on them has not received enough attention [10]. Dowling et al. [11] studied the arsenic releasing mechanisms in the Bengal Basin based on the statistical methods and correlation analysis. Kawawa et al. [12] used the hydrochemical methods and isotopic techniques to study the mechanism of salinity changes and hydrochemical evolution of groundwater in the Machile–Zambezi Basin, and concluded that high groundwater salinity was associated with pre–Holocene environmental changes and was restricted to a stagnant saline zone. Nipada et al. [13] took the Western Lampang Basin as the study area and researched the arsenic contamination in groundwater based on the PHREEQC software. Endorheic basins are widely distributed in China and nitrate contamination occurs. Many studies have been performed by domestic scholars [5,14,15], but these studies were mainly focused on special indices, such as arsenic and fluorine [16,17]. As regards nitrate concentration, some scholars [6,8,18] pointed out that the nitrate concentrations in their studied basin increased due to the human activities. Mukherjee et al. [19] indicated that the ingestion of untreated nitrate contaminated groundwater in the lower Ganga Basin caused a risk of methemoglobinemia. Avilés et al. [20] concluded that the nitrate content in the Titicaca Basin was influenced by manure piles, synthetic N fertilizers, and sewage collector pipes based on the $\delta^{15}N(NO_3)$–$\delta^{18}O(NO_3)$ isotopic technique. One major limitation of these studies is that the hydrochemical characteristics and identification of nitrate sources were separately researched in general. In fact, the comprehensive analysis of hydrochemical characteristics is conducive to revealing the variability of nitrate and identifying its source, and the two parts should be combined and comprehensively researched.

The northern Huangqihai Basin, located in the northern China, is a typical endorheic basin and plays an important role in the Beijing–Tianjin–Hebei region. Huangqihai Lake is one of the eight well–known lakes in Inner Mongolia, but its area shrank in the past two decades [21]. The groundwater level obviously declines [22]. Nitrate and other indices in the groundwater in some areas exceed the standard and are not suitable for drinking. Excessive exploitation and groundwater quality deterioration aggravate the contradiction between the supply and demand of groundwater resources. What was worse, the ecosystem reliant on groundwater resources has become increasingly fragile. Previous research on the northern Huangqihai Basin mainly focused on ecology [23–25], and pointed out that the wetland degeneration and the ecological deterioration were mainly controlled by a series of human activities, such as the unreasonable exploitation of groundwater, river closure and the increase of the building land. Regarding water quality, a few scholars [26,27] evaluated the trophic level of the surface water. However, few studies on the hydrochemistry and nitrate source of groundwater in the Huangqihai Basin have been reported. The northern Huangqihai Basin is an endorheic basin typical in arid and semiarid regions, and the hydrochemical research on it is expected to well develop the research system of endorheic basin. The study objective was the Quaternary phreatic water which often constitutes the most important source of drinking water in semiarid and arid regions but easily influenced by external factors, human health is closely associated with its hydrochemical evolution. In addition, the Huangqihai Basin plays an ecological significant role in the Beijing–Tianjin–Hebei region. Therefore, research on the hydrochemistry and tracing nitrate contamination in the northern Huangqihai Basin is not only significant for developing the theoretical research on endorheic basin, but also has great practical meanings of the sustainable utilization of regional groundwater and ecological protection.

Currently, there are few reports that have systematically and comprehensively analyzed the hydrochemical characteristics and nitrate contamination of groundwater, especially in endorheic basins. This paper intends to comprehensively analyze the hydrochemical characteristics and seeks to highlight the effectiveness of the combined use of hydrochemical and isotopic data for tracing nitrate source. Therefore, the main objectives of this study are: (1) to analyze the hydrochemical characteristics of groundwater and the spatial distributions of the main indices; (2) to reveal the hydrochemical evolution of groundwater; (3) to identify the source of groundwater and the way through which artificial contamination may enter groundwater; and (4) to integrate the dual isotope technique and hydrochemical analyses to identify nitrate contamination. It is expected that this research can enrich hydrogeochemical research on endorheic basins in arid and semiarid regions and provide an effective way to identify the nitrate source.

2. Materials and Methods

2.1. Study Area

The northern Huangqihai Basin (113°2′–113°28′ E, 40°43′–41°3′ N), is located in Right Chahaer County and the Jinning District of Wulanchabu city in Inner Mongolia, China (Figure 1a). The study area has a continental monsoon climate with an annual average temperature of 5.23 °C. The annual average rainfall is 359.30 mm and mainly concentrated in summer. The basin is surrounded by mountains on three sides; the terrain is generally high in the north and low in the south. Surface water resource is rare, the main rivers have dried up in their middle and downstream regions in recent years, and the other rivers are seasonal. The development of the regional social economy is highly dependent on groundwater, especially for agriculture.

The entire study area is covered by the unconsolidated Quaternary sediments and the main aquifer is the Quaternary phreatic aquifer. Based on the lithologic characteristics, the phreatic aquifer is further divided into two aquifers: the Quaternary Holocene lacustrine aquifer (Q_4^l) and the Quaternary Upper Pleistocene alluvial–diluvial aquifer (Q_3^{al+pl}) (Figure 1b). The Q_4^l aquifer is distributed around Huangqihai Lake and consists of medium and fine sand, while the Q_3^{al+pl} aquifer is distributed around the Q_4^l aquifer and mainly composed of sandy gravel, pebbles and coarse sand. According to previous research [22], the dynamic type of the groundwater level is the rainfall infiltration–artificial exploitation type. In summer, rainfall is abundant, but the groundwater level does not immediately rise and even declines due to the high consumption of irrigation. After irrigation, the groundwater level recovers due to the hysteresis recharge of rainfall and reaches a high level in spring. The groundwater levels and depths may change over time, but the overall flow direction of the Quaternary phreatic water does not obviously change during a hydrological year, that is, the groundwater generally flows from the north to the south following the topography. As for the Quaternary phreatic water (Figure 1b), the Q_3^{al+pl} aquifer is located upstream of the hydraulic head field and the Q_4^l aquifer is located downstream of that. Influenced by the terrain, geomorphic type and hydrological conditions, the hydraulic gradient upstream is steeper than that downstream. As seen from Figure 1c, the groundwater depth changes from deep to shallow from north to south. The groundwater depths near Huangqihai Lake and rivers are usually shallower than 5 m, and the depths in other areas are deeper than 5 m. According to previous studies [22], the extreme evaporation depth of the groundwater is 5 m, in other words, the depths of groundwater in most areas exceed the extreme evaporation depth.

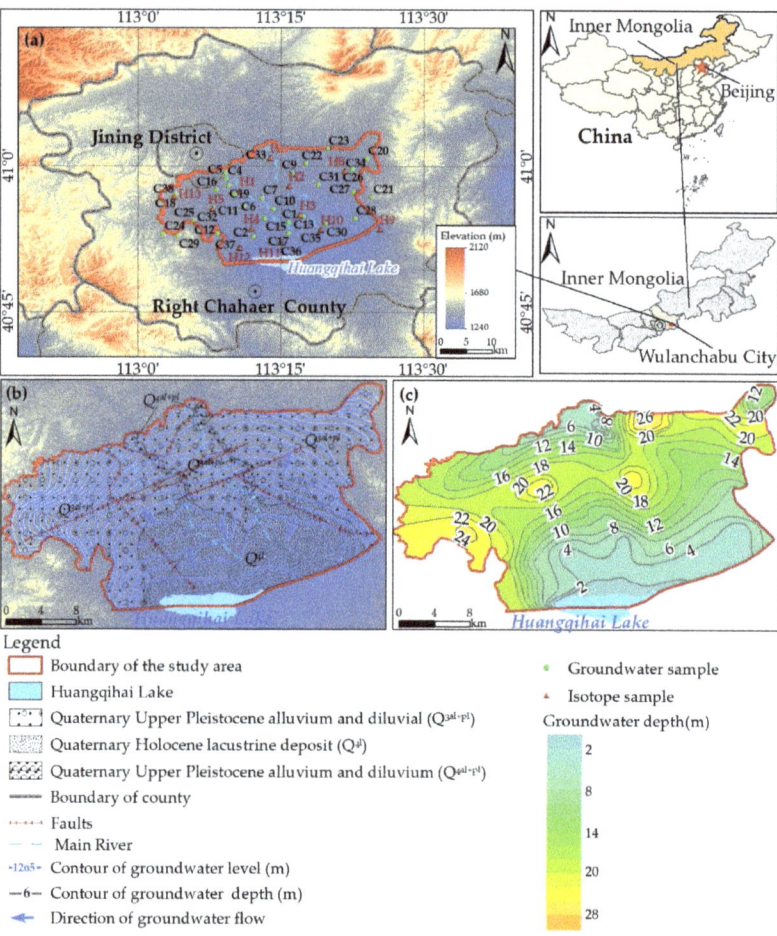

Figure 1. Location of the study area and water samples (**a**), hydraulic head field (**b**) and the burial depth of groundwater (**c**).

2.2. Data Preparation and Methods

2.2.1. Data Preparation

The groundwater level and depth data were measured in late September 2021. Thirty-eight groundwater samples were collected according to the Groundwater Quality Standard (GB/T 14848-2017) [28]. Before sampling, the wells were pumped for thirty minutes to obtain fresh groundwater. All groundwater samples were collected from the Quaternary phreatic aquifer and evenly distributed in different hydrogeological units (Figure 1a). Groundwater samples were sealed and stored in 5 L PVC bottles that were carefully cleaned before sampling. After collection, the samples were kept at 4 °C and later sent to the Inner Mongolia Mineral Resources Experimental Research Institute for analyzing. Hydrochemical indices were analyzed by using the standard methods as suggested by Analysis Methods of Groundwater quality (DZ/T 0064.1–2021) [29]. pH was analyzed by an ion meter (PXJ–1B, Jiangsu Electric Analysis Instrument Factory, Jiangyan, China). The concentrations of cations (Mg^{2+}, Ca^{2+}, Na^+ and K^+) and some trace elements (Fe, Mn, Cu, Pb, Cd, etc.) were analyzed by a PerkinElmer Optima 8300 with a detection accuracy of 0.001 mg/L. The concentrations of anions (SO_4^{2-}, Cl^-, F^- and NO_3^-) were measured by ion chromatography (IC850). The concentrations of NH_4^+, NO_2^- and H_2SiO_3

were analyzed by a visible spectrophotometer (7200, Tianmei Scientific Instument Co., LTD, Shanghai, China). The concentrations of HCO_3^-, total hardness (TH) and chemical oxygen demand of manganese (COD_{Mn}) were analyzed by the titration method. The total dissolved solids (TDS) were determined by the weighing method (Electronica scales JA31001). The accuracy of the testing results was checked using an ionic error equilibrium, and the relative error was controlled below 3%, which meant that the analyzing results were reliable [30]. Twenty–five chemical indices of the groundwater samples were analyzed. The concentrations of some indices, such as Cu, Cd, Hg, and Cr^{6+}, were low, even below the detection limit. Based on the previous study and the real conditions of the study area, the analysis was focused on the main ions and the overstandard indices.

The samples for testing δD, $\delta^{18}O$, $\delta^{15}N(NO_3)$ and $\delta^{18}O(NO_3)$ were collected in late September 2021 and early May 2022 based on the Handbook of Hydrogeology [30], and twelve D–^{18}O isotope samples and twelve ^{15}N–^{18}O(NO_3) isotope samples were collected in each phase. These samples were collected form the Quaternary phreatic aquifer, and numbered H1–H6 and H8–H13. D–^{18}O isotope samples were sealed in 10 mL EP plastic tubes, and ^{15}N–^{18}O(NO_3) isotope samples were sealed in 50 mL EP plastic tubes after filtering with a 0.45 μm filter, and then stored at a low temperature. All isotope samples were analyzed by the LICA United Technology Limited. The D–^{18}O isotope test machine was a Liquid water isotope analyzer (912–0050, Los Gatos Research, Inc., San Jose, CA, USA), and the ^{15}N–^{18}O(NO_3) isotope test machine was a Thermo Fisher MT253 and Flash 2000HT. Three parallel samples (H10′, H11′ and H12′) for analyzing those isotopes were collected and sent to another testing organization (Institute of Hydrogeology and Environmental Geology, Chinese Academy of Geological Sciences), and the relative errors between the results testing from the two testing organizations were all less than 3%.

The main ions (Mg^{2+}, Ca^{2+}, Na^+, K^+, HCO_3^-, SO_4^{2-}, Cl^-) and other indices (NO_3^-, NO_2^-, NH_4^+, TDS, TH, COD_{Mn} and H_2SiO_3) were used to reflect the hydrochemical characteristics, reveal the hydrochemical evolution and evaluate groundwater quality. The D and ^{18}O isotopes were conducted to analyze the source of groundwater. The relationship analysis among NO_3^-, SO_4^{2-}, Cl^-, Na^+ and K^+ and the $^{15}N(NO_3)$ and $^{18}O(NO_3)$ isotopes were combined to accurately identify nitrate contamination.

2.2.2. Methods

In this study, the hydrochemical characteristics of groundwater were analyzed from four aspects: the concentration characteristics of hydrochemical indices, spatial distribution, hydrochemical facies and correlation analysis among hydrochemical indices. Second, the hydrochemical evolution mechanisms were further studied by applying hydrochemical methods. Then, the main source of groundwater was recognized by using the D–^{18}O isotope technique. Based on the evaluation of groundwater quality, the main sources of nitrate in groundwater were identified by integrating the hydrochemical and isotopic data. The methodology flowchart (Figure 2) is shown below.

(1) Mole fraction

The mole fraction is the ratio of the amount of substance in a solution to the sum of the amounts of substance in each component, and the equation is listed below [31]. The mole fraction can reflect the relative amount of a substance in a solution and is the basis of the classification of hydrochemical facies and ion proportional coefficient method.

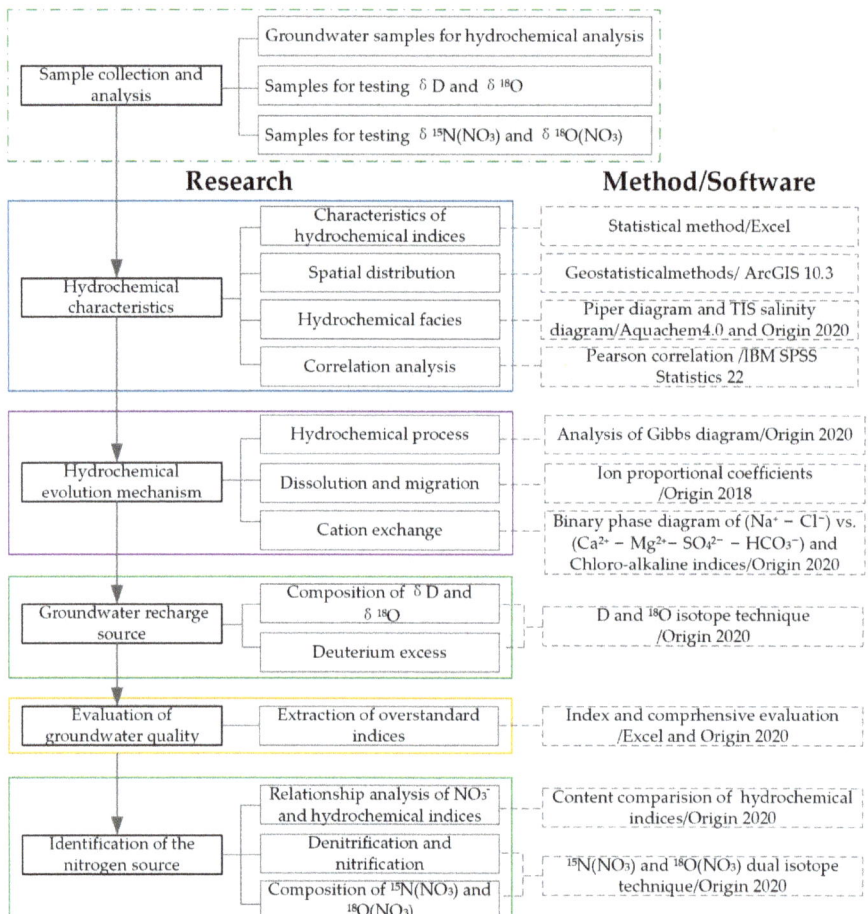

Figure 2. Methodology flowchart. At present, many methods are used to proceed the hydrochemcial research, such as Piper diagram, correlation analysis, Gibbs diagram, ion proportional coefficients, cluster–based methods [14,31–37]. According to previous studies and the study objectives, we used Excel to complete the statistical analysis. Statistics of the minimum, maximum, average and medium were used to reflect the concentration characteristics of the indices. The standard deviation (SD) was applied to reflect the variation degree between the average value and the actual value, and the coefficient of variation (C_v) was used to indicate the dispersion degree [38]. Based on the geostatistical methods, the spatial distributions of the main hydrochemical indices were obtained by using the Kriging interpolation with the help of ArcGIS 10.3. The Piper diagram drawn by Aquachem 4.0 and the iso–ionic–salinity (TIS) diagram were applied to classify the hydrochemical facies and the salinity distribution. Pearson correlation was undertaken to analyze the relationships among the hydrochemical indices via IBM SPSS Statistics 22. Gibbs diagram, ion proportional coefficients were conducted to reveal the hydrochemical evolution mechanism. Based on the binary phase diagram and chloro – alkaline indices, the cation exchange was judged. Integration of the $\delta^{15}N(NO_3) - \delta^{18}O(NO_3)$ dual isotope technique and hydrochemical analyses was designed to trace the nitrate contamination. Origin 2020 was used to plot Gibbs diagram, TIS diagram, binary phase diagrams, and isotope distribution figures. The details of the methods are listed below.

$$C_i = \frac{\rho_i / M_{ri}}{\sum_{i=1}^{n} \rho_i / M_{ri}} \times 100\% \tag{1}$$

where C_i is the mole fraction of the i ion, %; ρ_i is the mass concentration, mg/L; and M_{ri} is the relative molecular mass, 1.

(2) Shukalev classification

Shukalev classification is a common method of classifying hydrochemical facies and mainly based on the mole fractions of the main ions (Ca^{2+}, Mg^{2+}, $Na^+ + K^+$, HCO_3^-, SO_4^{2-} and Cl^-). Ions with a mole fraction greater than 25% should participate in the classification of the hydrochemical facies. Based on this, there are 49 hydrochemical facies [30].

Noticeably, the NO_3^- ion is not considered according to the Shukalev classification. This is mainly because the NO_3^- content is lower than 25% in the natural groundwater environment. However, due to the disturbance of human activities, the content of NO_3^- may increase and influence the hydrochemical evolution. Kpa o et al. [39] pointed out that the new hydrochemical types (NO_3–Ca and NO_3–Na type) were widely found in many agricultural areas of the former Soviet Union and the United States. Huang et al. [40] improved the Shukalev classification and took NO_3^- into consideration when determining the hydrochemical facies. Due to this, the NO_3^- ion was taken into consideration in this study to indicate nitrate contamination.

(3) Cation exchange

Through cation exchange, Ca^{2+} and Mg^{2+} in groundwater were replaced by Na^+, which may affect the cation concentrations and hydrochemical facies [16]. The binary phase diagram of ($Na^+ - Cl^-$) vs. ($Ca^{2+} + Mg^{2+} - SO_4^{2-} - HCO_3^-$) can indicate whether cation exchange occurs [16] based solely on the milligram equivalent ratios, this method is simple and not explained in detail here. Chlor–alkali indices can reflect the direction of cation exchange analysis based on the two indices (CAI_1 and CAI_2), and the equations were listed below [41].

$$CAI_1 = [Cl^- - (Na^+ + K^+)]/Cl^- \tag{2}$$

$$CAI_2 = [Cl^- - (Na^+ + K^+)]/(SO_4^{2-} + HCO_3^- + CO_3^{2-} + NO_3^-) \tag{3}$$

where the units of the ions involved in the equations are meq/L. When the values of both CAI_1 and CAI_2 were negative, the forward reaction of cation exchange occurred in groundwater; when the values of both CAI_1 and CAI_2 were positive, the backward reaction occurred in groundwater.

(4) Isotope values

The isotope concentration analyses are usually expressed as the sample deviations from the standard [42]. The calculated equation is shown below.

$$\delta = \frac{R_{sample} - R_{standard}}{R_{sample}} \times 1000\text{‰} \tag{4}$$

where R is the isotope ratio, such as $D/^1H$, $^{18}O/^{16}O$ and $^{15}N/^{14}N$, the H and O isotopes take Vienna Standard mean ocean water (V–SMOW) as the reference standard, the N isotope takes the atmospheric N_2 as the reference standard; δ is the sample deviation from the standard, such as δD, $\delta^{18}O$ and $\delta^{15}N$, and its unit is ‰.

(5) Evaluation of groundwater quality

According to the Groundwater Quality Standard (GB/T 14848-2017) [28], groundwater quality is divided into five levels (I~V). Groundwater at levels I~III can be used for domestic drinking, groundwater at level IV can be used for drinking after proper treatment, and groundwater at level V is not suitable for drinking. Taking level III as the standard (Table 1), if the concentration of the evaluated index is inferior to level III, it means that the index is over the standard. When comprehensively evaluating the groundwater sample, the determination of the evaluated level obeys the inferior principle [28].

Table 1. Evaluated standard of the main indices of groundwater.

Index	Level III	Index	Level III
pH	8.5	F^-	1
TH	450	COD_{Mn}	3
TDS	1000	Fe	0.3
SO_4^{2-}	250	Mn	0.1
Cl^-	250	I	0.08
Na^+	200	Hg	0.001
$NO_3(N)$	20	As	0.01
$NO_2(N)$	1	Cr^{6+}	0.05
$NH_4(N)$	0.5	Cd	0.005

Level III: Level III of Standard for Groundwater Quality (GB/T 14848-2017); Units: pH is unitless, the other indices have units of mg/L; the nitrogen species were expressed as N to be consistent with the Groundwater Quality Standard; Fe represents the total iron; Mn represents the total manganese.

(6) Denitrification and nitrification

Due to denitrification, nitrate and nitrite are reduced to gaseous nitrides and nitrogen in an anaerobic environment, which may change the composition of $\delta^{15}N(NO_3)$ and $^{18}O(NO_3)$ of different nitrogen sources. Therefore, an important prerequisite for using $\delta^{15}N(NO_3)$ and $\delta^{18}O(NO_3)$ isotopic values to identify the nitrogen source is that no significant denitrification occurs [43]. If denitrification occurs, the residual NO_3^- would enrich $^{15}N(NO_3)$, and the content of NO_3^- decreases [44]. The occurrence of denitrification can also be judged by the enrichment coefficient ($\varepsilon N/\varepsilon O$). According to the research results of Bottcher et al. [45] and Fukada et al. [46], the enrichment coefficient ($\varepsilon N/\varepsilon O$) should range between 1.3–2.1.

(7) $\delta^{15}N(NO_3) - \delta^{18}O(NO_3)$ dual isotope technique

When using the $^{15}N(NO_3)$ alone, the $\delta^{15}N$ value ranges between different sources overlap, which leads to multiple solutions. The $^{15}N(NO_3) - ^{18}O(NO_3)$ dual isotope technique provides a useful and powerful tool to identify nitrate contaminations by using the stable isotopes ^{15}N and ^{18}O of nitrate together [8]. Combined with hydrochemical methods, the source of nitrate can be accurately identified, which is of great significance to the prevention and control of nitrogen contamination.

Nitrate in groundwater may come from atmospheric deposition, inorganic fertilizer, soil, manure and sewage, and different nitrate sources have specific $\delta^{15}N(NO_3)$ and $\delta^{18}O(NO_3)$ value ranges [18,47]. Based on previous research [47–49], the ranges of $\delta^{15}N(NO_3)$ and $\delta^{18}O(NO_3)$ originating from different sources were classified and are shown in Table 2. Then, the sources of nitrate can be identified by using the $\delta^{15}N(NO_3) - \delta^{18}O(NO_3)$ dual isotope technique.

Table 2. Ranges of $\delta^{15}N(NO_3)$ and $\delta^{18}O(NO_3)$ originating from different sources (unit: ‰).

	Precipitation	Synthetic NO_3 Fertilizer	NH_4 Fertilizer	Soil Organic Nitrogen	Manure	Sewage
$\delta^{15}N(NO_3)$	−13~+13	−6~+6	−7~+5	0~+8	+5~+25	+4~+19
$\delta^{18}O(NO_3)$	+23~+75	+17~+25	+15~+25	−5~+14	−5~+10	

3. Results

3.1. Hydrochemical Characteristics Analysis

The groundwater in the study area was weakly alkaline with low salinity overall. The pH values ranged from 7.38 to 8.50, with an average value of 7.82. The TDS values ranged from 287.03 to 3426.54 mg/L, with an average of 744.56 mg/L. The SD value of TDS was 545.50 mg/L, and the C_v value was 73.22%, meaning that the spatial dispersion degree of TDS was high. The concentrations of TH were higher, ranging from 362.83 to 1851.67 mg/L, with an average of 451.54 mg/L. The SD value of TH was as high as 1170.65 mg/L, and the C_v value was 63.22%. Among the main anions in groundwater, the content of HCO_3^- was the highest, followed by Cl^-, SO_4^{2-} and NO_3^-. Among the main cations, Mg^{2+} and Ca^{2+} were the dominant ions, while the concentrations of Na^+ and K^+ were relatively lower. The concentrations of F^- ranged from 0.31 to 2.39 mg/L, with an average of 0.91 mg/L. The NO^{3-} concentrations fluctuated widely, ranging from 0.43 to 419.71 mg/L, with an average of 79.77 mg/L. The concentrations of COD_{Mn} ranged from 0.12 to 5.07 mg/L, with an average of 1.44 mg/L. According to Table 3, the SD values of Mg^{2+}, Na^+, HCO_3^-, SO_4^{2-}, Cl^-, NO_3^-, TDS, TH and COD_{Mn} were high, and their C_v values were correspondingly high, which indicated that the dispersion degrees of them were high. The probable reasons for the high C_v values are discussed below.

Table 3. Statistics of hydrochemical parameters of groundwater (n = 38).

Index	Maximum	Minimum	Average	Median	SD	C_v
pH	8.58	7.38	7.82	7.81	0.24	3.03
Ca^{2+}	160.32	39.08	80.33	69.14	30.46	37.92
Mg^{2+}	413.27	13.37	60.95	43.51	64.39	105.64
Na^+	577.60	16.66	91.51	53.35	98.86	108.03
HCO_3^-	1677.97	213.56	386.78	340.17	237.28	61.35
SO_4^{2-}	392.20	14.41	102.75	56.90	94.21	91.69
Cl^-	744.51	14.18	114.69	74.10	129.57	112.97
NO_3^-	419.71	0.43	79.77	53.87	94.27	118.18
NO_2^-	1.35	0.00	0.04	0.00	0.22	561.36
F^-	2.39	0.31	0.91	0.80	0.48	53.02
H_2SiO_3	50.00	16.67	26.67	26.04	5.86	21.97
TDS	3426.54	287.03	744.56	568.66	545.20	73.22
TH	451.54	362.83	1851.67	152.64	1170.65	63.22
COD_{Mn}	5.07	0.12	1.44	1.01	1.11	77.23

The units of all groundwater quality indices, except pH, are in mg/L. SD: mg/L. C_v: %.

Hydrochemical characteristics of groundwater may be influenced by natural factors and human activities. Some natural factors, such as earthquakes and volcanic eruptions, may lead to the remarkable increase in the ion contents (K^+, Na^+, HCO^{3-}, and SO_4^{2-}), and the influence areas of these factors are large [50,51], while the influences caused by other natural factors are relatively slight. Loose Quaternary sediment is widely distributed in the study area without obvious changes in lithology, no obvious tectonic movement has occurred in recent years, and the extreme values of the analytic indices are distributed discontinuously. Hence, it was inferred that natural factors were not the main cause of the high dispersion and high C_v values. According to previous studies [52], the hydrochemical indices, influenced by human activities, were characterized by high dispersion and high fluctuation, and their C_v values were correspondingly high. Thus, the C_V value can reflect the influence of human activities to some extent, and the disturbance of human activities can also lead to a high value of C_V. In general, there is a variation if the C_V value is higher than 30%, and there is a great variation if the C_V value is higher than 60%. The greater C_V is, the greater the difference, and the greater the influence of external factors on the groundwater index. As shown in Table 3, the C_v values of pH and H_2SiO_3 were low, meaning that the distributions of the two indices were spatially steady, while the C_v values of other indices were high, the C_v values of Mg^{2+}, Na^+, Cl^- and NO_3^- were even greater than 100%. The spatial variations were also reflected in Figure 3, the concentrations of these indices in some areas were much higher than those in other areas. In summary, the intensive variations were mainly influenced by human activities.

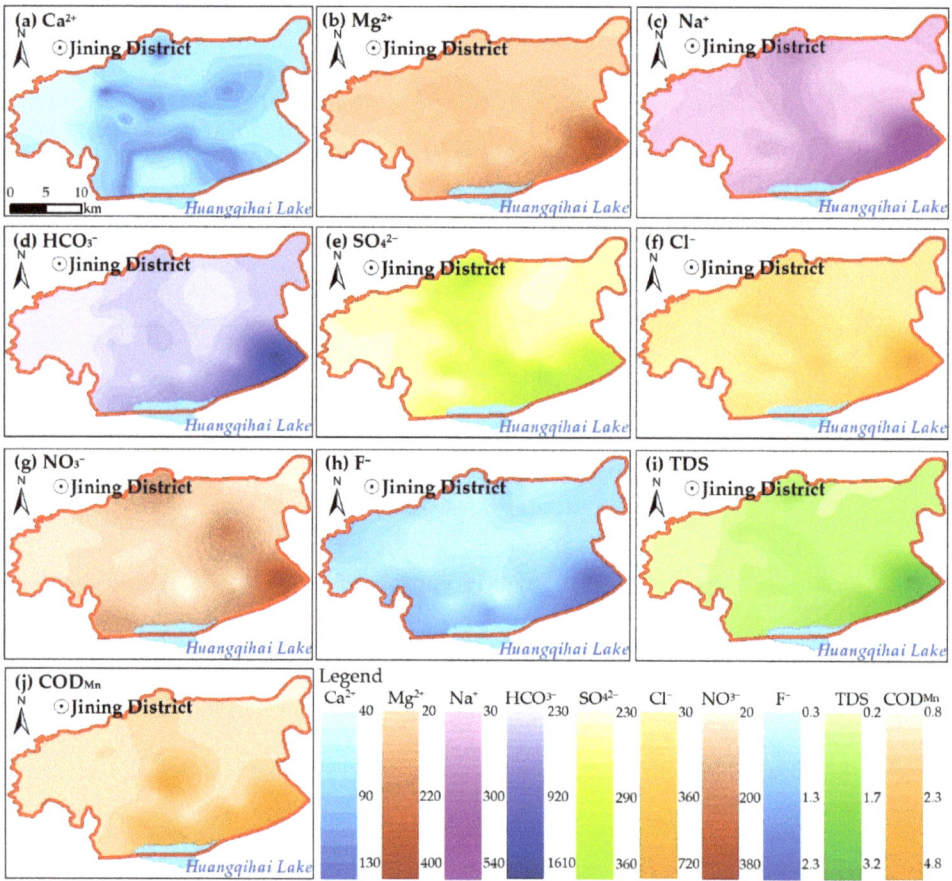

Figure 3. Spatial distribution of hydrochemical indices (a–j). The unit of TDS is g/L and the units of other indices are mg/L.

As shown in Figure 3, the concentrations of TDS, Ca^{2+}, Mg^{2+}, Cl^- and SO_4^{2-} in groundwater were generally high in the central region of the study area, low in the east and west, low in the north and high in the south. The terrain of the study area gently slopes toward Huangqihai Lake, and the groundwater runs off slowly; the water – rock interactions were fully reacted, and groundwater ions accumulated from upstream to downstream. Furthermore, the groundwater depths in most areas downstream were shallower than the extreme evaporation depth, and the evaporation of groundwater was intensive. Due to the above factors, the contents of TDS, Ca^{2+}, Mg^{2+}, Cl^- and SO_4^{2-} were generally high in the downstream area. The concentration of F^- was low in the middle and relatively high in the periphery. The concentrations of NO_3^- and COD_{Mn} were relatively low in most areas but high in area nearby the Huangqihai Lake.

Based on the Piper diagram (Figure 4a), the anions were mainly distributed closer to the HCO_3^- side for the groundwater samples of the upstream aquifer (Q_3^{al+pl}), while the cations were mainly distributed close to the Mg^{2+} and Ca^{2+} sides, indicating that the chemical facies upstream were mainly HCO_3–Mg·Ca or HCO_3–Ca·Mg. For the groundwater samples of the downstream aquifer(Q_4^l), the cations were obviously biased toward the side of Na^+, and the concentrations of anions (Cl^- and SO_4^{2-}) increased. Therefore, the anion types gradually transitioned from HCO_3 upstream to HCO_3·Cl (Cl·HCO_3) and Cl·SO_4 downstream, and the types of cations changed from Mg·Ca (Ca·Mg) to Mg·Ca·Na

(Ca·Mg·Na), and even Na·Mg·Ca (Na·Mg·Ca) and Na type. Noticeably, the NO_3^- mole fractions of the C29 and C30 samples were 36.25% and 33.18%, respectively. To reflect nitrate contamination [39,40], the NO_3^- ion was taken into account when determining the hydrochemical facies. Therefore, two new hydrochemical types, except the 49 types, appeared in the C29 and C30 samples, namely, the $NO_3 \cdot HCO_3$–Ca·Mg and $HCO_3 \cdot NO_3 \cdot Cl$–Ca·Mg facies. According to studies by Apollaro et al. [53], the iso–ionic–salinity (TIS) lines were added in the correlation plot of $(Na^+ + K^+)$ vs. $(Ca^{2+} + Mg^{2+})$ to reflect the ionic salinity. As shown in Figure 4b, the content of Ca^{2+} and Mg^{2+} was higher than that of $Na^+ + K^+$ on the whole, and the groundwater had low ionic salinity which ranged from 10.43 to 56.21 meq/L. The TDS values were comparatively low with an average of 744.56 mg/L which were in the range of fresh water (TDS < 1000 mg/L).

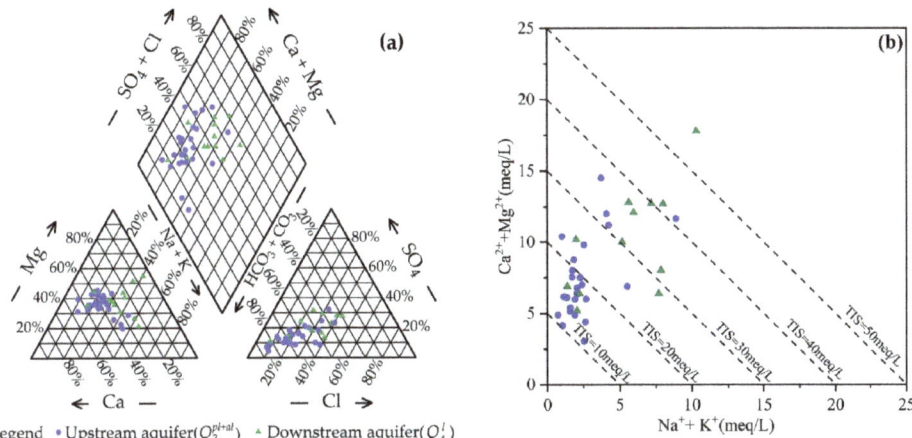

Figure 4. Piper diagram of groundwater (**a**) and correlation plot of $(Na^+ + K^+)$ vs. $(Ca^{2+} + Mg^{2+})$, also showing TIS salinity diagram for reference (**b**).

In this study, the Pearson correlations of groundwater indices were analyzed to reflect the possible sources and chemical reactions related to the hydrochemical indices (Table 4).

Table 4. Correlation matrices of hydrochemical indices (n = 38).

	pH	Ca^{2+}	Mg^{2+}	Na^+	HCO_3^-	SO_4^{2-}	Cl^-	F^-	H_2SiO_3	TDS	TH	COD_{Mn}
pH	1											
Ca^{2+}	−0.114	1										
Mg^{2+}	−0.225	0.135	1									
Na^+	−0.204	0.185	0.908 **	1								
HCO_3^-	−0.226	0.040	0.968 **	0.915 **	1							
SO_4^{2-}	−0.105	0.564 **	0.666 **	0.816 **	0.639 **	1						
Cl^-	−0.294	0.308	0.912 **	0.951 **	0.869 **	0.794 **	1					
F^-	0.290	−0.229	0.550 **	0.487 **	0.563 **	0.267	0.359 *	1				
H_2SiO_3	−0.188	0.000	0.596 **	0.477 **	0.528 **	0.205	0.565 **	0.079	1			
TDS	−0.224	0.352 *	0.949 **	0.962 **	0.917 **	0.825 **	0.961 **	0.455 **	0.524 **	1		
TH	−0.239	0.392 *	0.965 **	0.892 **	0.910 **	0.768 **	0.929 **	0.450 **	0.553 **	0.976 **	1	
COD_{Mn}	−0.040	0.086	0.671 **	0.711 **	0.708 **	0.584 **	0.679 **	0.331 *	0.345 *	0.699 **	0.646 **	1

* and ** represent significant levels of 0.05 and 0.01, respectively.

According to Table 4, the correlation coefficient (r) of TDS and TH was 0.976, indicating that there was a good positive correlation between them. The two indices had good positive correlations with Mg^{2+}, Na^+, Ca^{2+}, Cl^-, HCO_3^- and SO_4^{2-}, reflecting the significant contribution of these elements in mineralization of groundwater. r (Ca^{2+} vs· Mg^{2+}) did not reach a significance level of 0.05, meaning that there is no obvious relationship between Ca^{2+} and Mg^{2+}. This was mainly because the sources of Ca^{2+} and Mg^{2+} or the reactions related to the two ions in groundwater were different. Both Ca^{2+} and Mg^{2+} had a positive relationship with SO_4^{2-}, which indicated that sulfates rich in calcium and magnesium were dissolved in groundwater. r (Mg^{2+} vs. HCO_3^-) was as high as 0.968; that is, the dissolution of carbonate rich in Mg^{2+} in groundwater, such as dolomite, was the main source of Mg^{2+}. Meanwhile, r (Ca^{2+} vs. HCO_3^-) was small and did not reach the significance level of 0.05, indicating that the dissolution of carbonate was not the main source of Ca^{2+} in groundwater or that the relationship between Ca^{2+} and HCO_3^- was weaker due to other reactions, such as cation exchange and crystallization. r (Na^+ vs. Cl^-) was as high as 0.951, indicating that the dissolution of salt rocks in groundwater was the main source of Na^+ and Cl^-. According to r (Na^+ vs. HCO_3^-) and r (Na^+ vs. SO_4^{2-}), both of which were higher than 0.8, it was inferred that there were other sources of Na^+ in addition to the dissolution of salt rock. H_2SiO_3 was positively correlated with Mg^{2+} and Na^+, illustrating the dissolution of silicate containing Mg^{2+} and Na^+ in groundwater. F^- had a relatively good correlation with pH, HCO_3^-, Na^+ and Cl^-, that is, high F^- groundwater was generally accompanied by a distinctive hydrochemical characteristic: Ca-poor and Na-rich with alkaline conditions and high HCO_3^- concentration. These results were consistent with previous studies [16,54,55]. The dissolution of fluorite and some silicate minerals, such as micas, was the main source of F^- in groundwater [3].

3.2. Hydrochemical Evolution Mechanism
3.2.1. Hydrochemical Process

According to the Gibbs diagram, the main evolution mechanism of groundwater was classified into three types: evaporation, rock weathering and atmospheric precipitation [5,33,41]. As shown in Figure 5, the ratio of $Na^+/(Na^+ + Ca^{2+})$ ranged from 0.14 to 0.59, and the ratio of $Cl^-/(Cl^- + HCO_3^-)$ ranged between 0.05 and 0.59, indicating that the ions' concentrations of groundwater in the study area were mainly affected by rock weathering. The $Na^+/(Na^+ + Ca^{2+})$ and $Cl^-/(Cl^- + HCO_3^-)$ ratios of the downstream aquifer (Q_4^l) were larger than those of the upstream aquifer (Q_3^{al+pl}). The burial depth of groundwater gradually became shallow from upstream to downstream, evaporation strengthened, some Ca^{2+} ions were precipitated as $CaCO_3$, and the contents of Na^+ and Cl^- were further concentrated.

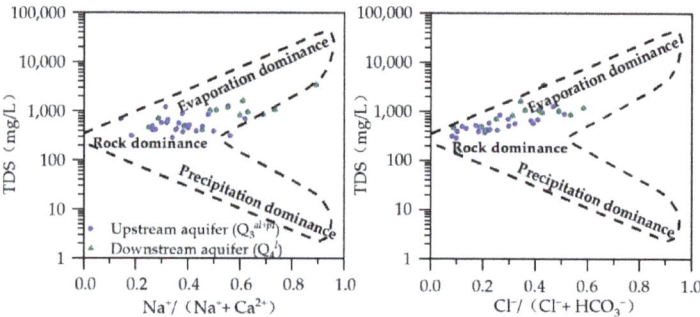

Figure 5. Gibbs diagram of groundwater.

3.2.2. Analysis of the Main Dissolution and Migration

Based on the above analysis, the probable dissolutions and migrations were further judged by using the ion proportional coefficients method [16,56].

(1) $(Ca^{2+} + Mg^{2+})/(HCO_3^- + SO_4^{2-})$

In general, Ca^{2+} and Mg^{2+} in groundwater mainly come from the dissolution of carbonate, silicate and evaporite, so the $(Ca^{2+} + Mg^{2+})/(HCO_3^- + SO_4^{2-})$ ratio was used to determine the main sources of Ca^{2+} and Mg^{2+} [16]. As shown in Figure 6a, the samples were mostly located above the 1:1 line, led by the groundwater samples in the upstream aquifer (Q_3^{al+pl}). Combined with Table 4, r (Ca^{2+} vs. SO_4^{2-}) and r (Mg^{2+} vs. SO_4^{2-}) were high, which indicated that Ca^{2+} and Mg^{2+} in the groundwater were derived not only from the dissolution of carbonate and silicate minerals, but also from the dissolution of sulfate minerals. The ratios of $(Ca^{2+} + Mg^{2+})/(HCO_3^- + SO_4^{2-})$ in some groundwater samples were less than 1:1, which may be caused by carbonate precipitation and cation exchange.

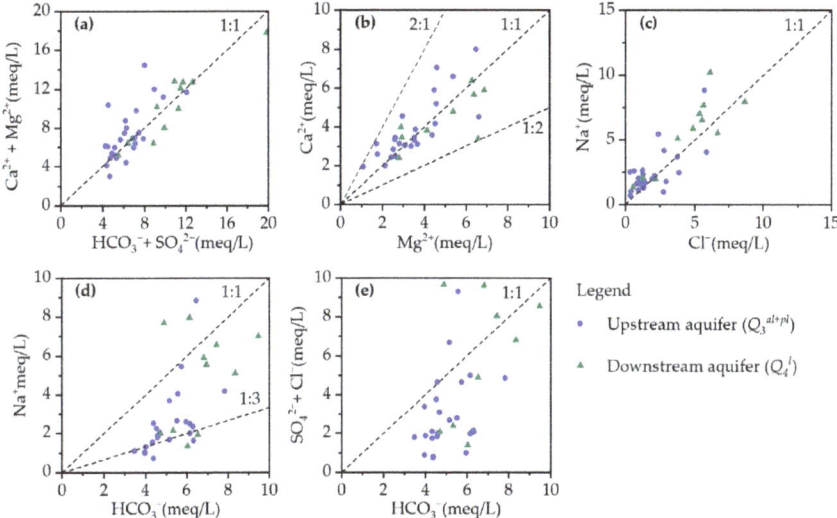

Figure 6. Ion proportional coefficient diagrams: $(Ca^{2+} + Mg^{2+})$ vs. $(HCO_3^- + SO_4^{2-})$ (a), Ca^{2+} vs. Mg^{2+} (b), Na^+ vs. Cl^- (c), Na^+ vs. HCO_3^- (d), and $(SO_4^{2-} + Cl^-)$ vs. HCO_3^- (e).

(2) Ca^{2+}/Mg^{2+}

The ratio of Ca^{2+} to Mg^{2+} was used to reflect the dissolution of calcite and dolomite [41]. As shown in Figure 6b, the samples were generally distributed near the 1:1 line. The samples in the upstream aquifer (Q_3^{al+pl}) were mostly distributed above the 1:1 line; that is, the Ca^{2+} content was higher than the Mg^{2+} content, indicating that the main dissolved carbonate in the groundwater was calcite. Meanwhile, the samples in the downstream aquifer (Q_4^l) were mostly distributed below 1:1; that is, the Ca^{2+} content was lower than the Mg^{2+} content in the downstream aquifer (Q_4^l). Dolomite dissolution produces 1:1 Ca^{2+}/Mg^{2+}, and calcite only produces Ca^{2+}. If calcite or dolomite was dissolved in groundwater, the content of Ca^{2+} should be higher than that of Mg^{2+}. In fact, the Ca^{2+} contents were lower than the Mg^{2+} contents, indicating that cation exchange may occur and that the Ca^{2+} ions in groundwater were adsorbed on the surface particles of the aquifer.

(3) Na^+/Cl^- and Na^+/HCO_3^-

The Na^+/Cl^- ratio can indicate the dissolution of salt rocks and silicates in groundwater [16]. Most of the samples were located above the 1:1 line (Figure 6c). In the process of hydrochemical evolution, the content of Cl^- was steady and participated less in the reaction, and its main source was the dissolution of the salt rocks. The Na^+ content was higher than the Cl^- content, which may be due to silicate dissolution as well as salt rock dissolution during the flow process of groundwater [57].

Plagioclase exists in the study area [22]. Plagioclase minerals generally include albite, labradorite (intermediate), anorthite and so on. Groundwater in the study area was in a slightly alkaline environment, in which plagioclase dissolved. The dissolution reactions are listed below. According to (5) and (6), the albite released 1:1 Na^+/HCO_3^- and 1:3 Na^+/SiO_2, and the labradorite produced 1:3 Na^+/HCO_3^- and 1:3 Na^+/SiO_2 [4]. Figure 6d shows that the ratio of Na^+/HCO_3^- mostly ranged between 1:3 and 1:1, which meant that the content of HCO_3^- was higher than that of Na^+. Combined with Table 4, r (Na^+ vs. HCO_3^-) and

r (Na$^+$ vs. H$_2$SiO$_3$) were positive, indicating the dissolution of albites and labradorites in groundwater. If the dissolution degrees of albite and labradorite were the same, both would release 1:3 Na$^+$/SiO$_2$. However, r (Na$^+$ vs. HCO$_3^-$) was higher than r (Na$^+$ vs. H$_2$SiO$_3$), meaning that the reaction degree of labradorites was more intensive than that of albites. Furthermore, cation exchange also affected the content of Na$^+$.

Albite: NaAlSi$_3$O$_8$ + CO$_2$ + 2H$_2$O = Na$^+$ + 3SiO$_2$ + Al(OH)$_3$ + HCO$_3^-$ (5)

Labradorites: 2NaCaAl$_3$Si$_5$O$_{16}$ + 3CO$_2$ + 9H$_2$O = 2Na$^+$ + 2 Ca^{2+} + 2SiO$_2$ + 3Al$_2$Si$_2$O$_5$(OH)$_4$ +6HCO$_3^-$ (6)

(4) (SO$_4^{2-}$ + Cl$^-$)/HCO$_3^-$

The dissolution of carbonate and silicate was the main source of HCO$_3^-$ in the study area, while the weathering and dissolution of salt rocks and the oxidation of sulfide minerals were the main sources of Cl$^-$ and SO$_4^{2-}$. Most of the samples were distributed below the 1:1 line (Figure 6e), which meant that the content of HCO$_3^-$ was higher than that of SO$_4^{2-}$ and Cl$^-$. This result indicated that the dissolution of carbonate and silicate minerals played a dominant role in the hydrochemical process, while the dissolution of salt rock and oxidation of sulfur minerals was relatively weak.

3.2.3. Cation Exchange

Cation exchange influenced the main cations concentration, and it was judged by using the binary phase diagram of (Na$^+$ − Cl$^-$) vs. (Ca^{2+} + Mg^{2+} − SO$_4^{2-}$ − HCO$_3^-$) and Chlor–Alkali indices [4,16].

(1) (Na$^+$ − Cl$^-$) vs. (Ca^{2+} + Mg^{2+} − SO$_4^{2-}$ − HCO$_3^-$)

If cation exchange was the dominant process influencing the contents of Na$^+$, Ca^{2+} and Mg^{2+}, the relationship between the two parameters was negative linear, with a slope of −1.0. As shown in Figure 7a, there was a certain linear relationship between (Na$^+$ − Cl$^-$) and (Ca^{2+} + Mg^{2+} − SO$_4^{2-}$ − HCO$_3^-$) in groundwater, but the correlation coefficient was low. It was indicated that cation exchange existed in the hydrogeochemical process, but it did not play a dominant role in the changes in the contents of Na$^+$, Ca^{2+} and Mg^{2+}.

(2) Chloro–alkaline indices

According to Figure 7b, the CAI$_1$ and CAI$_2$ values of most groundwater samples were smaller than 0, indicating that Ca^{2+} and Mg^{2+} in groundwater were replaced by Na$^+$, and the reactions occurred: Ca^{2+} + 2NaX $\rightleftarrows$ 2Na$^+$ + CaX; Mg^{2+} + 2NaX $\rightleftarrows$ 2Na$^+$ + MgX. This was mainly because the sorptive abilities of Ca^{2+} and Mg^{2+} are higher than that of Na$^+$ [31]. The CAI$_1$ and CAI$_2$ values of a few samples were greater than 0, which indicated that the Na$^+$ ions in groundwater in some areas were exchanged by Ca^{2+} or Mg^{2+} ions, and the reactions might occur: 2Na$^+$ + CaX $\rightleftarrows$ Ca^{2+} + 2NaX; 2Na$^+$ + MgX $\rightleftarrows$ Mg^{2+} + 2NaX. Previous studies [31,58,59] have shown that cation exchange is also influenced by other factors, such as the sediment granularity in the aquifer, pH and concentrations of the ions. Taking C1 and C7 as examples, the Na$^+$ contents of the two samples were much higher than the average content, and the high Na$^+$ content might cause the Na$^+$ ions in groundwater to be exchanged with the Ca^{2+} or Mg^{2+} ions in aquifer media.

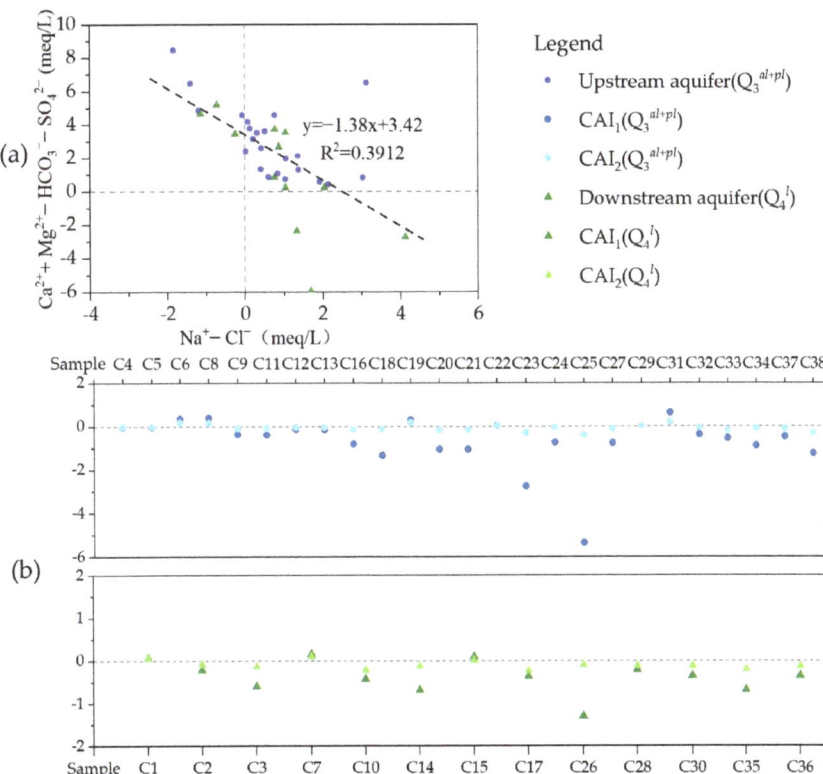

Figure 7. Coefficient diagrams of ($Na^+ - Cl^-$) vs. ($Ca^{2+} + Mg^{2+} - SO_4^{2-} - HCO_3^-$) for analyzing cation exchange reaction (**a**) and distribution diagram of Chloro–alkaline indices (**b**).

3.3. Groundwater Source

The global meteoric water line (GMWL) was $\delta D = 8\delta^{18}O + 10$ [60]. Due to the lack of the precipitation isotopic data of the study area, the local meteoric water line (LMWL) of Hohhot, which was located in the same climate zone as the study area and near the study area, was chosen to reflect the D–^{18}O isotopic composition of precipitation in the study area, and the LMWL equation was $\delta D = 7.68\delta^{18}O - 0.72$ ($R^2 = 0.8964$) [61]. The study area is situated inland with less precipitation and intensive evaporation, and is comprehensively influenced by the East Asian summer monsoon and westerly circulation. The water vapor of the precipitation, which was caused by westerly circulation, originates from the North Atlantic and is transported from Xinjiang to inland China. It is characterized by low humidity and obvious secondary evaporation, which causes the temperature effect [61,62]. The water vapor brought by the East Asian summer monsoon has a high humidity and is slightly influenced by evaporation; along with transportation inland, the heavy isotopes are preferentially condensed, and δD and $\delta^{18}O$ values in precipitation decrease with increasing of precipitation, which was called the rainfall effect [61,62]. For the above reasons, the slope of the LMWL was less than that of the GMWL (Figure 8).

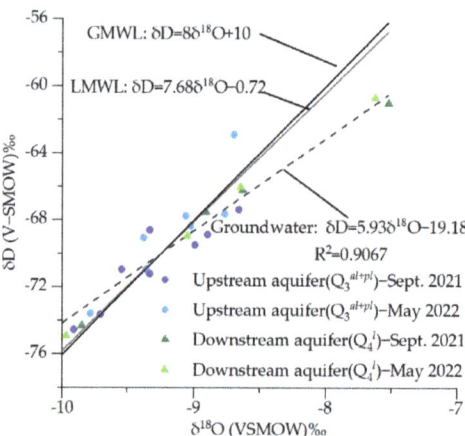

Figure 8. Relationship between δD and δ^{18}O of groundwater in the study area. Sampling sites: Upstream aquifer (Q$_3^{al+pl}$) and downstream aquifer (Q$_4^l$); sampling times: September 2021 and May 2022.

The δD values of groundwater ranged from −74.93 to −61.01‰, the δ^{18}O values of groundwater ranged from −10.08 to −7.39‰, and the averages of δD and δ^{18}O were 69.14 and 9.07‰, respectively. Samples were mostly distributed near the LMWL (Figure 8), and the linear equation of δD and δ^{18}O of groundwater was δD = 5.93δ^{18}O − 19.18 (R^2 = 0.9067), indicating that local atmospheric precipitation was the main source of groundwater. Contaminants were very likely to enter groundwater along with the precipitation infiltration. The linear slope of δD and δ^{18}O of groundwater was lower than that of GMWL (8) and LMWL (7.68), which meant that heavy isotopes were further enriched by evaporation during the runoff process of groundwater. The isotopic composition (δD and δ^{18}O) was influenced to some extent by the regional climate and local processes (evaporation, vegetation distribution, anthropogenic activities) [48]; thus, the δD and δ^{18}O values of groundwater may be different in different areas and different times. According to Table 5, the averages of δD and δ^{18}O downstream were higher than those upstream which was caused by the intensification of evaporation in the shallow groundwater depth area. Comparing the δD and δ^{18}O values in different seasons (Table 5), the δ^{18}O averages in May 2022 were slightly higher than those in September 2021 regardless of whether the groundwater samples were distributed upstream and downstream; the δD average downstream in May 2022 was just 0.30‰ higher than that in September 2021, and the average upstream in May 2022 was −0.55‰ lower than that in September 2021. The reasons for the difference between different times need further research based on monthly isotopic data over years, and measurement uncertainty should be taken into consideration.

Table 5. Statistics of δD, δ^{18}O and d–excess in different seasons (Unit: ‰).

Sample	Sampling Site	September 2021			May 2022		
		δD	δ^{18}O	d–Excess	δD	δ^{18}O	d–Excess
H1	Upstream aquifer (Q_3^{al+pl})	−69.51	−8.99	2.38	−69.06	−9.31	5.43
H2		−67.41	−8.65	1.79	−67.61	−8.64	1.50
H5		−74.58	−9.91	4.70	−74.93	−10.08	5.73
H6		−70.97	−9.54	5.35	−62.86	−8.56	5.65
H8		−68.60	−9.33	6.01	−67.74	−8.97	3.98
H9		−71.20	−9.33	3.40	−70.95	−9.29	3.36
H12		−68.89	−8.89	2.23	−68.32	−8.92	3.04
H13		−73.68	−9.70	3.95	−73.57	−9.76	4.52
Average		−70.60	−9.29	3.73	−69.38	−9.19	4.15
H3	Downstream aquifer (Q_4^l)	−66.19	−8.63	2.84	−66.01	−8.50	2.02
H4		−74.31	−9.85	4.49	−74.89	−9.96	4.82
H10		−61.01	−7.53	−0.81	−60.68	−7.39	−1.60
H11		−67.53	−8.91	3.74	−68.94	−8.95	2.68
Average		−67.26	−8.73	2.57	−67.63	−8.70	1.98

The deuterium excess values (d–excess = δD − 8δ^{18}O) were used to analyze the intensity of groundwater evaporation. The stronger the evaporation was, the more negative the d–excess value was [63]. The d–excess values of groundwater ranged from −1.60 to 6.01‰, with an average value of 3.38‰. The average of d–excess in the study area was positive but smaller than the d–excess average of the global meteoric water (10‰) [60]. It was indicated that the ion contents of groundwater in the study area were controlled by water–rock interactions and influenced by evaporation. Due to intensive evaporation, the contents of ions are generally concentrated in the groundwater. The solubilities of some salts (such as NaCl) are high, and the contents of their ions in groundwater (such as Na^+ and $Cl^−$) can increase to a high level. On the other hand, the solubilities of some salts (such as $CaCO_3$) are low. Taking $CaCO_3$ as an example, the content of Ca^{2+} increases due to evaporation, but once its content is saturated, the Ca^{2+} ion in groundwater precipitates in the form of $CaCO_3$, $Ca^{2+} + CO_3^− = CaCO_3\downarrow$, and the Ca^{2+} content in the groundwater decreases. This may change the ion compositions of groundwater, and further influence the hydrochemical facies. In general, intensive evaporation in the shallow aquifer can cause an increase in the TDS and TH contents [64].

By comparing the d–excess values of groundwater in different seasons (Table 5), the d–excess average in May (3.42‰) was slightly higher than that in September (3.40‰). Due to the lack of time series data of isotopes, the possible reasons were preliminarily inferred based on previous studies [61,65]: (1) the difference in the isotopic composition of precipitation (the main source of groundwater) between different seasons and (2) the different influences of evaporation on groundwater between different seasons. The measurement uncertainty should be considered. Continuous measurements of stable isotopes both in precipitation and groundwater and further research are needed. From the spatial distribution, the d–excess values decreased from the upstream aquifer (Q_3^{al+pl}) to the downstream aquifer (Q_4^l). The main reason was that the water–rock interactions continuously proceeded along the direction of groundwater runoff, and an oxygen shift occurred. The burial depth of groundwater became shallow from north to south, and evaporation intensified; thus, the heavy isotopes were enriched. From the time perspective, the content of $Cl^−$ would be enriched due to evaporation. According to the Pearson correlation analysis, r ($Cl^−$ vs. δ^{18}O) was 0.611, but it did not reach the significance level of 0.05, indicating that evaporation slightly affected the isotopic compositions.

3.4. Evaluation of Groundwater Quality

According to the Groundwater Quality Standard (GB/T 14848-2017), the comprehensive overstandard rate of groundwater in the study area was 60.53% by following the inferior principle, and the overstandard indices were TH, F, $NO_3(N)$, TDS, COD_{Mn}, SO_4^{2-}, Na^+, Cl^-, $NO_2(N)$ and As. The statistical data of the overstandard indices are shown in Table 6 and Figure 9.

Table 6. Statistics of the overstandard indices.

Index	Na^+	Cl^-	SO_4^{2-}	$NO_3(N)$	$NO_2(N)$	$NH_4(N)$
Level III	200	250	250	20	1.00	0.50
R_{over}	7.89	5.26	10.53	23.68	2.63	0.00
M	2.89	2.98	1.57	4.74	1.35	/
Index	F^-	TDS	TH	As	COD_{Mn}	
Level III	1	1000	450	0.01	3	
R_{over}	28.95	21.05	36.84	2.63	10.53	
M	2.39	3.43	4.11	1.24	1.69	

R_{over}: Overstandard ratio based on the level III standard, %; M: Multiple of the maximum concentration to a multiple of the level III standard; Level III: Level III of Standard for Groundwater Quality (GB/T 14848-2017), mg/L.

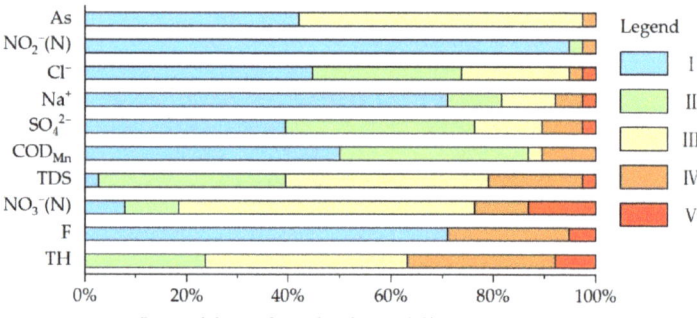

Figure 9. Evaluated results of groundwater, the division of the I~V levels was based on the Standard for Groundwater Quality (GB/T 14848-2017) [28].

The indices (TH, TDS, F^-, Na^+ and Cl^-) in groundwater exceeded the standard, mainly due to natural factors. The climate in the study area is arid, evaporation is intensive, the groundwater flows slowly due to the gentle terrain, and the background contents of F^- in some areas are high. All these factors led to the overstandard of these indices. The concentrations of $NO_3(N)$ and $NO_2(N)$ in some areas exceeded the standard. According to Table 6, the highest concentration of $NO_3(N)$ was 4.74 times that of Level III, and the concentration of $NO_2(N)$ was 1.35 times that of Level III. Combining Table 3 and Figure 3, the spatial variations in $NO_3(N)$ and $NO_2(N)$ in the study area were strong, indicating that the contents of $NO_3(N)$ and $NO_2(N)$ were influenced by human activities. COD_{Mn} is a common index to reflect the pollution of organic oxidizable substances in groundwater [66]. COD_{Mn} in the study area had a strong spatial variability. Concentrations of COD_{Mn} in some zones were higher than 3 mg/L, indicating that the synthetic organic compounds caused the deterioration of the groundwater environment by human factors.

3.5. Identification of Nitrogen Contamination

Based on the evaluated groundwater quality results, the $NO_3(N)$ concentrations in some parts of the study area were overstandard, and the highest concentration far exceeded the level III standard. Thus, the main source of nitrate contaminations should be scientifically identified.

3.5.1. Relationship between NO_3^- and Other Ions (Cl^-, SO_4^{2-}, Na^+ and K^+)

As shown in Figure 10a, the Cl^-/Na^+ ratios of most samples were close to 1, indicating that the Cl^-/Na^+ ratios were mainly affected by salt rock dissolution [16]. The NO_3^-/Na^+ ratios ranged from 0.00 to 3.69 with an average of 0.49, and the C_v was 1.34 (>1), indicating that there was a strong spatial variation. Combined with Figure 10b, the ratios of SO_4^{2-}/Na^+ were smaller than those of NO_3^-/Na^+ overall, and the groundwater samples were located near the agricultural side, meaning that the nitrate content of groundwater was mainly affected by agricultural activities. The distribution of the points in Figure 10a was scattered and the C_v of NO_3^-/Na^+ was higher than 1, meaning that agricultural activities in different zones had different impacts on groundwater. The discussed results were consistent with the field survey results. The farmland in the study area was widely distributed, and the main crops were potatoes, oats and vegetables. Fertilizers and pesticides were very likely to enter groundwater along with the water from irrigation and precipitation, resulting in excessive nitrogen in groundwater.

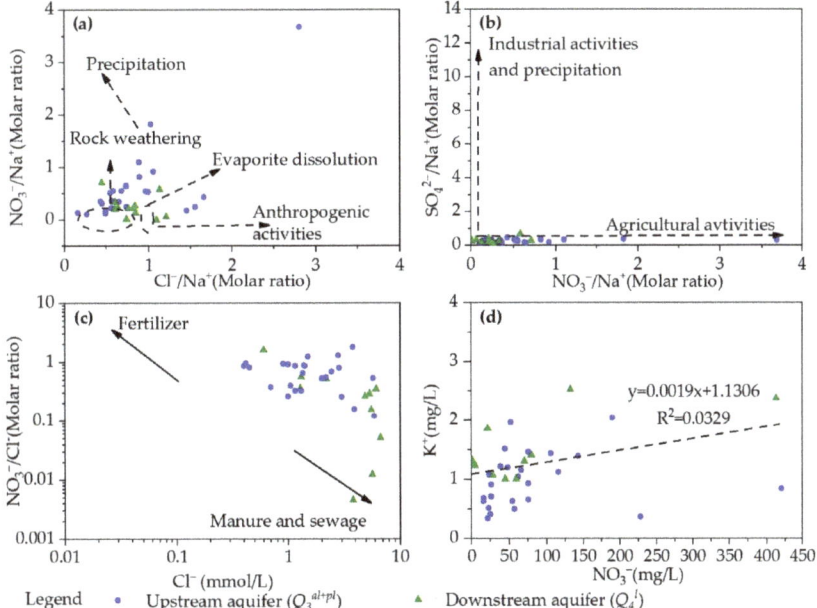

Figure 10. Relationships of (NO_3^-/Na^+) vs. (Cl^-/Na^+) (**a**), (SO_4^{2-}/Na^+) vs. (NO_3^-/Na^+) (**b**), (NO_3^-/Cl^-) vs. Cl^- (**c**), and NO_3^-/K^+ (**d**).

Contaminations coming from manure and sewage would show a higher Cl^- content and a lower NO_3^-/Cl^- ratio. The Cl^- molar concentration should usually be greater than 1 mmol/L, while the NO_3^-/Cl^- molar concentration ratio should range between 0.001 and 0.1 [44,67]. The contaminations originating from synthetic NO_3 fertilizer via agricultural activities should show characteristics of low Cl^- concentrations and high NO_3^-/Cl^- ratios, the Cl^- content should be less than 0.1 mmol/L, and the NO_3^-/Cl^- molar concentration ratio should range between 0.1 and 10 [18]. As shown in Figure 10c, the Cl^- contents were generally more than 1 mmol/L, and the molar concentration ratios of NO_3^-/Cl^- ranged between 0.1 and 10, with an average value of 0.58. The points were mostly distributed in the upper–right of Figure 10c, indicating that the nitrate content in groundwater was affected by a variety of factors such as manure, sewage and fertilizers. To further explore the probability of synthetic NO_3 fertilizer as a probable source of nitrate, the bivariate relationship between NO_3^- and K^+ was examined. There will be a strong correlation between NO_3^- and K^+ if NO_3 originates from synthetic NO_3 fertilizer [7]. In fact, there was a weak correlation between NO_3^- and K^+ (Figure 10d), indicating that synthetic NO_3 fertilizer is not the dominant source of NO_3^- in the study area.

3.5.2. Denitrification and Nitrification

As shown in Figure 11a, the relationship between NO_3^- and $\delta^{15}N(NO_3)$ was not negative, indicating that denitrification in groundwater was not obvious. Previous studies [45,46] have shown that the enrichment coefficient ($\varepsilon N/\varepsilon O$) should range between 1.3–2.1 if intensive denitrification occurs in groundwater, in other words, the slope of $\delta^{15}N(NO_3)$ and $\delta^{18}O(NO_3)$ should range from 0.48 to 0.77. As shown in Figure 11c, the slope of $\delta^{15}N(NO_3)$ and $\delta^{18}O(NO_3)$ was −0.09, far lower than 0.48 and 0.77, respectively. It was indicated that no obvious denitrification occurred during the transforming process of nitrate in the study area.

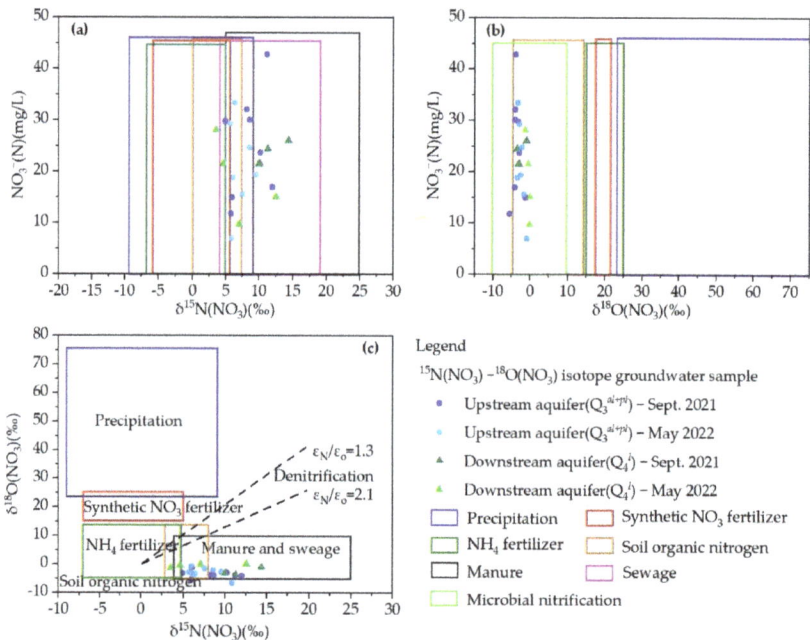

Figure 11. Relationships of NO_3^- vs. $\delta^{15}N(NO_3)$ (a), NO_3^- vs. $\delta^{18}O(NO_3)$ (b) and $\delta^{15}N(NO_3)$ vs. $\delta^{18}O(NO_3)$ (c).

Under an oxidative environment, NH_4^+ is transformed into NO_3^- by Nitrobacter, which is called nitrification. According to the experimental research of Kendall [68], one O atom of NO_3 comes from the atmosphere, and the other two O atoms come from H_2O in environment, i.e., $\delta^{18}O(NO_3) = 2/3\delta^{18}O(H_2O) + 1/3\delta^{18}O(O_2)$. The $\delta^{18}O(H_2O)$ average of groundwater in the study area was −9.07‰, and the $\delta^{18}O(O_2)$ value was +23.5‰ [69]. Based on this, the $\delta^{18}O(NO_3)$ theoretical value was calculated as 1.79‰. The actual values of $\delta^{18}O(NO_3)$ in the study area ranged from −6.47 to +1.24‰, and the average was −2.40‰. There was a difference between the actual values and theoretical value, the main possible reasons were: (1) the real conditions were different from the laboratory cultures; (2) the $\delta^{18}O(O_2)$ value was cited from previous study not from actual measurement. According to the previous research, the $\delta^{18}O(NO_3)$ value produced by microbial nitrification ranged from −10 to +10‰ in general [47,68,70]. As shown in Figure 11b, the $\delta^{18}O(NO_3)$ values all fell within the above range and indicated that nitrification was the main process of nitrogen transformation.

3.5.3. δ15. N(NO$_3$) – δ^{18}O(NO$_3$) Dual Isotope Technique

As shown in Figure 11a,b, the distribution of samples was an irregular plane, indicating that NO^3 contamination was a mixture of point sources and nonpoint sources. The δ^{15}N(NO$_3$) values were within the overlap range of multiple sources (Figure 11a), such as precipitation, soil organic nitrogen, manure and sewage, while the δ^{18}O(NO$_3$) values were in the range of soil organic nitrogen (Figure 11b). The results based on single isotope to identify the source of nitrogen were not accurate enough due to the overlaps of the δ^{15}N(NO$_3$) values among multiple sources.

To overcome this problem, the ^{15}N(NO$_3$) –^{18}O(NO$_3$) dual isotopes technique has been proposed [8] (Figure 11c). The δ^{15}N(NO$_3$) values ranged from +0.29 to +14.39‰, and the δ^{18}O(NO$_3$) values ranged from −6.47 to +1.24‰. 50% of the groundwater samples were in the range of manure and sewage, 41.67% of the groundwater samples were in the overlap area of soil organic nitrogen and manure and sewage, and 8.33% of the samples were located below the range of manure and sewage. High concentration of nitrate in groundwater indicated that soil organic nitrogen was not the main source of nitrate contamination. The mixed contamination of manure, sewage and NH$_4$ fertilizers increased the nitrate contents in groundwater, and the contribution of manure and sewage was greater than that of NH$_4$ fertilizers. The δ^{15}N(NO$_3$) and δ^{18}O(NO$_3$) data further affirmed the hydrochemical interpretation that precipitation, industrial activities and synthetic NO$_3$ were unlikely to be the main sources of NO$_3$ in the study area.

4. Conclusions

Hydrochemical characteristics and evolution process of groundwater in the northern Huangqihai Basin were comprehensively analyzed by using multiple hydrochemical methods. The groundwater in the study area was generally weakly alkaline with low salinity. The relative anionic abundance of groundwater samples was in the order of HCO$_3$$^-$ > Cl$^-$ > SO$_4$$^{2-}$ > NO$_3$$^-$, whereas the cationic abundance was Mg^{2+} > Ca^{2+} > Na$^+$ > K$^+$. The distributions of Mg^{2+}, Na$^+$, Cl$^-$ and NO$_3$$^-$ showed significant spatial variations (C_v > 100%), indicating the influence of human activities on groundwater. The main chemical facies of groundwater were HCO$_3$–Mg·Ca and HCO$_3$–Ca·Mg. The mole fractions of NO$_3$$^-$ in C29 and C30 samples were higher than 25%, and two new hydrochemical facies (NO$_3$·HCO$_3$–Ca·Mg and HCO$_3$·NO$_3$·Cl–Ca·Mg) appeared based on the improved Shukalev classification method. The hydrochemical evolution of groundwater was predominantly affected by rock weathering and also by cation exchange.

The main source of groundwater was precipitation by means of the D–^{18}O isotope technique, and the relationship between δD and δ^{18}O of groundwater was δD = 5.93δ^{18}O − 19.18 (R^2 = 0.9067). The *d–excess* range was −1.60 to +6.01‰ with an average value of 3.38‰ (<10‰), indicated that groundwater was controlled by water–rock interactions and influenced by evaporation. Contaminants were very likely to enter groundwater along with precipitation infiltration, resulting in excessive nitrogen in groundwater.

The NO$_3$(N) contents in some parts of the study area were far exceeded the level III standard. The NO$_3$$^-$/Na$^+$ ratios ranged from 0.00 to 3.69 with an average of 0.49, and the C_v was 1.34 (>1). The ratios of SO$_4$$^{2-}$/Na$^+$ were smaller than those of NO$_3$$^-$/Na$^+$ overall, indicating that the nitrate content of groundwater was mainly affected by agricultural activities. The Cl$^-$ contents were generally more than 1 mmol/L, and the NO$_3$$^-$/Cl$^-$ ratios ranged between 0.1 and 10 with an average value of 0.58. The correlation between NO$_3$$^-$ and K$^+$ was weak. The hydrochemical analysis showed that precipitation, industrial activities and synthetic NO$_3$ were unlikely to be the main sources of nitrate contamination.

The relationship between NO$_3$$^-$ and δ^{15}N(NO$_3$) was not negative, and the slope of δ^{15}N(NO$_3$) vs. δ^{18}O(NO$_3$) in groundwater was −0.09, far lower than 0.48 and 0.77, respectively. The δ^{18}O(NO$_3$) values fell within the range (−10 to +10‰). These analyses indicated that no obvious denitrification occurred, and that nitrification was the main process during nitrogen transformation in the study area. The δ^{15}N(NO$_3$) values ranged from +0.29 to +14.39‰, and the δ^{18}O(NO$_3$) values ranged from −6.47 to +1.24‰. The

δ^{15}N(NO$_3$) and δ^{18}O(NO$_3$) data further affirmed the hydrochemical interpretation that precipitation, industrial sewage and synthetic NO$_3$ were unlikely sources of NO$_3$, and the main sources were manure, sewage and NH$_4$ fertilizers.

The integration of hydrochemical analysis and dual isotope technique provides a further insight into the identification of nitrate contamination from multiple perspectives: hydrochemical characteristics, evolution mechanism, probable pathway of contaminants entering groundwater, bivariate relationships between NO$_3^-$ and other indices, and isotope composition. It is recommended that extensive and sustained monitoring of the D–^{18}O isotopes in groundwater and the ^{15}N(NO$_3$) and ^{18}O(NO$_3$) isotopes in the potential nitrate sources (fertilizers, manure and sewage) should be performed in further studies.

Author Contributions: Conceptualization, J.J.; methodology, J.J. and Z.W.; investigation, H.D. and J.Z.; data curation, Z.W. and Y.Z.; writing—original draft preparation, J.J.; writing—review and editing, J.J. All authors have read and agreed to the published version of the manuscript.

Funding: This research was funded by the Basic Scientific Research Foundation Special Project of the China Institute of Water Resources and Hydropower Research (No. MK2021J07 and MK2020J10), Project of Collaborative Innovation Center for Grassland Ecological Security (Ecohydrological Characteristics and Ecosystem Services Assessment in Tabu River Watershed, No. MK0143A032021) and Science and Technology Planning Project of Inner Mongolia (No. MK0143A012022).

Institutional Review Board Statement: Not applicable.

Informed Consent Statement: Not applicable.

Data Availability Statement: Not applicable.

Acknowledgments: The authors are grateful to all the editors and anonymous reviewers for their helpful comments that greatly improved the quality of the manuscript.

Conflicts of Interest: The authors declare no conflict of interest.

References

1. He, X.D.; Li, P.Y.; Ning, J.; He, S.; Yang, N.N. Geochemical processes during hydraulic fracturing in a tight sandstone reservoir revealed by field and laboratory experiments. *J. Hydrol.* **2022**, *612*, 128292. [CrossRef]
2. Fuoco, I.; Rosa, D.R.; Barca, D.; Figoli, A.; Gabriele, B.; Apollaro, C. Arsenic polluted waters: Application of geochemical modelling as a tool to understand the release and fate of the pollutant in crystalline aquifers. *J. Environ. Manag.* **2022**, *301*, 113796. [CrossRef] [PubMed]
3. Fuoco, I.; Marini, L.; De Rosa, R.; Figoli, A.; Gabriele, B.; Apollaro, C. Use of reaction path modelling to investigate the evolution of water chemistry in shallow to deep crystalline aquifers with a special focus on fluoride. *Sci. Total Environ.* **2022**, *830*, 154566. [CrossRef]
4. Shi, X.Y.; Wang, Y.; Jiao, J.J.; Zhong, J.L.; Wen, H.G.; Dong, R. Assessing major factors affecting shallow groundwater geochemical evolution in a highly urbanized coastal area of Shenzhen City, China. *J. Geochem. Explor.* **2018**, *184*, 17–27. [CrossRef]
5. Ren, C.B.; Zhang, Q.Q. Groundwater chemical characteristics and controlling factors in a region of northern China with intensive human activity. *Int. J. Env. Res. Public Health* **2020**, *17*, 9126. [CrossRef] [PubMed]
6. Su, F.; Wu, J.; Wang, D.; Zhao, H.; Wang, Y.; He, X. Moisture movement, soil salt migration, and nitrogen transformation under different irrigation conditions: Field experimental research. *Chemosphere* **2022**, *300*, 134569. [CrossRef] [PubMed]
7. Anornu, G.; Gibrilla, A.; Adomako, D. Tracking nitrate sources in groundwater and associated health risk for rural communities in the White Volta River basin of Ghana using isotopic approach (δ^{15}N, δ^{18}O–NO$_3$ and ^{3}H). *Sci. Total Environ.* **2016**, *603–604*, 687–698. [CrossRef] [PubMed]
8. Hu, M.M.; Wang, Y.C.; Du, P.C.; Shui, Y.; Cai, A.M.; Lu, C.; Bao, Y.F.; Li, Y.H.; Li, S.Z.; Zhang, P.W. Tracing the sources of nitrate in the rivers and lakes of the southern areas of the Tibetan Plateau using dual nitrate isotopes. *Sci. Total Environ.* **2019**, *658*, 132–140. [CrossRef]
9. Li, P.Y.; He, X.D.; Guo, W.Y. Spatial groundwater quality and potential health risks due to nitrate ingestion through drinking water: A case study in Yan'an City on the Loess Plateau of northwest China. *Hum. Ecol. Risk Assess.* **2019**, *25*, 11–31. [CrossRef]
10. Mayo, A.L.; Tingey, D.G. Shallow groundwater chemical evolution, isotopic hyperfiltration, and salt pan formation in a hypersaline endorheic basin: Pilot Valley, Great Basin, USA. *Hydrogeol. J.* **2021**, *29*, 2219–2243. [CrossRef]
11. Dowling, C.B.; Poreda, R.J.; Basu, A.R.; Peters, S.L.; Aggarwal, P.K. Geochemical study of arsenic release mechanisms in the Bengal Basin groundwater. *Water Resour. Res.* **2002**, *38*, 1173. [CrossRef]
12. Banda, K.E.; Wilson, M.; Rasmus, J.; Jason, O.; Imasiku, N.; Flemming, L. Mechanism of salinity change and hydrogeochemical evolution of groundwater in the Machile–Zambezi Basin, South–western Zambia. *J. Afr. Earth Sci.* **2019**, *153*, 72–82. [CrossRef]

13. Nipada, S.; Saowani, S.; Schradh, S. Arsenic contamination in groundwater and potential health risk in western Lampang Basin, northern Thailand. *Water* **2022**, *14*, 465. [CrossRef]
14. Wen, Y.; Qiu, J.H.; Cheng, S.; Xu, C.C.; Gao, X.J. Hydrochemical evolution mechanisms of shallow groundwater and its quality assessment in the estuarine coastal zone: A case study of Qidong, China. *Int. J. Env. Res. Pub. He.* **2020**, *17*, 3382. [CrossRef]
15. Zhang, Y.H.; Xu, M.; Li, X.; Qi, J.H.; Zhang, Q.; Guo, J.; Yu, L.L.; Zhao, R. Hydrochemical characteristics and multivariate statistical analysis of natural water system: A case study in Kangding county, southwestern China. *Water* **2018**, *10*, 80. [CrossRef]
16. Su, H.; Kang, W.D.; Kang, N.; Liu, J.T.; Li, Z. Hydrogeochemistry and health hazards of fluoride–enriched groundwater in the Tarim Basin, China. *Environ. Res.* **2021**, *200*, 111476. [CrossRef] [PubMed]
17. Feng, F.; Jia, Y.F.; Yang, Y.; Huan, H.; Lian, X.Y.; Xu, X.J.; Xia, F.; Han, X.; Jiang, Y.H. Hydrogeochemical and statistical analysis of high fluoride groundwater in northern China. *Environ. Sci. Pollut. Res. Int.* **2020**, *27*, 34840–34861. [CrossRef]
18. Chen, Z.X.; Yu, L.; Liu, W.G.; Lam, M.H.W. Nitrogen and oxygen isotopic compositions of water–soluble nitrate in Taihu Lake water system, China: Implication for nitrate sources and biogeochemical process. *Env. Earth Sci.* **2014**, *71*, 217–223. [CrossRef]
19. Mukherjee, I.; Singh, U.K. Characterization of groundwater nitrate exposure using Monte Carlo and Sobol sensitivity approaches in the diverse aquifer systems of an agricultural semiarid region of Lower Ganga Basin, India. *Sci. Total Environ.* **2021**, *787*, 147657. [CrossRef]
20. Avilés, F.; Patricia, G.; Lorenzo, S.; Elisa, S.; Yvan, R.; Joel, S.; Eduardo, R.O.; Céline, D. Hydrogeochemical and nitrate isotopic evolution of a semiarid mountainous basin aquifer of glacial–fluvial and paleolacustrine origin (Lake Titicaca, Bolivia): The effects of natural processes and anthropogenic activities. *Hydrogeol. J.* **2022**, *30*, 181–202. [CrossRef]
21. Chen, J.Q.; Lv, J.M.; Wang, Q.W.; Wang, J. External Groundwater alleviates the degradation of closed lakes in semi–arid regions of China. *Remote Sens.* **2019**, *12*, 45. [CrossRef]
22. Han, X.Q.; Lin, J.; Yang, Y.; Sun, X.M.; Zhu, X.L.; Wu, J.F.; Wu, J.C. Three–dimensional modeling and visualization of hydrogeological structure in water supply well fields of Jining district, Inner Mongolia. *J. Inn. Mong. Agric. Univ.* **2017**, *38*, 45–52. (In Chinese) [CrossRef]
23. Chen, M.M.; Liu, J.G. Historical trends of wetland areas in the agriculture and pasture interlaced zone: A case study of the Huangqihai Lake Basin in northern China. *Ecol. Model.* **2015**, *318*, 168–176. [CrossRef]
24. Tian, F.; Wang, Y.; Zhao, Z.L.; Dong, J.; Liu, J.; Ling, Y.; Yuan, Y.P.; Ye, M.N. Holocene vegetation and climate changes in the Huangqihai Lake region, Inner Mongolia. *Acta Geol. Sin.* **2020**, *94*, 1178–1186. [CrossRef]
25. Fu, Y.C.; Zhao, J.Y.; Peng, W.Q.; Zhu, G.P.; Quan, Z.J.; Li, C.H. Spatial modelling of the regulating function of the Huangqihai Lake wetland ecosystem. *J. Hydro.* **2018**, *564*, 283–293. [CrossRef]
26. Guo, Z.X.; Hao, W.G.; Zhang, S.; Zhang, Z.G. Evaluation on present situation of water environmental quality in Huangqihai Lake. *Chin. Agric. Sci. Bull.* **2007**, *23*, 346–350. (In Chinese) [CrossRef]
27. Hao, W.G.; Bao, X.Q.; Wei, Y.F.; Guo, Z.X.; Zhuang, J.; Liang, J. Eutrophication and its dynamic changes in Huangqihai lake. *Yellow River* **2008**, *30*, 74–75. (In Chinese) [CrossRef]
28. *GB/T14848-2017*; Standard for Groundwater Quality. Standards Press of China: Beijing, China, 2017. (In Chinese)
29. *DZ/T 0064-2021*; Analysis Method of Groundwater Quality. Standards Press of China: Beijing, China, 2021. (In Chinese)
30. China Geological Survey. *Handbook of Hydrogeology*, 2nd ed.; Geological Publishing House: Beijing, China, 2018; pp. 108–110.
31. Qian, H.; Ma, Z.Y. *Hydrogeochemistry*; Geology Publishing House: Beijing, China, 2005. (In Chinese)
32. Piper, A.M. A graphic procedure in the geochemical interpretation of water analysis. *Trans. Am. Geophys. Union* **1944**, *25*, 914–928. [CrossRef]
33. Gibbs, R.J. Mechanisms controlling world water chemistry. *Science* **1970**, *170*, 1088–1090. [CrossRef]
34. Lachaal, F.; Messaoud, R.B.; Jellalia, D.; Chargui, S.; Chekirbane, A.; Mlayah, A.; Massuel, S.; Leduc, C. Impact of water resources management on groundwater hydrochemical changes: A case of Grombalia shallow aquifer, NE of Tunisia. *Arab. J. Geosci.* **2018**, *11*, 304. [CrossRef]
35. Lachache, S.; Nabou, M.; Merzouguui, T.; Amroune, A. Hydrochemistry and origin of principal major elements in the groundwater of the Béchar–Kénadsa basin in arid zone, south–west of Algeria. *J. Water Land Dev.* **2018**, *36*, 77–87. [CrossRef]
36. Eskandari, E.; Mohammadzadeh, H.; Nassery, H.; Vadiati, M.; Zadeh, A.M.; Kisi, O. Delineation of isotopic and hydrochemical evolution of karstic aquifers with different cluster–based (HCA, KM, FCM and GKM) methods. *J. Hydrol.* **2022**, *609*, 127706. [CrossRef]
37. Apollaro, C.; Di Curzio, D.; Fuoco, I.; Buccianti, A.; Dinelli, E.; Vespasiano, G.; Castrignanò, A.; Rusi, S.; Barca, D.; Figoli, A.; et al. A multivariate non–parametric approach for estimating probability of exceeding the local natural background level of arsenic in the aquifers of Calabria region (Southern Italy). *Sci. Total Environ.* **2022**, *806*, 150345. [CrossRef] [PubMed]
38. Deepika, B.V.; Ramakrishnaiah, C.R.; Naganna, S.R. Spatial variability of ground water quality: A case study of Udupi district, Karnataka State, India. *J. Earth Syst. Sci.* **2020**, *129*, 221. [CrossRef]
39. Kpao, C.P.; φo, ..; Wang, Y.X. Geochemical and ecological consequences of changes in groundwater chemical composition under the influence of pollutants. *J. Geosci. Transl.* **1992**, *19*, 73–80. (In Chinese) [CrossRef]
40. Huang, G.X.; Liu, C.Y.; Sun, J.C.; Zhang, M.; Jing, J.H.; Li, L.P. A regional scale investigation on factors controlling the groundwater chemistry of various aquifers in a rapidly urbanized area: A case study of the Pearl River Delta. *Sci. Total Environ.* **2018**, *625*, 510–518. [CrossRef]

41. Hu, C.P.; Liu, Z.Q.; Xiong, K.N.; Lyu, X.X.; Li, Y.; Zhang, R.K. Characteristics of and influencing factors of hydrochemistry and carbon/nitrogen variation in the Huangzhouhe river basin, a World Natural Heritage Site. *Int. J. Environ. Res. Public Health* **2021**, *18*, 13169. [CrossRef]
42. Heaton, T.H.E. Isotopic studies of nitrogen pollution in the hydrosphere and atmosphere: A review. *Chem. Geol. Isot. Geosci. Sect.* **1986**, *59*, 87–102. [CrossRef]
43. Adebowale, T.; Surapaneni, A.; Faulkner, D.; McCance, W.; Wang, S.Q.; Currell, M. Delineation of contaminant sources and denitrification using isotopes of nitrate near a wastewater treatment plant in peri-urban settings. *Sci. Total Environ.* **2019**, *65*, 2701–2711. [CrossRef]
44. Widory, D.; Petelet, G.E.; Négrel, P.; Ladouche, B. Tracking the sources of nitrate in groundwater using coupled nitrogen and boron isotopes: A synthesis. *Environ. Sci. Tech.* **2005**, *39*, 539–548. [CrossRef]
45. Bottcher, J.; Sreebel, O.; Voerkelius, S.; Schmidt, H.L. Using isotope fractionation of nitrate–nitrogen and nitrate–oxygen for evaluation of microbial denitrification in a sandy aquifer. *J. Hydrol.* **1990**, *114*, 413–424. [CrossRef]
46. Fukada, T.; Hiscock, K.M.; Dennis, P.F.; Grischek, T. A dual isotope approach to identify denitrification in groundwater at a river–bank infiltration site. *Water Res.* **2003**, *37*, 3070–3078. [CrossRef]
47. Xue, D.M.; Botte, J.; De, B.B.; Accoe, F.; Nestler, A.; Taylor, P.; Van, C.O.; Berglund, M.; Boeckx, P. Present limitations and future prospects of stable isotope methods for nitrate source identification in surface and groundwater. *Water Res.* **2009**, *43*, 1159–1170. [CrossRef] [PubMed]
48. Andrew, A.A.; Jun, S.; Takahiro, H.; Makoto, K.; Akoachere, R.; George, E.N.; Aka, F.T.; Ono, M.; Eyong, G.E.T.; Tandia, B.K. Flow dynamics and age of groundwater within a humid equatorial active volcano (Mount Cameroon) deduced by δD, $\delta^{18}O$, ^3H and chlorofluorocarbons (CFCs). *J. Hydrol.* **2013**, *502*, 156–176. [CrossRef]
49. Liao, H.W.; Jiang, Z.C.; Zhou, H.; Qin, X.Q.; Huang, Q.B. Isotope–based study on nitrate sources in a karst wetland water, southwest China. *Water* **2022**, *14*, 1533. [CrossRef]
50. Raj, P.S.; Prasad, B.S.; Amrita, W.; Friedemann, F.T. Earthquake chemical precursors in groundwater: A review. *J. Seismol.* **2018**, *22*, 1293–1314. [CrossRef]
51. Skelton, A.; Liljedahl-Claesson, L.; Wästeby, N.; Andrén, M.; Stockmann, G.; Sturkell, E.; Mörth, C.M.; Stefansson, A.; Tollefsen, E.; Siegmund, H.; et al. Hydrochemical changes before and after earthquakes based on long-term measurements of multiple parameters at two sites in northern Iceland–A review. *J. Geophys. Res. Sol. Ea.* **2019**, *124*, 2702–2720. [CrossRef]
52. Peng, C.; He, J.T.; Liao, L.; Zhang, Z.G. Research on the influence degree of human activities on groundwater quality by the method of geochemistry: A case study from Liujiang Basin. *Earth Sci. Front.* **2017**, *24*, 321–331. (In Chinese) [CrossRef]
53. Apollaro, C.; Vespasiano, G.; De Rosa, R.; Marini, L. Use of mean residence time and flowrate of thermal waters to evaluate the volume of reservoir water contributing to the natural discharge and the related geothermal reservoir volume. Application to Northern Thailand hot springs. *Geothermics* **2015**, *58*, 62–74. [CrossRef]
54. Li, J.X.; Zhou, H.L.; Qian, K.; Xie, X.J.; Xue, X.B.; Yang, Y.J.; Wang, Y.X. Fluoride and iodine enrichment in groundwater of north China plain: Evidences from speciation analysis and geochemical modeling. *Sci. Total Environ.* **2017**, *598*, 239–248. [CrossRef]
55. Singh, G.; Rishi, M.S.; Herojeet, R.; Kaur, L.; Sharma, K. Evaluation of groundwater quality and human health risks from fluoride and nitrate in semi–arid region of northern India. *Environ. Geochem. Health* **2020**, *42*, 1833–1862. [CrossRef] [PubMed]
56. An, Y.K.; Lu, W.X. Hydrogeochemical processes identification and groundwater pollution causes analysis in the northern Ordos Cretaceous Basin, China. *Environ. Geochem. Health* **2017**, *40*, 1209–1219. [CrossRef] [PubMed]
57. Farid, I.; Zouari, K.; Rigane, A.; Beji, R. Origin of the groundwater salinity and geochemical processes in detrital and carbonate aquifers: Case of Chougafiya basin (Central Tunisia). *J. Hydrol.* **2015**, *530*, 508–532. [CrossRef]
58. Capuano, R.M.; Jones, C.R. Cation exchange in groundwater-chemical evolution and prediction of paleo-groundwater flow: A natural-system study. *Water Resour. Res.* **2020**, *56*, e2019WR026318. [CrossRef]
59. Eaman, S.; De Louw, P.G.B.; Van der Zee, S.E.A.T.M. Cation exchange in a temporally fluctuating thin freshwater lens on top of saline groundwater. *Hydrogeol. J.* **2017**, *25*, 223–241. [CrossRef]
60. Craig, H. Isotopic variations in meteoric waters. *Science* **1961**, *133*, 1702–1703. [CrossRef]
61. Guo, X.; Li, W.B.; Du, L.; Jia, D.B.; Liu, T.X. Characteristics and influence factors for the hydrogen and oxygen isotopic of precipitation in inner Mongolia. *China Environ. Sci.* **2022**, *42*, 1088–1096. (In Chinese) [CrossRef]
62. Dansgaard, W. Stable isotopes in precipitation. *Tellus* **1964**, *16*, 436–468. [CrossRef]
63. Claus, K.; Rolf, F.R.; Fernando, R.B.; Iñaki, V. Vapor source and spatiotemporal variation of precipitation isotopes in southwest Spain. *Hydrol. Process.* **2021**, *35*, e14445. [CrossRef]
64. Chen, X.; Jiang, C.L.; Zheng, L.G.; Zhang, L.Q.; Fu, X.J.; Chen, S.G.; Chen, Y.C.; Hu, J. Evaluating the genesis and dominant processes of groundwater salinization by using hydrochemistry and multiple isotopes in a mining city. *Environ. Pollut.* **2021**, *283*, 117381. [CrossRef]
65. Xia, C.; Liu, G.; Mei, J.; Meng, Y.; Hu, Y. Characteristics of hydrogen and oxygen stable isotopes in precipitation and the environmental controls in tropical monsoon climatic zone. *Int. J. Hydrog. Energy.* **2019**, *44*, 5417–5425. [CrossRef]
66. Wang, C.; Zhang, H.; Lei, P.; Xin, X.K.; Zhang, A.J.; Yin, W. Evidence on the causes of the rising levels of CODMn along the middle route of the South-to-North Diversion Project in China: The role of algal dissolved organic matter. *J. Environ. Sci.* **2022**, *113*, 281–290. [CrossRef] [PubMed]

67. Koba, K.; Tokuchi, N.; Wada, E.; Nakajima, T.; Iwatsubo, G. Intermittent denitrification: The application of a ^{15}N natural abundance method to a forested ecosystem. *Geochim. Cosmochim Ac.* **1997**, *61*, 5043–5050. [CrossRef]
68. Kendall, C. Tracing nitrogen sources and cycling in catchments. *Isot. Tracers Catchment Hydrol.* **1998**, *1*, 519–576. [CrossRef]
69. Amberger, A.; Schmidt, H.L. Natürliche isotopengehalte von nitrat als indikatoren für dessen herkunft. *Geochim. Et Cosmochim. Acta* **1987**, *51*, 2699–2705. [CrossRef]
70. Kent, R.; Landon, M.K. Trends in concentrations of nitrate and total dissolved solids in public supply wells of the Bunker Hill, Lytle, Rialto, and Colton groundwater sub basins, San Bernardino county, California: Influence of legacy land use. *Sci. Total Environ.* **2013**, *452–453*, 125–136. [CrossRef]

Article

Stimulating Nitrate Removal with Significant Conversion to Nitrogen Gas Using Biochar-Based Nanoscale Zerovalent Iron Composites

Siyuan Liu, Xiao Han, Shaopeng Li, Wendi Xuan and Anlei Wei *

College of Urban and Environmental Sciences, Northwest University, Xi'an 710127, China
* Correspondence: alwei@nwu.edu.cn

Abstract: For efficient and environmentally friendly removal of nitrate from groundwater, biochar-based nanoscale zerovalent iron composites were prepared, where biochar was derived from pine sawdust at 4 different pyrolysis temperatures. The results show that biochar with different pyrolysis temperatures played a great role in both nitrate removal efficiency and nitrate conversion rate to nitrogen gas for the prepared composites. Specifically, the composite with biochar pyrolyzed at 500 °C, ZB12-500, showed the best performance in both nitrate removal and conversion to nitrogen gas. With an initial solution pH from 5 to 10, ZB12-500 maintained high removal efficiencies varying from 97.29% to 89.04%. Moreover, the conversion of nitrate to nitrogen gas increased with the initial nitrate concentration, and it reached 31.66% with an initial nitrate concentration of 100 mg/L. Kinetics analysis showed that the nitrate removal process fit well with a two-compartment first-order kinetic model. Meanwhile, the test of nitrate removal by ZB12-500 in synthetic groundwater showed that HCO_3^- and SO_4^{2-} limited nitrate removal but improved nitrate conversion to nitrogen gas. Furthermore, the nitrate removal mechanism suggested that biochar could facilitate electron transfer from zero valent iron to nitrate, which led to high nitrate removal efficiency. In addition, the interaction of ferrous ions and the quinone group of biochar could increase the nitrate conversion to nitrogen gas. Therefore, this study suggests that ZB12-500 is a promising alternative for the remediation of nitrate-contaminated groundwater.

Keywords: nano zero-valent iron; biochar; nitrate removal; nitrogen gas

Citation: Liu, S.; Han, X.; Li, S.; Xuan, W.; Wei, A. Stimulating Nitrate Removal with Significant Conversion to Nitrogen Gas Using Biochar-Based Nanoscale Zerovalent Iron Composites. *Water* 2022, *14*, 2877. https://doi.org/10.3390/w14182877

Academic Editor: Jianhua Wu

Received: 24 August 2022
Accepted: 11 September 2022
Published: 15 September 2022

Publisher's Note: MDPI stays neutral with regard to jurisdictional claims in published maps and institutional affiliations.

Copyright: © 2022 by the authors. Licensee MDPI, Basel, Switzerland. This article is an open access article distributed under the terms and conditions of the Creative Commons Attribution (CC BY) license (https://creativecommons.org/licenses/by/4.0/).

1. Introduction

Groundwater plays an important role in human freshwater resources. About 20% of the world's fresh water supply is provided by groundwater. Therefore, ensuring the safety of groundwater is of great importance to human beings. However, as a result of rapid expansion in modern agriculture and industry, inorganic nitrogen pollution has become a major problem in the world. The main types of nitrogen present in water are nitrate, nitrite, and ammonia, but the most common pollutant is nitrate, and nitrate nitrogen pollution has been reported as a worldwide pollution problem [1–3]. Nitrate nitrogen entering water may lead to eutrophication or other negative effects on water quality [4–7]. On the other hand, nitrate can also end up in the human body through groundwater and cause methemoglobinosis, also known as "blue baby syndrome," and can be converted into carcinogenic nitrite amine preforms [8]. The WHO limited the concentration of NO_3^--N to 10 mg/L [9]. Therefore, advanced treatment technologies are needed to remove nitrate from groundwater in an economical and environmentally friendly manner.

At present, mainstream nitrate removal technologies in groundwater include chemical catalysis of nitrate reduction, anion exchange, low-pressure reverse osmosis, and microbial methods. Among them, the anion-exchange method and reverse osmosis method both have the disadvantage of requiring frequent regeneration medium and producing secondary pollutants [10,11], while biological denitrification requires a long repair time and generates sludge, which requires a large amount of maintenance cost [12]. Compared with other methods, the chemical reduction method provides the benefits of quick effect, lower cost, and no secondary pollution, and is more suitable for nitrate remediation in groundwater [13].

Nanoscale zero-valent iron (nZVI) has been reported to effectively remedy groundwater contamination in recent years because of its finer particle size, higher reactivity, and larger specific surface area to remove pollutants. Although nZVI can effectively reduce nitrate, it also has some disadvantages, such as easy agglomeration, easy corrosion, poor electronic selectivity, low reactivity, and poor product selectivity [14]. For these reasons, a variety of modification methods for nZVI have been studied. The bimetallic method is to dope some high-potential metals into the nZVI so that nZVI and the doped metals can make up micro-macroscopic coupled electrode systems to get rid of more nitrate and product selectivity of the material. Sparis et al. used ZVI-5%Cu particles doped with Cu ions into ZVI to reduce more than 80% of nitrate in 20 min and completely remove it in 1 h, whereas ZVI alone removed only 74.5% of nitrate in 1 h [15]. Lubphoo et al. found that a trimetal (Pd-Cu)-ZVI material using Cu and Pd as catalysts was more efficient in reducing nitrate than pure nZVI, as well as an increased ratio of N_2 production [16]. Zhang et al. found that the addition of palladium increased the gas production of nitrate, nitrite, and ammonia recovered from aqueous and solid-phase supports [17]. Although the dopped metal materials could be beneficial for nitrate removal and its selective conversion to nitrogen gas, they also mean increased material cost and environmental risk. Pre-magnetization could also boost nZVI's reactivity due to its superiority in magnetic memory. The nitrate removal rate of the pre-magnetized Fe^0 system was 1.99 times better than that of the non-pre-magnetized one [18]. During nitrate reduction, the magnetic field gradient force drove nitrate gathering at the surface of the pre-magnetized Fe^0 system, and meanwhile led to more nitrate conversion to nitrogen gas [19]. However, it is difficult to apply the pre-magnetization method on a large scale in the field of groundwater treatment engineering. In fact, nZVI prepared by the loading method can prevent aggregation of ZVI in the reaction, provide more active sites, and have a wider range of pH application conditions [20]. The application scenario of the material is more suitable for practical groundwater remediation.

The carrier materials commonly used in the loading method include organic materials such as alginate matrix [21], polymeric styrene anion exchanger [22], and inorganic materials such as activated carbon [23], biochar [24], zeolite [20], etc. Biochar often presents with better porosity and specific surface area, which makes it a promising carrier for nanoscale materials [25]. Namasivayam et al. used a waste coconut shell to prepare $ZnCl_2$ activated carbon to recover nitrate from water. The results showed that pH had a great impact on the recovery of nitrate, and the desorption rate could reach 58% and 92% at pH 2 and 11, while the desorption rate was negligible at pH 3–10 [26]. Kamyar et al. prepared TBC by impregnating magnetic nanoparticles on tea biochar to remove heavy metals and nutrients in water, up to 147.84 mg/g of Ni^{2+}, 160.00 mg/g of Co^{2+}, 49.43 mg/g of NH_4^+ and 112.61 mg/g of PO_4^{3-} could be adsorbed onto tested biochar [27]. For nZVI, nitrate was mainly conversed to NH_4^+ (93.5%) instead of N_2 (5.7%), while the N_2 conversion ratio of ZVI/BC composite can reach 60.1% [28]. Oh et al. used straw as raw material to prepare biochar-loaded nZVI material at 900 °C and reaction results showed that NO_3^--N was almost completely removed and the selectivity of the N_2 product was also very high [29]. Gao et al. produced biochar-loaded ZVI at 400 °C to remove Cr^{6+} from aqueous solutions and reached a maximum removal capacity of 126 mg/g at pH 2.5, whereas ZVI was highly agglomerated at the same pH [30]. Wei et al. prepared BC/nZVI composites with different mass ratios from straw to remove nitrate nitrogen from water. The prepared composite presented superiority to nZVI, and its removal capacity was 229 mg NO_3^--N/g [28]. Some studies have concluded that the process of nitrate removal by ZVI composite materials can be explained by Equations (1)–(10) [31–34]. Obviously, biochar-supported nZVI composites have great potential for remediation of nitrate contamination. As a low-cost and environmentally friendly natural material, biochar could also mediate environmentally related abiotic redox processes [35,36], so it is desirable to use composite materials for nitrate treatment in groundwater. When biochar is used as the carrier of composite materials, pyrolysis temperature affects the properties of composite materials by changing the physicochemical properties of biochar (including specific surface area, functional group, hydrophobicity, and graphitization) [37–39]. However, little is known about the effect of the pyrolysis temperature of biochar support on the product selectivity for nitrate removal by nZVI/BC composites.

$$Fe^0 + 2H^+ \rightarrow Fe^{2+} + H_2 \uparrow \tag{1}$$

$$2Fe^0 + O_2 + 2H_2O \rightarrow 2Fe^{2+} + 4OH^- \tag{2}$$

$$Fe^0 + NO_3^- + 2H^+ \rightarrow Fe^{2+} + NO_2^- + H_2O \tag{3}$$

$$4Fe^0 + NO_3^- + 10H^+ \rightarrow 4Fe^{2+} + NH_4^+ + 3H_2O \tag{4}$$

$$3Fe^0 + NO_2^- + 8H^+ \rightarrow 3Fe^{2+} + NH_4^+ + 3H_2O \tag{5}$$

$$5Fe^0 + 2NO_3^- + 12H^+ \rightarrow 5Fe^{2+} + N_2(g) + 6H_2O \tag{6}$$

$$8Fe^0 + 3NO_3^- + 9H_2O \xrightarrow{BC} 4Fe_2O_3 + 3NH_4^+ + 6OH^- \tag{7}$$

$$10Fe^0 + 6NO_3^- + 3H_2O \xrightarrow{BC} 5Fe_2O_3 + 3N_2(g) + 6OH^- \tag{8}$$

$$3Fe^0 + NO_3^- + 3H_2O \xrightarrow{BC} Fe_3O_4 + NH_4^+ + 2OH^- \tag{9}$$

$$18Fe^0 + 3Fe^{2+} + 10NO_3^- + 2H_2O \xrightarrow{BC} 7Fe_3O_4 + 5N_2(g) + 4OH^- \tag{10}$$

Our previous study have proved the feasibility of the nZVI/BC for nitrate removal from groundwater [28]. In order to explore the effect of biochar prepared at different pyrolysis temperatures on the removal of nitrate from groundwater by nano zero-valent iron/biochar composite here, nZVI/BC composites with a mass ratio of 1:2 (ZB12) were prepared by using biochar at different pyrolysis temperatures. The prepared composites were characterized using surface analysis techniques, including a scanning electron microscope, X-ray diffraction pattern, Fourier transform infrared spectroscopy, X-ray photoelectron spectroscopy, and Brunauer–Emmett–Teller (BET) specific surface area measurement. Moreover, nitrate removal efficiencies and their product selectivity were evaluated under different conditions of groundwater, including dosage, initial pH, initial nitrate concentration, and co-existing ions. In addition, the nitrate removal kinetics were investigated, and a mechanism of nitrate removal was proposed.

2. Materials and Methods

2.1. Materials

Anhydrous ethanol (CH_3CH_2OH, 99.7%) was obtained from Fuyu Chemical Co., Ltd. (Tianjin, China). Sodium borohydride ($NaBH_4$, 97.0%), potassium nitrate (KNO_3, 99.0%), potassium persulfate ($K_2S_2O_8$, 99.0%), ammonium chloride (NH_4Cl, 99.8%), and sodium nitrite ($NaNO_2$, 99.0%) were acquired from Kermel Chemical Reagent (Tianjin, China). Ferrous sulfate heptahydrate ($FeSO_4 \cdot 7H_2O$, 99.0%) was purchased from Sheng Ao Chemical Reagent (Tianjin, China). The raw pine sawdust was collected from Weinan, Shaanxi Province, China. The sawdust was rinsed three times with deionized (DI) water and then dried overnight in an oven at 80 °C. nZVI used for comparison were purchased from Xiangtian Nanomaterials Co., LTD., Shanghai, China.

2.2. Synthesis of ZB12

Biochar was prepared by pyrolysis. Briefly, the crucible was filled with pre-treated pine sawdust, which was then put into a muffle furnace for carbonization. The carbonization was under a nitrogen atmosphere with a nitrogen flow rate of 80 mL/min and a heating rate of 10 °C/min. The carbonization process lasted 4 h at target temperatures of 350, 500, 650 and 800 °C, respectively. Then the heating stopped, and the muffle furnace was cooled down to room temperature. Next, the biochar was removed and pulverized through a 150-mesh sieve for later use.

The synthesis of ZB12 was conducted based on a sodium borohydride reduction method, as in Equation (11) [40]. According to our previous study [28], 1:2 was the optimal mass ratio of iron content to biochar, so this ratio was used in the sample preparation in this study. Briefly, 1.39 g FeSO$_4$·7H$_2$O was dissolved in 50 mL deionized water, then 0.56 g of the prepared biochar sample was added, and next, the mixture was placed in an anaerobic flask for 1 h under ultrasound to make full contact with the iron solution. Then, nitrogen was purged for 30 min to create an anoxic environment. According to the reaction formula, NaBH$_4$ with a slightly excess amount was dissolved in deionized water of 20 mL, and then added dropwise. The Fe^{2+} was completely reduced to Fe^0 after reaction at 120 rpm in a shaker for half an hour. The prepared composite was rinsed using deionized water and then anhydrous ethanol, and such rinsing was repeated three times and then dried in a constant temperature water bath under nitrogen protection. The composites prepared from biochar pyrolysis at 350, 500, 650 and 800 °C were labeled ZB12-350, ZB12-500, ZB12-650, and ZB12-800, respectively. The preparation process for ZB12 is shown in Figure 1.

$$Fe^{2+} + 2BH_4^- + 6H_2O \rightarrow Fe^0 + 2B(OH)_3 + 7H_2 \uparrow \qquad (11)$$

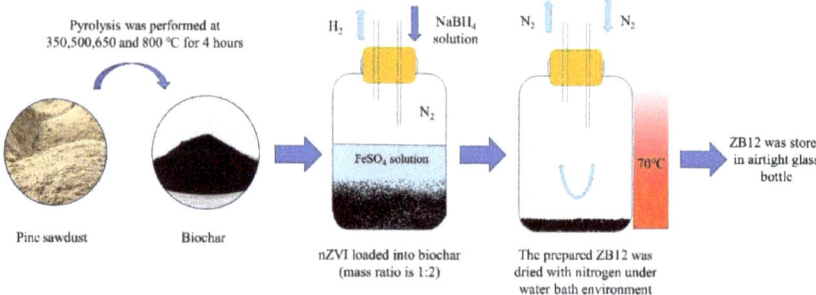

Figure 1. Preparation process of ZB12.

2.3. Characterization

According to previous studies, the nitrate removal reaction can reach about half of the reaction process after 1 h, and the reaction is almost complete after 24 h [41]. Therefore, the sample that reacted for 1 h is selected as the sample during the reaction, and the sample that is reacted for 24 h is the sample after the reaction.

The morphology of the samples before and after the reaction was investigated using a scanning electron microscope (SEM). The specific surface area (SSA) was measured using a surface area BET analyzer. X-ray diffraction (XRD) patterns were determined using an X-ray diffractometer. The chemical composition of the samples before, during, and after the reaction was analyzed according to the diffraction peaks. The scanning range 2θ was 10°~70°. In addition, the surface function groups were analyzed using Fourier transform infrared spectroscopy (FTIR). Meanwhile, the surface composition of the samples was investigated using X-ray photoelectron spectroscopy (XPS).

2.4. Experiments for Chemical Reduction of Nitrate by ZB12

Initial concentrations of nitrate were investigated in the range of 30–100 NO$_3^-$-N mg/L, which was set according to a survey of nitrate-contaminated groundwater in China [42]. In addition, nitrate removal experiments were conducted in 40 mL nitrate solution (30 NO$_3^-$-N mg/L) with 0.2 g of the prepared samples (nZVI or ZB12 samples). Moreover, the role of dosage in a 40 mL solution (30 NO$_3^-$-N mg/L) was investigated with 0.08, 0.12, 0.16, 0.2, 0.24, and 0.28 g of ZB12 samples. The impact of the solution pH was studied between

5 and 10, which was adjusted with HCl or NaOH solutions. On the other hand, the effects of co-existing ions were also investigated according to Table 1, which was obtained from a survey of groundwater samples in Weinan City, Shaanxi Province, China. In the above experiments, all the reactions happened in anaerobic bottles, which were sealed with butyl rubber stopper.

Table 1. Water quality data of groundwater and experimental configuration water.

Water Type	Monitoring Index	Maximum Value	Minimum Value	Average Value	Experimental Value
Ground water	K^+	9.18	0.87	2.90	0
	Na^+	953.74	12.98	286.12	250
	Mg^{2+}	274.62	14.56	61.93	60
	Ca^{2+}	140.98	13.05	43.66	40
	Cl^-	358.32	15.07	68.87	60
	SO_4^{2-}	604.65	12.26	42.30	40
	HCO_3^-	388.64	28.81	78.05	70
	NO_3^-	141.49	0.81	32.96	30

Note: K^+, Na^+, Mg^{2+}, Ca^{2+}, Cl^-, SO_4^{2-}, HCO_3^-, NO_3^- are in mg/L.

Kinetic and mechanistic experiments were carried out as follows: 0.2 g of ZB12 samples was reacted with nitrate with an initial concentration of 30 NO_3^--N mg/L in 40 mL solution in anaerobic bottles. Samples in each bottle were measured in turn at presetting time points. The measurement included pH, DO, oxidation-reduction potential (ORP), and the concentrations of nitrogen species (NO_3^--N, NO_2^--N, NH_4^+-N and total nitrogen (TN)). The values of DO, pH, and ORP were monitored by a portable DO meter and a pH meter, respectively. The concentrations of NO_3^--N, NO_2^--N, NH_4^+-N and TN were determined by UV-spectrophotometric method, spectrophotometric method, Nessler's reagent spectrophotometry, and alkaline potassium persulfate digestion spectrophotometry, and the instrument for determination was spectrophotometer. To explore the selectivity of nitrate reduction, gas samples were gathered using a microsyringe. N_2 and N_2O in the gas samples were analyzed by GC-TCD with a Molecular Sieve 5A column and a Porapak Q column, respectively [43]. NO_x (NO_2 and NO) were analyzed using the chemiluminescence detection method [44]. The models of all the instruments used in the experiment are shown in Table 2. All the experiments were conducted at room temperature (25 ± 2 °C) with 120 rpm shaking.

Table 2. Models of experimental instruments.

Instruments	Models
Scanning electron microscope	Vega-3XMU, Tescan, Czech
Surface area BET analyzer	Micromeritics ASAP 2020, Norcross, USA
X-ray diffractometer	TTR-III, Rigaku, Japan
Fourier transform infrared spectroscopy	Equinox 55, Bruker Banner Lane, Germany
X-ray photoelectron spectroscopy	Escalab 250, Thermo Fisher Scientifific, USA
DO meter	HQ30d, Hach, USA
pH meter	MP220, Mettler Toledo, Switzerland
spectrophotometer	UV-1802, BeifenRuili, China
GC-TCD	GC-8A, Shimadzu, Japan

2.5. Analysis Method

2.5.1. Calculation Method

Many previous studies have shown that nitrite, ammonia, and nitrogen are the main products in the process of nitrate reduction, while NO_x and N_2O are basically not found [45,46], the same result was also found in the preliminary experiment of this study. Therefore, in this experiment, only nitrite nitrogen, ammonia nitrogen, and nitrogen were considered in the reduction products. The removal efficiency of NO_3^--N (η) and the product conversion ratio ($S_{product}$) are calculated as follows:

$$\eta = \frac{C_0 - C_{NO_3^-}}{C_0} \times 100\% \qquad (12)$$

$$S_{NH_4^+} = \frac{C_{NH_4^+}}{C_0 - C_{NO_3^-}} \times 100\% \qquad (13)$$

$$S_{NO_2^-} = \frac{C_{NO_2^-}}{C_0 - C_{NO_3^-}} \times 100\% \qquad (14)$$

$$S_{N_2} = \frac{C_0 - C_{NO_3^-} - C_{NO_2^-} - C_{NH_4^+}}{C_0 - C_{NO_3^-}} \times 100\% \qquad (15)$$

where, C_0 is the initial concentration of NO_3^--N, mg/L; $C_{NO_3^-}$, $C_{NH_4^+}$ and $C_{NO_2^-}$ are the concentration of NO_3^--N, NH_4^+-N and NO_2^- after reaction, mg/L.

2.5.2. Kinetic Studies

Many studies have shown that the removal process of nitrate from solution by ZVI composites conforms to first-order or second-order kinetics [47]. The removal process of NO_3^--N from the solution by the ZB12-500 composite can be divided into two stages: the fast removal stage and the slow removal stage. Therefore, the first-order kinetic model, second-order kinetic model, and two-compartment first-order kinetic model were used for analysis and fitting in this experiment, and the fitting equations were as follows:

$$\frac{C_t}{C_0} = e^{-k_1 t} \qquad (16)$$

$$\frac{1}{C_t} - \frac{1}{C_0} = k_2 t \qquad (17)$$

$$\frac{C_t}{C_0} = f_f * e^{-k_f * t} + f_s * e^{-k_s * t} \qquad (18)$$

where C_t (mg/L) is the residual concentration of nitrate, C_0 (mg/L) is the initial concentrations of nitrate; k_1 and k_2 (1/h) are the reaction rate constants of first and second order reaction kinetics, respectively; f_f and f_s are the proportion of fast and slow compartment removal in the total removal, respectively, and $f_f + f_s = 1$; k_f and k_s (1/h) are the fast and slow compartment reaction rate constants, respectively.

3. Results

3.1. Characterization of ZB12-500

The SEM analysis results of ZB12-500 before, during, and after the reaction are shown in Figure 2a–c. Figure 2a shows the ZB12-500 material before the reaction. The particle size of the composite material is about 30 μm, roughly spherical, smooth surface, and rich honeycomb channels. nZVI particles are evenly distributed on the surface and pores of biochar particles, such as a spider web, and nZVI particles in different pores are separated by a carbon skeleton. This is because of the rich porous structure and large specific surface area of biochar [48]. This image shows that the ZB12-500 composite was well prepared by the sodium borohydride method. Figure 2b shows ZB12-500 during the reaction. Compared with before the reaction, the nZVI particles on the surface of the ZB12-500 were obviously corroded, and the corrosion of nZVI particles was mostly surface corrosion. Some of the porous structures on the surface of the ZB12-500 were blocked because the nZVI lost electrons, and the iron ions diffused from the inner core of the nZVI to the outer core, forming iron oxides on the surface of the particles. In this process, there may be mechanisms such as iron dissolution, migration, and reagglomeration of nZVI particles, and migration of iron ions between different particles [49]. Figure 2c shows ZB12-500 after the reaction. The surface morphology of the material was highly crystallized, and the porous structure on the surface of the composite material completely disappeared and was covered by iron oxide.

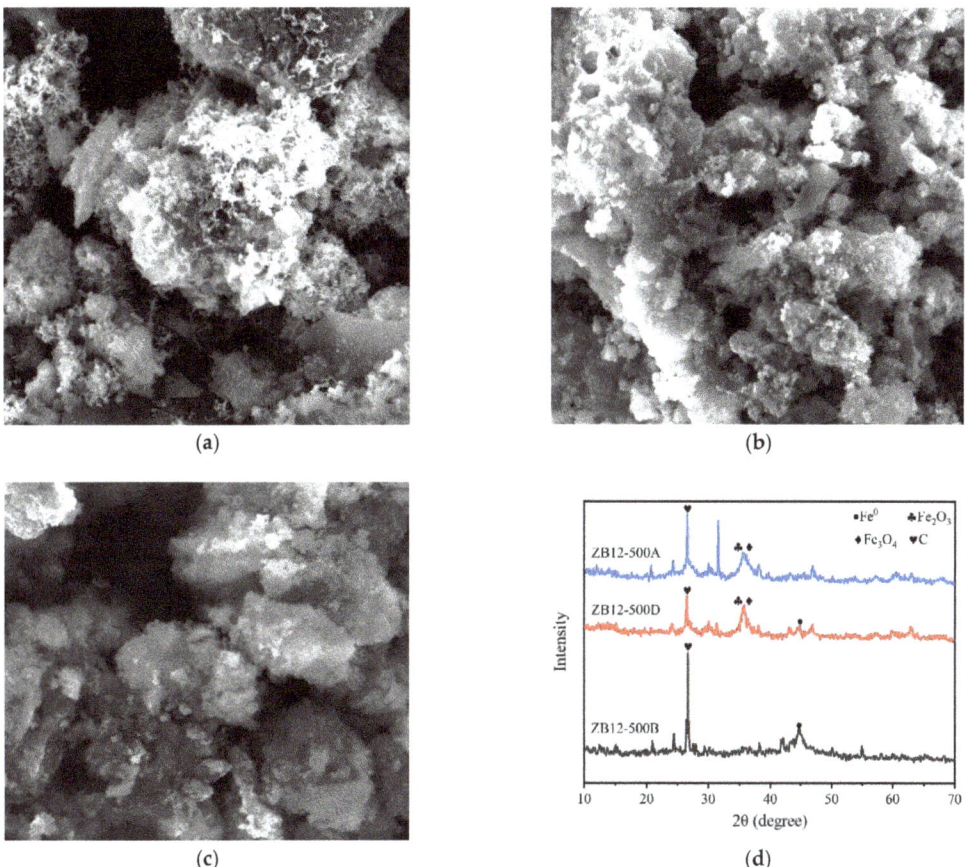

Figure 2. SEM images of ZB12-500 (**a**) before, (**b**) during, and (**c**) after the reaction. (**d**) XRD patterns of ZB12-500 before, during, and after the reaction.

The SSA analysis results of ZB12-500 before, during, and after the reaction are shown in Table 3. After the reaction, the SSA of the material increased. The possible reason is that the iron oxides generated by the nZVI reaction were redistributed on the biochar support, which may cover the original nZVI particles or other active sites, resulting in a larger gap between the iron oxides and an increase in SSA.

Table 3. SSA of ZB12-500 before, during, and after the reaction.

ZB12-500 Samples	SSA/(m^2/g)
before reaction	10.69
during the reaction	63.47
after reaction	83.32

The XRD analysis results of ZB12-500 before, during, and after the reaction are shown in Figure 2d. ZB12-500B, ZB12-500D, and ZB12-500A represent ZB12-500 samples before, during, and after the reaction, respectively. The characteristic peaks at 2θ = 44.8°, 65.32°, and 82.60° represent planes (110), (200), and (211) of the body-centered cubic crystal structure in nZVI particles, respectively [50]. Before the reaction, the ZB12-500B sample showed a diffraction peak at 2θ = 44.8°, which corresponds to the body-centered cubic α-Fe^0 (110) crystal plane, indicating that the prepared ZB12-500 material contains α-Fe^0, and nZVI particles are successfully loaded on the surface of biochar support [51,52]. The peak intensity of Fe^0 in the ZB12-500D sample decreases significantly, while the diffraction peak of Fe^0 in the ZB12-500A sample completely disappears, indicating that in the process of reducing NO_3^--N by ZB12-500, nZVI was all consumed or covered by reaction products. The characteristic peaks of Fe_2O_3 and Fe_3O_4 located at 2θ = 35.5° appear in sample ZB12-500D, which is widened to a certain extent in sample ZB12-500A, while these two peaks are not found in sample ZB12-500B, indicating that the main iron oxides generated after the reduction reaction are Fe_2O_3 and Fe_3O_4. Some scholars have found that Fe_3O_4 is formed at the interface between nZVI and iron oxide, while Fe_2O_3 is formed at the interface between iron oxide and water [53–55].

The results of FTIR analysis of ZB12-500 before, during, and after the reaction are shown in Figure 3. As can be seen from the figure, these three samples contain abundant functional groups, among which a characteristic peak with high strength appears near 3430 cm^{-1}, which corresponds to the tensile vibration of the –OH bond [56–59]. The characteristic peak intensity at 1620 cm^{-1} is slightly smaller and is related to the C=O and –OH bonds of carbonyl or carboxyl groups. The intensity of characteristic peaks at 1320 cm^{-1} and 1100 cm^{-1} is small, which represents the tensile vibration of the C–O bond. The characteristic peak intensity at 670 cm^{-1} is weak, which is the characteristic peak of the Fe–OH bond [60–62]. The number and types of functional groups will greatly affect the physical and chemical properties of materials, especially the –OH functional group, which will affect the absorption of other substances by biochar [63]. Compared with ZB12-500B, the intensity corresponding to the –OH characteristic peak of ZB12-500A and ZB12-500D at 3430 cm^{-1} increased, while the intensity of the characteristic peaks at 1620 cm^{-1}, 1320 cm^{-1} and 670 cm^{-1} decreased slightly, which may be due to the formation of C–O–Fe, Fe–O–Fe, and C–N–Fe.

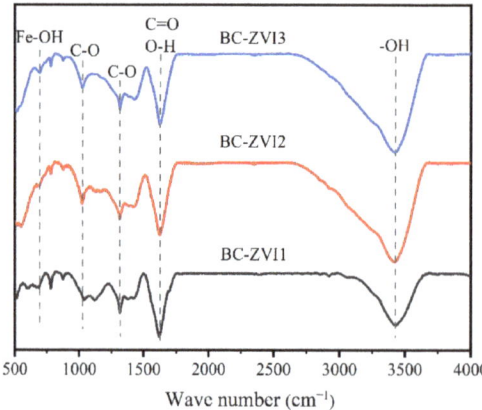

Figure 3. FTIR spectra of ZB12-500 before, during, and after the reaction.

Figure 4a–c shows the full spectrum scan of ZB12-500 material before, during, and after the reaction. The peaks of binding energies at 710 eV, 530 eV, and 284 eV correspond to the absorption peaks of Fe 2p, O 1s, and C 1s, respectively [60]. The main elements in the three samples are Fe, O, and C, but the content of the elements is different. Among them, the Fe element is mainly from nZVI and iron oxide, the O element is mainly from oxygen-containing functional groups in biochar, and C element is mainly from biochar. By comparing the scanning spectra, it can be found that the content of the Fe element increased continuously after the beginning of the reaction. This is because although nZVI was consumed during the reaction, the increased content of Fe oxide exceeded the consumed nZVI content. One hour after the reaction between ZB12-500 and NO_3^--N solution, the content of the Fe element increased greatly, and in the following 23 h, the content of the Fe element increased slightly, indicating that the reaction was nearly complete after one hour of reaction, which is consistent with the change of the iron element in XRD. The content of the O element in the ZB12-500 increased first and then decreased in the reaction process, which may be related to the consumption of nZVI and the formation of iron oxides in the reaction process. The change in the –OH group content in the FTIR spectra also showed the same trend. The content of element C decreased by 12.98% in 1 h after the reaction, and then decreased by 0.74% in the following 23 h, indicating that the functional group containing C participated in the reaction. This is consistent with the variation of the C element in the XRD and FTIR spectra.

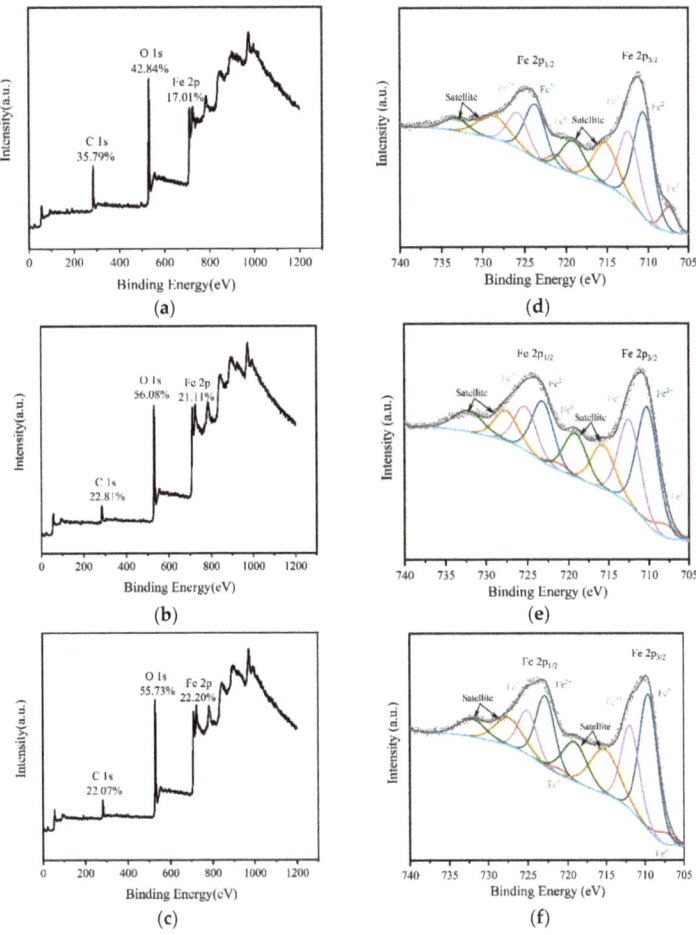

Figure 4. XPS survey spectra of ZB12-500 (a) before, (b) during, and (c) after the reaction. High-resolution XPS scan spectra over Fe 2p of ZB12-500 (d) before, (e) during, and (f) after the reaction.

Figure 4d–f shows the high-resolution XPS scan spectra over Fe 2p of ZB12-500 before, during, and after the reaction. The peaks at the binding energy of 707 eV, 710 eV, and 712 eV represent the absorption peaks of Fe^0, Fe^{2+} and Fe^{3+} in Fe 2p3/2, respectively. The peaks at the binding energy of 721 eV, 723 eV, and 725 eV represent the absorption peaks of Fe^0, Fe^{2+} and Fe^{3+} in Fe 2p1/2, respectively [64,65]. The relative contents of the three different valence iron elements changed obviously before, during, and after the reaction of ZB12-500, especially the peak of Fe^0 at 707 eV binding energy, and the intensity of this peak almost disappeared after the reaction. Table 4 shows the relative contents of Fe in different valence states in ZB12-500 before, during, and after the reaction. Compared with the ZB12-500 before the reaction, the content of Fe^0 in the composite during and after reaction decreased by 3.84% and 5.00%, respectively, indicating that nZVI was continuously consumed during the whole reaction. The increased of Fe^{3+} content mainly occurred in the first hour of the reaction process, and then decreased in the next 23 h. Meanwhile, the increased of Fe^{2+} content mainly occurred in the last 23 h of the reaction process, indicating that Fe^{3+} was converted to Fe^{2+} in the later reaction.

Table 4. The relative content of Fe valence states measured with XPS.

Fe Valence States	ZB12-500B	ZB12-500D	ZB12-500A
Fe^0	9.87%	6.03%	4.87%
Fe^{2+}	54.39%	54.36%	58.59%
Fe^{3+}	35.74%	39.61%	36.54%

3.2. Effect of the Pyrolysis Temperature of ZB12-500 on Biochar

Figure 5 shows ZB12-500 was the most effective reactant, followed, in decreasing order, by ZB12-350, ZB12-650, ZB12-800, and nZVI with removal efficiencies of 93.49%, 86.92%, 86.65%, 84.90%, and 40.14%, respectively. In the five different composites, the N_2 conversion ratios from large to small were ZB12-500 > ZB12-350 > ZB12-800 > ZB12-650 > nZVI, corresponding to 27.34%, 24.12%, 24.11%, 16.83%, and 5.71%, respectively.

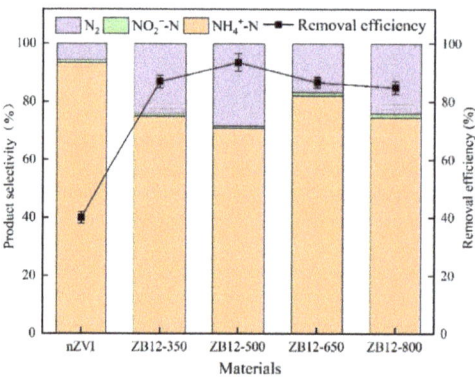

Figure 5. Nitrate removal by nZVI and ZB12 prepared with different pyrolysis temperatures (initial NO_3^--N concentration: 30 mg/L, pH 6, dosage: 5 g/L).

3.3. Effect of the Dosage of ZB12-500

Figure 6 shows how the dosage of ZB12-500 affected the removal efficiency of nitrate. When the dosages of ZB120-500 were 2, 3, 4, 5, 6, and 7 g/L, the removal efficiencies were 54.62%, 72.62%, 89.49%, 93.98%, 94.78%, and 96.04%, respectively. A positive correlation was observed between the dosage of ZB12-500 and the efficiency of nitrate removal. In terms of the selectivity of nitrogen products, when the dosage was less than 5 g/L, the proportion of various reduction products basically did not change, and the N_2 conversion ratio was about 27%, while when the dosage was more than 5 g/L, the N_2 conversion ratio decreased to about 24%, and the conversion ratio of NH_4^+-N and NO_2^- increased slightly with the increase in dosage.

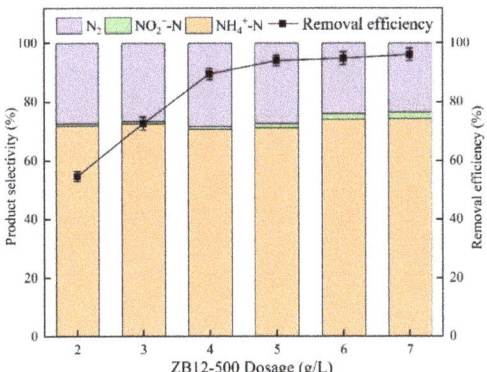

Figure 6. Effect of ZB12-500 dosage (initial NO_3^--N concentration: 30 mg/L, pH 6, ZB12 sample: ZB12-500).

3.4. Effect of pH

Figure 7 shows the effect of the initial pH on nitrate removal by ZB12-500. Obviously, the removal efficiency was negatively correlated with the initial pH of the solution. When the initial pH increased from 5 to 10, the removal efficiency decreased from 97.29% to 89.04%. The N_2 conversion ratio was the highest when pH = 5, followed by pH = 6, and the lowest when pH = 7, which were 27.13%, 26.38%, and 21.92%, respectively. In particular, ZB12-500 exhibited similar nitrate reduction results at initial pH values of 5 and 6.

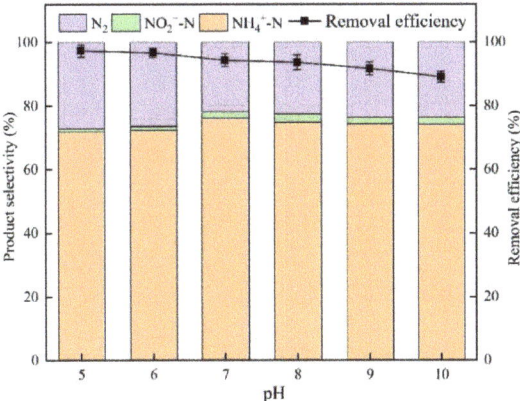

Figure 7. Effect of initial pH (initial NO_3^--N concentration: 30 mg/L, Dosage: 5 g/L, ZB12 sample: ZB12-500).

3.5. Effect of Initial Nitrate Concentration

Figure 8 shows the effect of the initial concentration of nitrate (NO_3^--N) on the removal of nitrate by ZB12-500. When the initial concentration of NO_3^--N were 30 mg/L, 50 mg/L, 70 mg/L and 100 mg/L, the removal efficiencies were 93.94%, 82.60%, 75.3%, and 51.42%, respectively, and the N_2 conversion ratios were 26.82%, 28.45%, 29.74%, and 31.66%, respectively.

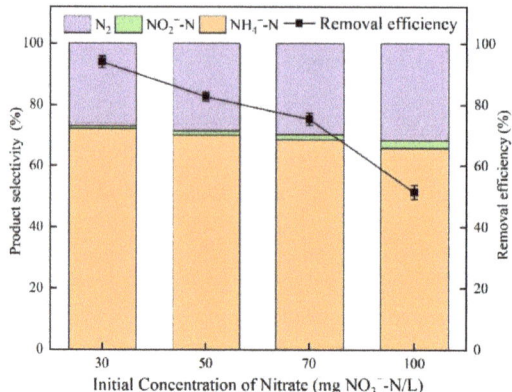

Figure 8. Effect of initial nitrate (NO_3^--N) concentration (pH 6, dosage: 5 g/L, ZB12 sample: ZB12-500).

3.6. Effect of Co-Existing Ions

Figure 9 shows the effect of the co-existing ions on nitrate removal by ZB12-500. Common ions (Na^+, Mg^{2+}, Ca^{2+}, Cl^-, SO_4^{2-}, and HCO_3^-) in groundwater were investigated. According to the experimental data in the previous sections, without adding co-existing ions, the removal efficiency of NO_3^--N exceeded 93%, and the conversion ratio of N_2 exceeded 25% under the same reaction conditions. However, after adding co-existing ions Na^+, Mg^{2+}, Ca^{2+}, Cl^-, SO_4^{2-}, and HCO_3^- into the solution, the removal efficiencies of NO_3^--N were 91.14%, 95.20%, 94.33%, 92.88%, 80.20%, and 57.00%, and the conversion ratios of N_2 were 26.30%, 26.01%, 25.02%, 25.82%, 29.49%, and 37.01%, respectively.

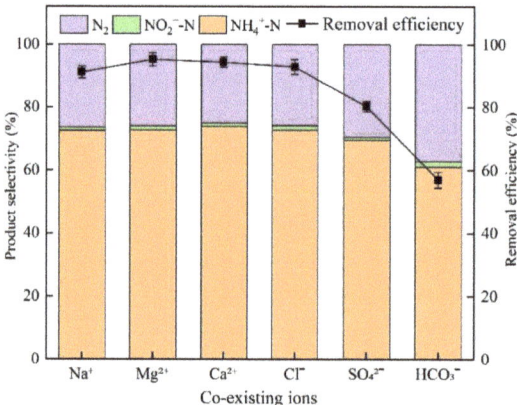

Figure 9. Effect of co-existing ions (initial NO_3^--N concentration: 30 mg/L, pH 6, Dosage: 5 g/L, ZB12 sample: ZB12-500).

3.7. Kinetics

The kinetics data were fitted using first-order kinetic, second-order kinetic, and two-compartment first-order kinetic equations, as shown in Figure 10. The whole removal process was obviously divided into two stages: 0–2 h for the rapid removal stage and 2–24 h for the slow removal stage. The kinetics parameters fitted by the kinetics model are shown in Table 5. The results show that the R^2 values of the three kinetics models were all above 0.99, indicating that both adsorption and reduction reactions existed in the removal process. The R^2 of the two-compartment first-order kinetic model was the highest, which was 0.997, indicating that it is more reasonable to use the two-compartment first-order kinetic model to explain the process of NO_3^--N removal by the ZB12-500 composite. From the two-compartment first-order kinetic parameters, the main stage in the whole removal process was the fast compartment reaction stage, accounting for 92.5%, while the slow compartment reaction stage only accounted for 7.5%. The fast compartment reaction rate constant was 3.093 h^{-1}, and the slow compartment reaction rate constant was 0.038 h^{-1}.

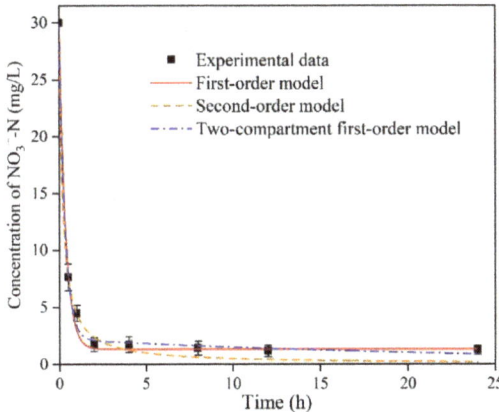

Figure 10. Kinetics of ZB12-500 for NO_3^--N. (initial NO_3^--N concentration: 30 mg/L, pH 6, Dosage: 5 g/L, ZB12 sample: ZB12-500).

Table 5. Kinetic parameters of NO_3^--N removal by ZB12-500.

Sample	Material	First-Order		Second-Order		Two-Compartment First-Order				
		k_1/(1/h)	R^2	k_2/(1/h)	R^2	f_f	k_f/(1/h)	f_s	k_s/(1/h)	R^2
ZB12-500	NO_3^--N	2.905	0.993	0.166	0.996	0.925	3.093	0.075	0.038	0.997

4. Discussion

4.1. Effect of the Pyrolysis Temperature of ZB12-500 on Biochar

For nZVI, the removal efficiency of nitrate is greatly affected by agglomeration. The low N_2 selectivity of pure nZVI is also affected by the natural defects of the materials. In the reduction process, after the surface active site of the aggregate is inactivated by the reaction, the internal active site is also inactivated by being covered. For ZB12, the removal efficiency of NO_3^--N firstly increased and then decreased with the increase of biochar pyrolysis temperature. However, the selectivity of the nitrogen products had no obvious rule with the pyrolysis temperature of the biochar carrier. Although increasing the temperature of pyrolysis increases the electronic conductivity and the degree of graphitization of biochar, it also leads to the loss of functional groups [37,66]. Some scholars have pointed out that with an increase in pyrolysis temperature, the number of functional groups of wood-based biochar (mainly –OH and aliphatic C–H functional groups) and grass-based biochar (mainly C–O functional groups) decreases [67]. The trend of nitrate removal efficiency and nitrogen product selectivity with pyrolysis temperature proves to some extent that when the pyrolysis temperature of biochar is 500 °C, the types and number of functional groups related to reduced nitrate in ZB12 reach an ideal equilibrium state with their electrical conductivity and graphitization degree. The NO_2^--N produced by the reaction of the five kinds of samples with nitrite nitrogen was very small, indicating that NO_2^--N was the intermediate product of the reaction. Briefly, ZB12-500 was the best reactant for nitrate removal and was selected for the following studies.

4.2. Effect of the Dosage of ZB12-500

When the dosage is low, the removal efficiency increased rapidly with increasing dosage because the increased dosage provided more Fe^0 active sites, which is the reason for the positive correlation. At the same time, there were sufficient reaction raw materials for ZB12-500 in the whole reaction process, and the active site of ZB12-500 could be fully utilized in the reduction process. Therefore, when the dosage was less than 5 g/L, ZB12-500 experienced the same reaction conditions and environment, resulting in the same nitrogen product selectivity. When the dosage was greater than 5 g/L, nitrate dilution significantly inhibited the reactivity of the residual nitrate, leading to incomplete reduction of more nitrate, i.e., an increased proportion of intermediates accumulated, resulting in a slight change in nitrate removal efficiency and a slight reduction in N_2 selectivity. Some previous studies have found similar results [68,69]. Accordingly, a dosage of 5 g/L is optimum for this study.

4.3. Effect of pH

From the changing trend of NO_3^--N removal efficiency with the initial pH, it can be seen that the reaction between the ZB12-500 material and NO_3^--N is an acidophilic reaction. Hao et al. also reached this conclusion in their research results [70]. The reasons are as follows: Under acidic conditions, H^+ in the solution is increased, the adsorption of NO_3^--N on the biochar surface is enhanced, and protons also participate in the reduction process of nitrate nitrogen [71]. Under alkaline conditions, the increase of OH^- leads to more metal hydroxide precipitation ($Fe(OH)_2$ and $Fe(OH)_3$) and metal carbonate ($FeCO_3$). This led to increased corrosion of the nZVI. These iron oxides limited the diffusion of nitrate ions and coated the zero-valent iron, reducing the active sites on the surface of ZB12-500 composite and reducing the removal efficiency of NO_3^--N [72]. Many studies have found that pH can affect the reduction of nitrate by ZVI particles [19,73,74]. However, in this section, the composite material can achieve good selectivity on N_2 (21.92–27.13%) within the initial pH range of 5–10. This may be related to the rich functional groups, huge specific surface area, and abundant active sites of biochar carriers with pyrolysis temperature of 500 °C [75]. The reason N_2 conversion ratio of ZB12-500 was the lowest under neutral conditions was that the content of active sites or functional groups on biochar surface decreases under neutral conditions.

4.4. Effect of Initial Nitrate Concentration

The initial nitrate concentration is negatively correlated with its removal efficiency. The study of Sparis et al. also showed that the higher the initial nitrate nitrogen concentration, the lower the reduction rate constant [15]. In fact, when nitrate concentration in solution is low, the active sites on the ZB12-500 composite are relatively abundant, which makes the utilization rate of ZVI very high. However, for higher concentrations of nitrate in solution, nitrate ions compete for limited active sites on ZB12-500 composites, which makes it easier for nitrate anions to squeeze onto the ZB12-500 surface and then rapidly oxidize nZVI particles on the ZB12-500 surface, eventually seriously blocking the porous structure of ZB12-500. As a result, the material loses more active sites more quickly, which inevitably hinders nitrate reduction. In the experiment, N_2 conversion ratio was positively correlated with the initial nitrate concentration, which might be the result of the influence of solution ion concentration on the surface charge density of the material. Mikami et al.'s study also showed that increased nitrate concentration was conducive to the transformation of reduced products into N_2 rather than NH_4^+-N [76]. As the density of N-species on the surface of the composite became high with the increased in NO_3^--N concentration, the selectivity to N_2 increased.

4.5. Effect of Co-Existing Ions

In this study, Na^+, Mg^{2+}, Ca^{2+} and Cl^- had insignificant effects on the removal efficiency of NO_3^--N, while SO_4^{2-} and HCO_3^- significantly reduced the removal efficiency of NO_3^--N. General cations and low concentrations of Cl^- have a negligible effect on the removal of anionic NO_3^-. However, the outer-spherically sorbing anions, especially SO_4^{2-}, have a significant interference effect on nitrate adsorption, and adsorption competition will occur on the limited adsorption sites on the surface of ZB12-500 [77,78]. HCO_3^- can be ionized to form CO_3^{2-} and H^+ or hydrolyzed to form OH^- and H_2CO_3. However, the degree of hydrolysis of HCO_3^- is greater than the degree of ionization, so more OH^- is generated by hydrolysis, which can form $Fe(OH)_2$, $Fe(OH)_3$ and $FeCO_3$ with iron ions in the system. Thus, the corrosion of nZVI was aggravated, and the formation of iron oxides will cover the outer surface of nZVI particles, preventing it from continuing to react with NO_3^--N, and ultimately reducing the reduction efficiency of NO_3^--N. The reason HCO_3^- obviously improved the product selectivity of N_2 may be related to the atomic structure of HCO_3^-. The atomic arrangement of HCO_3^- is planar, and the carbon in the center is bonded with three oxygen atoms (one C=O, one C–OH, and one C–O–). All of these functional groups are involved in the reduction process of NO_3^--N. If the ZB12-500 composite is applied to actual groundwater, the influence of HCO_3^- and SO_4^{2-} should be considered.

4.6. Kinetics

The kinetics of nitrate removal complied with the two-compartment first-order kinetic equations. In the rapid removal stage, that is, within 2 h of reaction, the concentration of NO_3^--N decreased almost in a straight line, indicating that the reaction was strong and rapid at this time. In the slow removal stage, that is, within 2 h to 24 h after the reaction, the concentration of nitrate nitrogen decreased extraordinarily little, indicating that the reaction in the system was extremely slow and that the reaction was almost complete. At the beginning of the removal reaction, ZB12-500 reacted rapidly with NO_3^--N in solution when it just entered the system. This is because the initial concentration of nitrate is relatively high and the mass transfer driving force is large, so NO_3^--N is easily sent to the active site on the surface of the composite materials. This is manifested by the film diffusion of nitrate ions from the bulk liquid phase to the external surface of ZB12. As the reaction produced a large amount of iron oxide, the active site of nZVI was blocked so that nZVI could not be exposed to the system to react with the remaining NO_3^--N. Moreover, as the concentration of nitrate in the solution decreased, the mass transfer driving force also decreased, resulting in a slow reaction in the later stage.

4.7. Nitrate Reduction Mechanism

During the nitrate removal process, ZVI nanoparticles in ZB12 also corroded. In the XRD pattern, the disappearance of Fe^0 peak and the appearance of Fe_3O_4 and Fe_2O_3 peaks in the composite also indicate the corrosion of nZVI and the appearance of iron oxides. The infrared spectrum shows that the peak strength and peak width of the –OH functional group increased significantly, indicating that –OH was involved in the reaction between ZB12 and NO_3^--N. XPS survey spectra shows that Fe^{2+} was consumed and generated during the reaction.

In order to deeply explore the reaction process between ZB12 and NO_3^--N, this section analyzes the nitrogen species (NO_3^--N, NO_2^--N, NH_4^+-N and TN), pH, ORP, and DO during the reaction process. The experimental results are shown in Figures 11 and 12. According to the change in the concentration of each substance in the reaction process, the reaction can be basically divided into three stages.

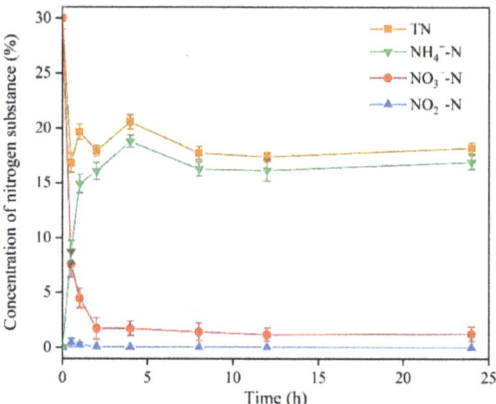

Figure 11. Evolution of the concentration of nitrogen species during the reaction (initial NO_3^--N concentration: 30 mg/L, pH 6, Dosage: 5 g/L, ZB12 sample: ZB12-500).

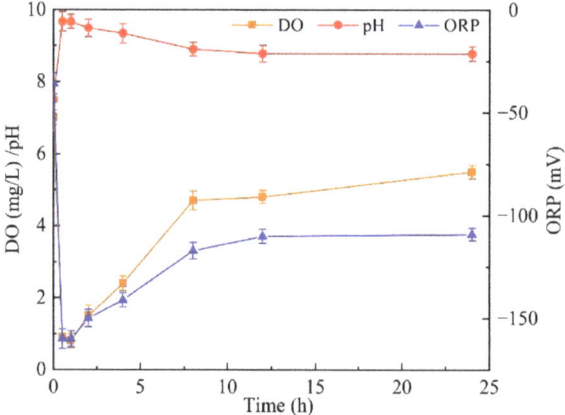

Figure 12. Evolution of DO, pH, and ORP during reaction (initial NO_3^--N concentration: 30 mg/L, pH 6, Dosage: 5 g/L, ZB12 sample: ZB12-500).

The first stage occurred in 0–0.5 h. Here, the concentration of NO_3^--N decreased to 8.85 mg/L in a straight line because the active sites of ZVI on ZB12 were abundant at this time, which could fully react with nitrate in the solution. The concentrations of NO_2^--N, NH_4^+-N and TN at the time of 0.5 h were 0.5 mg/L, 10.18 mg/L and 19.53 mg/L, respectively, indicating that nitrite and ammonia were generated during this stage (Equations (3) and (4)). The decrease in TN concentration indicated that N_2 was generated during the reaction (Equation (6)). Additionally, the pH increased from 7.5 to 9.67, indicating that H^+ in the system participated in the reaction (Equation (1)); DO decreased from 7 mg/L to 0.9 mg/L, indicating that dissolved oxygen in the system competed with NO_3^--N and participated in the reaction (Equation (2)); and ORP decreased, indicating that the system was transformed into a reducing environment and Fe^{2+} was generated.

The second stage occurred at 0.5–8 h. Here, the concentration of NO_3^--N decreased from 8.85 mg/L to 1.65 mg/L, which slowed the reaction speed. The concentration of NO_2^--N decreased from 0.50 mg/L to 0.05 mg/L, indicating that nitrite was an intermediate product of the reaction (Equation (5)). The concentrations of NH_4^+-N increased from 10.18 mg/L to 18.88 mg/L, becoming the main component of TN. The pH dropped from

9.67 to 8.9, probably buffered by oxygen-containing functional groups in the biochar. The rise in ORP indicated the consumption of Fe^{2+} in the system (Equation (10)). The rise in DO was due to the release of adsorbed oxygen or the buffering effect of functional groups from the biochar. Combined with the characterization analysis, the slow reaction rate was affected by the formation of iron oxide in the reaction process. In addition, functional groups in biochar were also involved in the reaction (Equations (7)–(9)).

In the last stage (8–24 h). The concentration of NO_3^--N, TN and NO_2^--N were still decreasing, and the concentration of NH_4^+-N was slightly increasing, indicating that the reduction reactions still occur. pH, ORP, and DO leveled off, indicating that the reaction proceeded at a very low rate.

The reaction mechanism of the ZB12 composite with NO_3^--N is shown in Figure 13. The nZVI particles loaded on the surface and pores of biochar can react with NO_3^--N or DO in solution. During this period, electrons escape from the core and shell of nZVI to form Fe^{2+} and generate NO_2^-, NH_4^+ and N_2, with NH_4^+ being the main product and NO_2^- being partially converted to NH_4^+ as an intermediate. nZVI is converted into iron oxide to cover the surface of biochar or exist in solution, and the Fe^{2+} generated by the reaction contributes to the reduction of nitrate [79]. Carboxyl groups on the surface of biochar will exist in the form of esters [80], while ester groups mainly undergo hydrolysis reaction, so ester groups in biochar will react with H_2O, nZVI, and Fe^{2+} to generate quinone groups (Equation (10)), which is consistent with the results of the FTIR spectrum. The generated quinone group can also provide H^+ for the reaction, which converts NO_3^--N to N_2 and itself into an ester group. In addition, ester groups on the surface of biochar can dissociate in a wide pH range, thus buffering the pH of the reaction and slowing down the reaction inhibition at a higher pH.

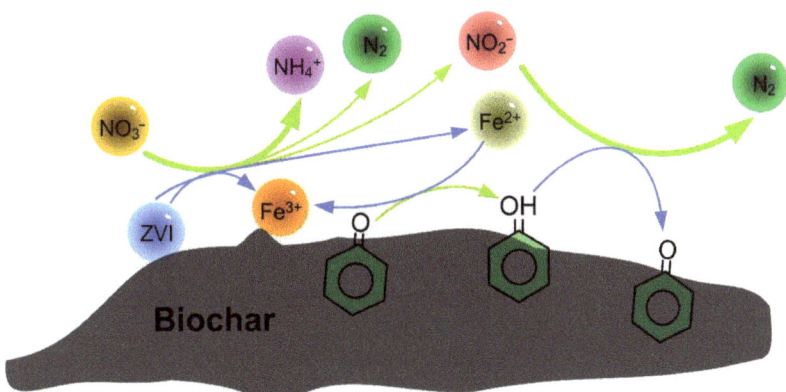

Figure 13. Reaction mechanism of ZB12 composite with NO_3^--N.

In addition to the role of biochar in regulating the pH of the system and increasing the reactivity, the conductivity of the carbon matrix and the electron-mediated ability of functional groups may also contribute to the reduction of nitrate nitrogen. The electrical conductivity of biochar contributes to the transfer of electrons from nZVI to nitrate, and the higher the pyrolysis temperature of biochar, the stronger the electrical conductivity of the carbon matrix [81]. The electron-mediating capacity of functional groups can be divided into electron donation capacity and electron acceptor capacity, among which the electron donation capacity is attributed to phenolic functional groups, and the electron acceptor capacity is attributed to quinone groups and concentrated aromatic hydrocarbons [82]. These REDOX functional groups can mediate the electron transfer process among nZVI, biochar, and nitrate.

In modified ZVI materials, carbon with high potential is used as the cathode and iron as the anode. When the two contact, many microscopic galvanic cells will be generated. The potential difference causes electrons to transfer from the ZVI core shell to the carbon. Some studies have shown that nitrate nitrogen reduction occurs primarily on the carbon surface with a higher electric potential, which reduces the hindering effect of iron oxides on nitrate reduction [15,83]. Therefore, for the ZB12 composite material in this paper, a huge number of miniature galvanic couples can be formed between biochar and nZVI. nZVI acts as the anode, losing electrons and being oxidized to ferrous ions, and biochar acts as the cathode for nitrate reduction.

5. Conclusions

On the basis of the 1:2 mass ratio of nZVI to biochar studied in our previous work [28], ZB12 was successfully prepared. The effects of different biochar pyrolysis temperatures on nitrate removal by the ZB12 composite were explored. It was found that the best biochar pyrolysis temperature was 500 °C. In addition, ZB12 has a higher N_2 conversion ratio (21.9~27.13%) in a wide pH range (5–10) under the premise of high nitrate removal efficiency (89.04–97.59%), which is more environmentally friendly in practical application. Increasing the initial concentration of nitrate would lead to a decrease in the removal efficiency, but a higher density of N-species on the surface of the composite might lead to an increase in the conversion ratio of N_2. The co-existence of HCO_3^- or SO_4^{2-} in the solution can reduce the removal efficiency of NO_3^--N to 57.00% or 80.20% and increase the conversion ratio of N_2 to 37.01% or 29.49%, respectively. The removal of nitrate by ZB12 was in accordance with the two-compartment first-order kinetics. Biochar plays a mediating role in the reduction of nitrate by nZVI. The pyrolysis temperature of biochar will affect its electrical conductivity and the electron-mediated ability of its surface functional groups, thus affecting the removal of nitrate by the ZB12 composite and the formation of products. Therefore, the results of this study provide further references for the eco-friendly removal of nitrate from groundwater.

Author Contributions: Conceptualization, A.W.; methodology, S.L. (Siyuan Liu) and A.W.; software, S.L. (Siyuan Liu); validation, S.L. (Shaopeng Li) and W.X.; formal analysis, S.L. (Siyuan Liu); investigation, W.X.; resources, X.H. and S.L. (Shaopeng Li); data curation, S.L. (Siyuan Liu); writing—original draft preparation, S.L. (Siyuan Liu) and X.H.; writing—review and editing, A.W.; visualization, X.H.; supervision, A.W. and X.H.; project administration, A.W.; funding acquisition, A.W. All authors have read and agreed to the published version of the manuscript.

Funding: This research work was funded by the National Natural Science Foundation of China (NSFC) (Grant No. 51208424).

Institutional Review Board Statement: Not applicable.

Informed Consent Statement: Not applicable.

Data Availability Statement: Not applicable.

Conflicts of Interest: All authors certify that they have no affiliations with or involvement in any organization or entity with any financial interest or non-financial interest in the subject matter or materials discussed in this manuscript. The authors declare no conflict of interest.

References

1. Zhai, Y.Z.; Lei, Y.; Wu, J.; Teng, Y.; Wang, J.; Zhao, X.; Pan, X. Does the groundwater nitrate pollution in China pose a risk to human health? A critical review of published data. *Environ. Sci. Pollut. Res.* **2017**, *24*, 3640–3653. [CrossRef] [PubMed]
2. Allaire, M.; Wu, H.W.; Lall, U. National trends in drinking water quality violations. *Proc. Natl. Acad. Sci. USA* **2018**, *115*, 2078–2083. [CrossRef] [PubMed]
3. Han, D.M.; Currell, M.J.; Cao, G.L. Deep challenges for China's war on water pollution. *Environ. Pollut.* **2016**, *218*, 1222–1233. [CrossRef] [PubMed]
4. Zghibi, A.; Tarhouni, J.; Zouhri, L. Assessment of seawater intrusion and nitrate contamination on the groundwater quality in the Korba coastal plain of Cap-Bon (North-east of Tunisia). *J. Afr. Earth Sci.* **2013**, *87*, 1–12. [CrossRef]

5. Kuang, P.J.; Natsui, K.; Einaga, Y. Comparison of performance between boron-doped diamond and copper electrodes for selective nitrogen gas formation by the electrochemical reduction of nitrate. *Chemosphere* **2018**, *210*, 524–530. [CrossRef] [PubMed]
6. Duan, W.J.; Li, G.; Lei, Z.C.; Zhu, T.H.; Xue, Y.Z.; Wei, C.H.; Feng, C.H. Highly active and durable carbon electrocatalyst for nitrate reduction reaction. *Water Res.* **2019**, *161*, 126–135. [CrossRef]
7. Fajardo, A.S.; Westerhoff, P.; Sanchez-Sanchez, C.M.; Garcia-Segura, S. Earth-abundant elements a sustainable solution for electrocatalytic reduction of nitrate. *Appl. Catal. B-Environ.* **2021**, *281*, 119465. [CrossRef]
8. Martinez, J.; Ortiz, A.; Ortiz, I. State-of-the-art and perspectives of the catalytic and electrocatalytic reduction of aqueous nitrates. *Appl. Catal. B-Environ.* **2017**, *207*, 42–59. [CrossRef]
9. Chen, J.; Wu, H.; Qian, H.; Gao, Y.Y. Assessing Nitrate and Fluoride Contaminants in Drinking Water and Their Health Risk of Rural Residents Living in a Semiarid Region of Northwest China. *Expo. Health* **2017**, *9*, 183–195. [CrossRef]
10. Vilardi, G.; Di Palma, L. Kinetic Study of Nitrate Removal from Aqueous Solutions Using Copper-Coated Iron Nanoparticles. *Bull. Environ. Contam. Toxicol.* **2017**, *98*, 359–365. [CrossRef]
11. Peng, L.; Liu, Y.W.; Gao, S.H.; Chen, X.M.; Xin, P.; Dai, X.H.; Ni, B.J. Evaluation on the Nanoscale Zero Valent Iron Based Microbial Denitrification for Nitrate Removal from Groundwater. *Sci. Rep.* **2015**, *5*, 12331. [CrossRef] [PubMed]
12. Cecconet, D.; Devecseri, M.; Callegari, A.; Capodaglio, A.G. Effects of process operating conditions on the autotrophic denitrification of nitrate-contaminated groundwater using bioelectrochemical systems. *Sci. Total Environ.* **2018**, *613*, 663–671. [CrossRef] [PubMed]
13. Della Rocca, C.; Belgiorno, V.; Meric, S. Overview of in-situ applicable nitrate removal processes. *Desalination* **2007**, *204*, 46–62. [CrossRef]
14. Guan, X.H.; Sun, Y.K.; Qin, H.J.; Li, J.X.; Lo, I.M.C.; He, D.; Dong, H.R. The limitations of applying zero-valent iron technology in contaminants sequestration and the corresponding countermeasures: The development in zero-valent iron technology in the last two decades (1994–2014). *Water Res.* **2015**, *75*, 224–248. [CrossRef] [PubMed]
15. Sparis, D.; Mystrioti, C.; Xenidis, A.; Papassiopi, N. Reduction of nitrate by copper-coated ZVI nanoparticles. *Desalination Water Treat.* **2013**, *51*, 2926–2933. [CrossRef]
16. Lubphoo, Y.; Chyan, J.M.; Grisdanurak, N.; Liao, C.H. Nitrogen gas selectivity enhancement on nitrate denitrification using nanoscale zero-valent iron supported palladium/copper catalysts. *J. Taiwan Inst. Chem. Eng.* **2015**, *57*, 143–153. [CrossRef]
17. Jiang, Z.M.; Zhou, H.G.; Chen, C.; Wei, S.Q.; Zhang, W.M. The Enhancement of Nitrate Reduction by Supported Pd-Fe Nanoscale Particle. *Sci. Adv. Mater.* **2015**, *7*, 1734–1740. [CrossRef]
18. Ren, Y.; Yang, J.H.; Li, J.; Lai, B. Strengthening the reactivity of Fe-0/(Fe/Cu) by premagnetization: Implications for nitrate reduction rate and selectivity. *Chem. Eng. J.* **2017**, *330*, 813–822. [CrossRef]
19. Liu, Y.; Wang, J.L. Reduction of nitrate by zero valent iron (ZVI)-based materials: A review. *Sci. Total Environ.* **2019**, *671*, 388–403. [CrossRef]
20. Zeng, Y.B.; Walker, H.; Zhu, Q.Z. Reduction of nitrate by NaY zeolite supported Fe, Cu/Fe and Mn/Fe nanoparticles. *J. Hazard. Mater.* **2017**, *324*, 605–616. [CrossRef]
21. Lee, C.S.; Gong, J.; Huong, C.V.; Oh, D.S.; Chang, Y.S. Macroporous alginate substrate-bound growth of Fe-0 nanoparticles with high redox activities for nitrate removal from aqueous solutions. *Chem. Eng. J.* **2016**, *298*, 206–213. [CrossRef]
22. Jiang, Z.M.; Zhang, S.J.; Pan, B.C.; Wang, W.F.; Wang, X.S.; Lv, L.; Zhang, W.M.; Zhang, Q.X. A fabrication strategy for nanosized zero valent iron (nZVI)-polymeric anion exchanger composites with tunable structure for nitrate reduction. *J. Hazard. Mater.* **2012**, *233*, 1–6. [CrossRef] [PubMed]
23. Zhu, H.; Jia, Y.; Wu, X.; Wang, H. Removal of arsenic from water by supported nano zero-valent iron on activated carbon. *J. Hazard. Mater.* **2009**, *172*, 1591–1596. [CrossRef]
24. Li, P.J.; Lin, K.R.; Fang, Z.Q.; Wang, K.M. Enhanced nitrate removal by novel bimetallic Fe/Ni nanoparticles supported on biochar. *J. Clean. Prod.* **2017**, *151*, 21–33. [CrossRef]
25. Yao, Y.; Gao, B.; Chen, J.J.; Zhang, M.; Inyang, M.; Li, Y.C.; Alva, A.; Yang, L.Y. Engineered carbon (biochar) prepared by direct pyrolysis of Mg-accumulated tomato tissues: Characterization and phosphate removal potential. *Bioresour. Technol.* **2013**, *138*, 8–13. [CrossRef]
26. Namasivayam, C.; Sangeetha, D. Removal and recovery of nitrate from water by $ZnCl_2$ activated carbon from coconut coir pith, an agricultural solid waste. *Indian J. Chem. Technol.* **2005**, *12*, 513–521.
27. Shirvanimoghaddam, K.; Czech, B.; Tyszczuk-Rotko, K.; Kończak, M.; Fakhrhoseini, S.M.; Yadav, R.; Naebe, M. Sustainable synthesis of rose flower-like magnetic biochar from tea waste for environmental applications. *J. Adv. Res.* **2021**, *34*, 13–27. [CrossRef]
28. Wei, A.L.; Ma, J.; Chen, J.J.; Zhang, Y.; Song, J.X.; Yu, X.Y. Enhanced nitrate removal and high selectivity towards dinitrogen for groundwater remediation using biochar-supported nano zero-valent iron. *Chem. Eng. J.* **2018**, *353*, 595–605. [CrossRef]
29. Oh, S.Y.; Seo, Y.D.; Kim, B.; Kim, I.Y.; Cha, D.K. Microbial reduction of nitrate in the presence of zero-valent iron and biochar. *Bioresour. Technol.* **2016**, *200*, 891–896. [CrossRef]
30. Gao, J.; Yang, L.Z.; Liu, Y.Y.; Shao, F.L.; Liao, Q.J.H.; Shang, J.G. Scavenging of Cr(VI) from aqueous solutions by sulfide-modified nanoscale zero-valent iron supported by biochar. *J. Taiwan Inst. Chem. Eng.* **2018**, *91*, 449–456. [CrossRef]
31. Yang, G.C.C.; Lee, H.L. Chemical reduction of nitrate by nanosized iron: Kinetics and pathways. *Water Res.* **2005**, *39*, 884–894. [CrossRef] [PubMed]

32. Huang, Y.H.; Zhang, T.C. Effects of low pH on nitrate reduction by iron powder. *Water Res.* **2004**, *38*, 2631–2642. [CrossRef] [PubMed]
33. Alowitz, M.J.; Scherer, M.M. Kinetics of nitrate, nitrite, and Cr(VI) reduction by iron metal. *Environ. Sci. Technol.* **2002**, *36*, 299–306. [CrossRef] [PubMed]
34. Huang, Y.H.; Zhang, T.C. Kinetics of nitrate reduction by iron at near neutral pH. *J. Environ. Eng.* **2002**, *128*, 604–611. [CrossRef]
35. Pignatello, J.J.; Mitch, W.A.; Xu, W.Q. Activity and Reactivity of Pyrogenic Carbonaceous Matter toward Organic Compounds. *Environ. Sci. Technol.* **2017**, *51*, 8893–8908. [CrossRef]
36. Nayyar, D.; Shaikh, M.A.N.; Nawaz, T. Remediation of Emerging Contaminants by Naturally Derived Adsorbents. In *New Trends in Emerging Environmental Contaminants*; Singh, S.P., Agarwal, A.K., Gupta, T., Maliyekkal, S.M., Eds.; Springer: Singapore, 2022; pp. 225–260.
37. Jiang, S.F.; Ling, L.L.; Chen, W.J.; Liu, W.J.; Li, D.C.; Jiang, H. High efficient removal of bisphenol A in a peroxymonosulfate/iron functionalized biochar system: Mechanistic elucidation and quantification of the contributors. *Chem. Eng. J.* **2019**, *359*, 572–583. [CrossRef]
38. Oh, S.Y.; Seo, Y.D.; Ryu, K.S. Reductive removal of 2,4-dinitrotoluene and 2,4-dichlorophenol with zero-valent iron-included biochar. *Bioresour. Technol.* **2016**, *216*, 1014–1021. [CrossRef]
39. Fakhrhoseini, S.M.; Czech, B.; Shirvanimoghaddam, K.; Naebe, M. Ultrafast microwave assisted development of magnetic carbon microtube from cotton waste for wastewater treatment. *Colloids Surf. A Physicochem. Eng. Asp.* **2020**, *606*, 125449. [CrossRef]
40. Yan, J.C.; Han, L.; Gao, W.G.; Xue, S.; Chen, M.F. Biochar supported nanoscale zerovalent iron composite used as persulfate activator for removing trichloroethylene. *Bioresour. Technol.* **2015**, *175*, 269–274. [CrossRef]
41. Wang, L.; Liu, S.Y.; Xuan, W.D.; Li, S.P.; Wei, A.L. Efficient Nitrate Adsorption from Groundwater by Biochar-Supported Al-Substituted Goethite. *Sustainability* **2022**, *14*, 7824. [CrossRef]
42. Qiu, J. China to Spend Billions Cleaning Up Groundwater. *Science* **2011**, *334*, 745. [CrossRef] [PubMed]
43. Jung, J.; Bae, S.; Lee, W. Nitrate reduction by maghemite supported Cu-Pd bimetallic catalyst. *Appl. Catal. B-Env.* **2012**, *127*, 148–158. [CrossRef]
44. Sakamoto, Y.; Kamiya, Y.; Okuhara, T. Selective hydrogenation of nitrate to nitrite in water over Cu-Pd bimetallic clusters supported on active carbon. *J. Mol. Catal. A-Chem.* **2006**, *250*, 80–86. [CrossRef]
45. Jung, S.; Bae, S.; Lee, W. Development of Pd-Cu/hematite catalyst for selective nitrate reduction. *Environ. Sci. Technol.* **2014**, *48*, 9651–9658. [CrossRef] [PubMed]
46. Shuai, D.; Choe, J.K.; Shapley, J.R.; Werth, C.J. Enhanced activity and selectivity of carbon nanofiber supported Pd catalysts for nitrite reduction. *Environ. Sci. Technol.* **2012**, *46*, 2847–2855. [CrossRef]
47. Choe, S.; Chang, Y.Y.; Hwang, K.Y.; Khim, J. Kinetics of reductive denitrification by nanoscale zero-valent iron. *Chemosphere* **2000**, *41*, 1307–1311. [CrossRef]
48. Greenlee, L.F.; Torrey, J.D.; Amaro, R.L.; Shaw, J.M. Kinetics of Zero Valent Iron Nanoparticle Oxidation in Oxygenated Water. *Environ. Sci. Technol.* **2012**, *46*, 12913–12920. [CrossRef]
49. Liu, C.-M.; Diao, Z.-H.; Huo, W.-Y.; Kong, L.-J.; Du, J.-J. Simultaneous removal of Cu2+ and bisphenol A by a novel biochar-supported zero valent iron from aqueous solution: Synthesis, reactivity and mechanism. *Environ. Pollut.* **2018**, *239*, 698–705. [CrossRef]
50. Xu, Y.; Schoonen, M.A.A. The absolute energy positions of conduction and valence bands of selected semiconducting minerals. *Am. Mineral.* **2000**, *85*, 543–556. [CrossRef]
51. Zhang, D.; Wang, B.; Gong, X.; Yang, Z.; Liu, Y. Selective reduction of nitrate to nitrogen gas by novel Cu2O-Cu0@Fe0 composite combined with HCOOH under UV radiation. *Chem. Eng. J.* **2019**, *359*, 1195–1204. [CrossRef]
52. Wang, T.; Jin, X.Y.; Chen, Z.L.; Megharaj, M.; Naidu, R. Green synthesis of Fe nanoparticles using eucalyptus leaf extracts for treatment of eutrophic wastewater. *Sci. Total. Environ.* **2014**, *466–467*, 210–213. [CrossRef] [PubMed]
53. Ritter, K.; Odziemkowski, M.S.; Gillham, R.W. An in situ study of the role of surface films on granular iron in the permeable iron wall technology. *J. Contam. Hydrol.* **2002**, *55*, 87–111. [CrossRef]
54. Kohn, T.; Livi, K.J.T.; Roberts, A.L.; Vikesland, P.J. Longevity of granular iron in groundwater treatment processes: Corrosion product development. *Environ. Sci. Technol.* **2005**, *39*, 2867–2879. [CrossRef]
55. Odziemkowski, M.S.; Simpraga, R.P. Distribution of oxides on iron materials used for emediation of organic groundwater contaminants-Implications for hydrogen evolution reactions. *Can. J. Chem.* **2004**, *82*, 1495–1506. [CrossRef]
56. Sajjadi, B.; Broome, J.W.; Chen, W.Y.; Mattern, D.L.; Egiebor, N.O.; Hammer, N.; Smith, C.L. Urea functionalization of ultrasound-treated biochar: A feasible strategy for enhancing heavy metal adsorption capacity. *Ultrason. Sonochem.* **2019**, *51*, 20–30. [CrossRef]
57. Oh, W.D.; Lisak, G.; Webster, R.D.; Liang, Y.N.; Veksha, A.; Giannis, A.; Moo, J.G.S.; Lim, J.W.; Lim, T.T. Insights into the thermolytic transformation of lignocellulosic biomass waste to redox-active carbocatalyst: Durability of surface active sites. *Appl. Catal. B-Environ.* **2018**, *233*, 120–129. [CrossRef]
58. Zhao, G.X.; Li, J.X.; Ren, X.M.; Chen, C.L.; Wang, X.K. Few-Layered Graphene Oxide Nanosheets As Superior Sorbents for Heavy Metal Ion Pollution Management. *Environ. Sci. Technol.* **2011**, *45*, 10454–10462. [CrossRef]
59. Li, J.; Chen, C.L.; Zhang, R.; Wang, X.K. Nanoscale Zero-Valent Iron Particles Supported on Reduced Graphene Oxides by Using a Plasma Technique and Their Application for Removal of Heavy-Metal Ions. *Chem. Asian J.* **2015**, *10*, 1410–1417. [CrossRef]

60. Qian, L.B.; Shang, X.; Zhang, B.; Zhang, W.Y.; Su, A.Q.; Chen, Y.; Ouyang, D.; Han, L.; Yan, J.C.; Chen, M.F. Enhanced removal of Cr(VI) by silicon rich biochar-supported nanoscale zero-valent iron. *Chemosphere* **2019**, *215*, 739–745. [CrossRef]
61. Zuo, X.J.; Liu, Z.G.; Chen, M.D. Effect of H2O2 concentrations on copper removal using the modified hydrothermal biochar. *Bioresour. Technol.* **2016**, *207*, 262–267. [CrossRef]
62. Xing, M.; Wang, J.L. Nanoscaled zero valent iron/graphene composite as an efficient adsorbent for Co(II) removal from aqueous solution. *J. Colloid Interface Sci.* **2016**, *474*, 119–128. [CrossRef] [PubMed]
63. Li, H.B.; Dong, X.L.; da Silva, E.B.; de Oliveira, L.M.; Chen, Y.S.; Ma, L.N.Q. Mechanisms of metal sorption by biochars: Biochar characteristics and modifications. *Chemosphere* **2017**, *178*, 466–478. [CrossRef] [PubMed]
64. Lua, A.C.; Yang, T.; Guo, J. Effects of pyrolysis conditions on the properties of activated carbons prepared from pistachio-nut shells. *J. Anal. Appl. Pyrolysis* **2004**, *72*, 279–287. [CrossRef]
65. Yamashita, T.; Hayes, P. Analysis of XPS spectra of Fe^{2+} and Fe^{3+} ions in oxide materials. *Appl. Surf. Sci.* **2008**, *254*, 2441–2449. [CrossRef]
66. Oh, S.Y.; Seo, Y.D. Factors Affecting Sorption of Nitro Explosives to Biochar: Pyrolysis Temperature, Surface Treatment, Competition, and Dissolved Metals. *J. Environ. Qual.* **2015**, *44*, 833–840. [CrossRef]
67. Keiluweit, M.; Nico, P.S.; Johnson, M.G.; Kleber, M. Dynamic Molecular Structure of Plant Biomass-Derived Black Carbon (Biochar). *Environ. Sci. Technol.* **2010**, *44*, 1247–1253. [CrossRef]
68. Salam, M.A.; Fageeh, O.; Al-Thabaiti, S.A.; Obaid, A.Y. Removal of nitrate ions from aqueous solution using zero-valent iron nanoparticles supported on high surface area nanographenes. *J. Mol. Liq.* **2015**, *212*, 708–715. [CrossRef]
69. Zhang, J.H.; Hao, Z.W.; Zhang, Z.; Yang, Y.P.; Xu, X.H. Kinetics of nitrate reductive denitrification by nanoscale zero-valent iron. *Process. Saf. Env.* **2010**, *88*, 439–445. [CrossRef]
70. Wilkin, R.T.; Su, C.M.; Ford, R.G.; Paul, C.J. Chromium-removal processes during groundwater remediation by a zerovalent iron permeable reactive barrier. *Environ. Sci. Technol.* **2005**, *39*, 4599–4605. [CrossRef]
71. Bock, E.; Smith, N.; Rogers, M.; Coleman, B.; Reiter, M.; Benham, B.; Easton, Z.M. Enhanced Nitrate and Phosphate Removal in a Denitrifying Bioreactor with Biochar. *J. Environ. Qual.* **2015**, *44*, 605–613. [CrossRef]
72. Huang, L.H.; Liu, G.F.; Dong, G.H.; Wu, X.Y.; Wang, C.; Liu, Y.Y. Reaction mechanism of zero-valent iron coupling with microbe to degrade tetracycline in permeable reactive barrier (PRB). *Chem. Eng. J.* **2017**, *316*, 525–533. [CrossRef]
73. Zhu, X.Y.; Chen, X.J.; Yang, Z.M.; Liu, Y.; Zhou, Z.Y.; Ren, Z.Q. Investigating the influences of electrode material property on degradation behavior of organic wastewaters by iron-carbon micro-electrolysis. *Chem. Eng. J.* **2018**, *338*, 46–54. [CrossRef]
74. Chen, S.S.; Hsu, H.D.; Li, C.W. A new method to produce nanoscale iron for nitrate removal. *J. Nanoparticle Res.* **2004**, *6*, 639–647. [CrossRef]
75. Mukherjee, A.; Zimmerman, A.R.; Harris, W. Surface chemistry variations among a series of laboratory-produced biochars. *Geoderma* **2011**, *163*, 247–255. [CrossRef]
76. Mikami, I.; Sakamoto, Y.; Yoshinaga, Y.; Okuhara, T. Kinetic and adsorption studies on the hydrogenation of nitrate and nitrite in water using Pd-Cu on active carbon support. *Appl. Catal. B-Environ.* **2003**, *44*, 79–86. [CrossRef]
77. Mehdinejadiani, B.; Amininasab, S.M.; Manhooei, L. Enhanced adsorption of nitrate from water by modified wheat straw: Equilibrium, kinetic and thermodynamic studies. *Water Sci. Technol.* **2019**, *79*, 302–313. [CrossRef]
78. Li, W.; Wang, L.J.; Liu, F.; Liang, X.L.; Feng, X.H.; Tan, W.F.; Zheng, L.R.; Yin, H. Effects of Al^{3+} doping on the structure and properties of goethite and its adsorption behavior towards phosphate. *J. Env. Sci* **2016**, *45*, 18–27. [CrossRef]
79. Suzuki, T.; Moribe, M.; Oyama, Y.; Niinae, M. Mechanism of nitrate reduction by zero-valent iron: Equilibrium and kinetics studies. *Chem. Eng. J.* **2012**, *183*, 271–277. [CrossRef]
80. Chen, Z.M.; Xiao, X.; Chen, B.L.; Zhu, L.Z. Quantification of Chemical States, Dissociation Constants and Contents of Oxygen-containing Groups on the Surface of Biochars Produced at Different Temperatures. *Environ. Sci. Technol.* **2015**, *49*, 309–317. [CrossRef]
81. Sun, T.R.; Levin, B.D.A.; Schmidt, M.P.; Guzman, J.J.L.; Enders, A.; Martinez, C.E.; Muller, D.A.; Angenent, L.T.; Lehmann, J. Simultaneous Quantification of Electron Transfer by Carbon Matrices and Functional Groups in Pyrogenic Carbon. *Environ. Sci. Technol.* **2018**, *52*, 8538–8547. [CrossRef]
82. Kluepfel, L.; Keiluweit, M.; Kleber, M.; Sander, M. Redox Properties of Plant Biomass-Derived Black Carbon (Biochar). *Environ. Sci. Technol.* **2014**, *48*, 5601–5611. [CrossRef] [PubMed]
83. Hosseini, S.M.; Ataie-Ashtiani, B.; Kholghi, M. Nitrate reduction by nano-Fe/Cu particles in packed column. *Desalination* **2011**, *276*, 214–221. [CrossRef]

Article

Geospatial Assessment of Groundwater Quality with the Distinctive Portrayal of Heavy Metals in the United Arab Emirates

Imen Ben Salem [1], Yousef Nazzal [1], Fares M. Howari [1], Manish Sharma [1,*], Jagadish Kumar Mogaraju [2] and Cijo M. Xavier [1]

[1] College of Natural and Health Sciences, Zayed University, Abu Dhabi 144534, United Arab Emirates; imen.bensalem@zu.ac.ae (I.B.S.); yousef.nazzal@zu.ac.ae (Y.N.); fares.howari@zu.ac.ae (F.M.H.); cijo.xavier@zu.ac.ae (C.M.X.)
[2] Agro-Ecosystems Specialist Group, IUCN-CEM, Varanasi 221456, India; jagadishmogaraju@gmail.com
* Correspondence: manish.sharma@zu.ac.ae; Tel.: +971-25993804

Abstract: Groundwater is a valuable resource, and its quality is critical to human survival. Optimal farming and urbanization degraded groundwater reserves. This research investigates and reports the spatial variability of selected heavy metals developed in the Liwa area of the United Arab Emirates. Forty water samples were collected from existing wells and analyzed for different elements. Principal components analysis was applied to a subgroup of the data set in terms of their usefulness for determining the variability of groundwater quality variables. Geographic information systems were used to produce contour maps to analyze the distribution of heavy metals. Ordinary kriging was used with Circular, Spherical, Tetraspherical, Pentaspherical-Bessel, K-Bessel, Hole effect, and Stable models for better representation. The water quality index was constructed using heavy metal concentrations and other variables. This yielded a value of 900 beyond the limit stated by WHO and US EPA. Nugget analysis showed that Cd (0), K (7.38%), and SO_4 (1.81%) variables exhibited strong spatial dependence. Al (27%), Ba (40.87%), Cr (63%), Cu (34%), EC (27%), HCO_3 (56%), NO_3 (36%), Pb (64%), and TDS (53%) represented moderate spatial dependence. As (76%), Mn (79%), Ni (100%), pH (100%), Temp (93%), and Zn (100%) exhibited weak spatial dependence.

Keywords: geostatistics; GIS; heavy metal pollution; groundwater; UAE

1. Introduction

Groundwater reaches the aperture of the landmass and its interior through natural, artificial, and indirect recharge. It was used by at least 2 billion people [1]. This infringement of this precious resource can be attributed to its nearest obtainability with partial efforts and spatial obstinacy. Aquifers at specific demand locations are being drained, and this state may lead to deprivation of this indispensable resource unless allayed [2]. The hydraulic conductivity and permeability of the rocks depend on the rocks' porosity [3]. Heavy metals are among the most significant pollutants of groundwater sources [4,5]. Nonetheless, the toxicity of heavy metals depends on their concentration levels in the environment. With increasing concentrations in the environment and decreasing soils' capacity toward retaining heavy metals, they leach into groundwater and soil solution [6,7]. Then, these toxic heavy metals can be accumulated and concentrated via the food chain in living tissues [8]. Some of these heavy metals are Arsenic (As), Lead (Pb), Nickle (Ni), Chromium (Cr), and Zinc (Zn). They can be categorized as critical heavy metals that pollute groundwater and affect human health [9]. One of the advanced techniques used in groundwater quality data interpolation is geostatistics [10]. The results obtained from the geostatistics can help a decision-maker to adopt suitable remedial measures to protect the quality of groundwater sources [11].

Focus has been devoted to studying groundwater quality and quantity to prevent groundwater contamination [12]. The native hydrogeological conditions and pollutant loads are dependent on lateral physicochemical interactions on the surface and sub-surface. Geostatistical analysis is used in subjecting data to interpolate by understanding the resemblances [13]. Radial functions and inverse distance weighted values are significant in data smoothing [14]. We can investigate uncertainties and produce surface predictions to get related information without subjecting data to erroneous functions and unnecessary manipulations. The geostatistical tools help us understand data in graphical forms while maintaining the native peaks and troughs, as witnessed by field reports [15]. Three-dimensional visualization evolved as an in-silico currency to reveal native information about the local geology and visualization thereof [16]. The sample points measured spatially can be used in the autocorrelation process in the ordinary kriging method [17]. The association between the transformation between the measured and predicted values can be described with a semi variogram [18]. Semi variance is the product of slope that appears within a fitted model and distance between the location pairs. Geostatistical techniques improved the spatial data distribution with delimited accuracy while maintaining portability [19]. Primarily, spatial analysis mandates the use of GIS and statistics. GIS apparatus is expected to deliver predictions and improved interpretation accuracy per se. Though technical constraints limit us, we can present a near-accurate portrayal of processes and features with GIS tools equipped with geostatistics [20].

Inverse distance weighting (IDW) can be stated as a simple interpolation technique. The weighted average is considered within a neighborhood in IDW [21]. The analyst can regulate a specific mathematical form of the weight function and the neighborhood size [22]. It is necessary to consider IDW with natural neighbor in analyzing substantial data sets. It uses cluster scatter points to identify datasets under investigation. This method is apt for discrete sample data [23].

The geostatistical analysis such as ordinary kriging provides insights into the groundwater situation. Kriging assumes that random processes with spatial autocorrelation can mimic at least some of the spatial variation observed in natural events and that the spatial autocorrelation must be explicitly modeled. However, it has much versatility as a simple prediction tool. An integrated approach to the assessment combining aquifer-based preselection criteria and multivariate non-parametric geostatistics was proposed to overcome the traditional approach's limitations and include the intrinsic hydrogeological and geochemical heterogeneity into the definition of groundwater water quality [24]. Arsenic (As) is one of the most harmful inorganic contaminants in water streams for the environment and human health [25]. The correlations between different groundwater quality indices and the causes and impacting variables of groundwater pollution can be tracked using statistical and multivariate techniques [26]. The meta-evaluation of the groundwater quality index was attempted [27]. The quality of groundwater that is being used for irrigation can be affected by several factors [28].

Groundwater is a vital resource for human life. Land use changes and urbanization harmed groundwater. This study examines the geographical variability of selected heavy metals in the UAE's Liwa region. There were limited studies that used geostatistics to explain the groundwater quality variability across this study area. This work is focused on the application of the geostatistical tools to explain the groundwater contamination. The objectives of this study are: (1) To describe the geospatial relationship between the observed groundwater variables using geostatistics, and multivariate analysis; (2) To determine the parameters' variability at various sample points spread over the study area. PCA was used to define the parameters controlling the groundwater chemistry and the information using significant variables was revealed; (3) To investigate the suitability of water for drinking (Using Weighted Arithmetic Water Quality Index (WA-WQI); (4) To employ ordinary kriging and select an appropriate model representing spatial distribution and spatial dependence variables.

2. General Characteristics of Study Area

The primordial settlements in Abu Dhabi traced to 7000 years ago and elsewhere are regarded as from the late stone age [29]. Agricultural hubs were established by the semi-nomadic peoples at Al Ain and Liwa regions [30]. It is an elapsed statistic that most of the populace endured on sustainable water management, attributed to rough climatic conditions that prevailed then. Stringent punishments were levied on anyone who accidentally or purposely threatened the water resources. The Liwa desert accommodates high temperatures ranging from 40 to 50 °C [31]. We can witness the highest dunes southwards of the Liwa crescent, prevalently known as the Moreeb dune [32]. Though dry scape appears on the surface, dunes can aid in excellent groundwater recharge with whatever rainfall is available over this zone. In the northern side of Liwa, i.e., Madinat Zayed, we can witness fresh water under the dunes [33].

Hydrogeology

The hydraulic conductivities measured at Liwa Crescent and Madinat Zayed ranged from 10–100 m/d. They were marked as a peak in Abu Dhabi. This property can be attributed to the sands that are unconsolidated and homogeneous. In the north of Liwa, water is of low salinity. It was classified as old water as it arrived at the aquifer long ago. Most of the groundwater levels are shown at the agricultural zones of Liwa with high nitrate concentrations [34]. The farm soils of Liwa are Torripsamments with no specific profile [35]. The agricultural activity in this area depends on desalinated water, and hence the native soils are less saline. The Liwa region can be considered an essential food production zone in the United Arab Emirates. Its water reserves are abundant for agriculture. The aquifer beneath this region has become vulnerable to pollution, attributed to several factors. The annual precipitation is confined to the winter season accounting for 100 mm/year [36]. The groundwater recharge is just 4%, and this area is devoid of surface water resources. The Liwa region is being degraded due to excess salinity. Its aquifer lodges increased levels of chromium derived from natural resources.

The Liwa aquifer can be categorized as lens-shaped and with a thickness of 121 mm. The average transmissivity was observed as 300 m^2/day [37]. The groundwater movement was observed in the north and south, especially in low-lying areas like sabkha physiographic regions, dunes, and sand salt flats [38]. Barchan dune complexes border the Liwa oasis at the south. There is a gradual incline passing to Oman's Hajjar massif mountain. The oil and gas-related activities injected brine into the groundwater zone. Ummer, Radhuma, Dammam, and Miocene are important aquifers underlying the Liwa region. Rus and Lower Fars were designated as confining units of this area [39]. Limestone dominates Ummer Radhuma and Dammam. Figure 1 shows the hydrogeological cross section of the Liwa region.

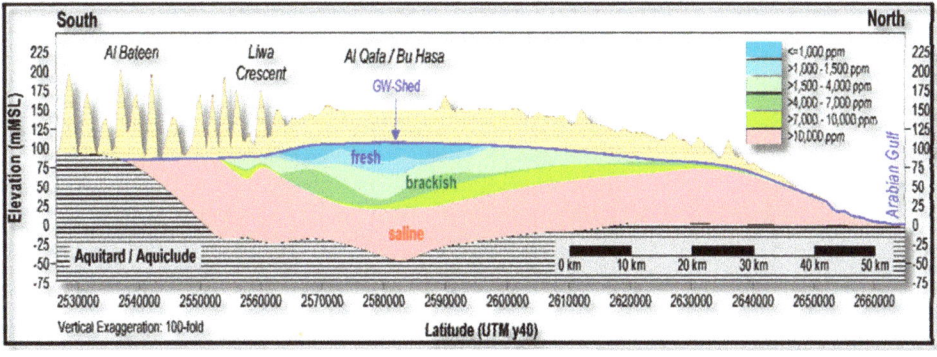

Figure 1. Hydrogeological cross-section of the northern Liwa area.

Miocene aquifer is dominated by sandstone. There is minimal silt and clay at the upper zones of the soil, resulting in high transmissivity and porosity. The Liwa area is hydrologically unique in the western region, making it a highly productive groundwater resource. Under the Liwa lens, it was observed that 38 bcm (billion cubic meters) of groundwater with TDS greater than 15,000 mg/L is present [40]. Since this aquifer is limited with its connectivity with adjoining aquifer systems, most of the solutes observed might be of atmospheric origins. The recharge of this shallow aquifer is negligible. The flow will be confined to the upper aquifer during intense precipitation especially in unsaturated zones [41]. Unmitigated irrigation-related activities are dumping nitrates into the aquifers. The freshwater lens is being depleted in the Liwa area due to several agricultural activities per se. Liwa oasis provided an agricultural base for the semi-nomadic Bani Yas community. This led to settlements on a large scale in Abu Dhabi [42]. The groundwater studies using geostatistics of the study area were attempted previously and concluded that the anthropogenic and natural processes affected the quality of the groundwater [43,44].

3. Materials and Method

3.1. Sampling and Analysis

The groundwater samples were collected from the Liwa region of the UAE. Forty samples were collected from the study area, and their hydrogeochemical parameters were analyzed. The water samples were analyzed using the American Public Health Association (APHA) standards [45]. The Inductively Coupled Plasma—optical emission spectroscopy (ICP-OES, Avio 200, Perkin Elmer, Waltham, MA, USA) was used to quantify the heavy metals present in groundwater samples. The correlation coefficient and PCA were determined to know the correlation between the elements in the sample. After analyzing the database, the spatial distribution of quality parameters was obtained using ArcGIS 10.8 software to create spatial and layered maps. Semi-variograms were prepared and an appropriate model selected based on the nugget analysis to establish spatial dependence.

Polyethylene bottles of 1 L capacity were used to collect the samples. In order to minimize the risk of contamination, the plastic bottles were rinsed with distilled water prior to being filled with sampled water. The samples were preserved with 65% nitric acid (HNO3) for a pH of 2, and bottles were kept cool at 4 °C. ICP-OES system was used to study the heavy and trace elements (As, Cr, Al, Mn, Ni, Cu, Pb, Zn, Cr, and Cu). Potassium (K), Calcium (Ca), Nitrate (NO_3), and Sulfate (SO_4) were analyzed using Ion Chromatography (ICS 5000+, Thermo Fisher, Waltham, MA, USA). Bicarbonate was determined by titration. Analyses were conducted in duplicate to minimize manual and instrumental errors.

3.2. Water Quality Index (WQI)

There have been many water quality assessment methods proposed by international scholars, such as set pair analysis [46–48], rough set and TOPSIS [49–52], entropy water quality index [53–55]. However, water quality index (WQI) is the most popular and widely adopted methods for overall water quality assessment [56,57]. In this study, the water quality index (WQI) was constructed using the weighted arithmetic average method as shown below [58].

$$Calculation\ for\ water\ quality\ rating (Q_n) = 100 \times \frac{(V_n - V_0)}{(S_n - V_0)} \qquad (1)$$

Q_n: Water quality rating for the nth parameter, V_n: Observed value of the nth parameter, V_0: Ideal value, S_n: Standard permissible value of nth parameter.

The unit weight of the corresponding parameter was an inverse proportional value to the recommended standard value of S_n

$$Calculation\ of\ unit\ weight\ (W_n) = \frac{K}{S_n} \qquad (2)$$

W_n: unit weight for the nth parameter, S_n: standard value of the n th parameter, K is the constant for proportionality: $K = \frac{1}{\sum \frac{1}{S_n}}$

The total water quality index was calculated linearly by adding the quality rating to the unit weight:

$$WQI = \sum Q_n W_n / \sum W_n \qquad (3)$$

3.3. Principal Component Analysis (PCA)

PCA is one of the popular statistical analysis techniques that can be used to investigate data patterns. The Principal Components Approach can be assumed as a comprehensive Factor Analysis method. The goal of principle component analysis (PCA) is to construct new variables, known as principal components, from a set of existing original variables [59,60]. The new variables are created by linearly combining the current variables. The PCA reduces an extensive data set of variables into a few elements known as the principal components, which can then be analyzed to show the underlying data structure. It is one of the features of primary components that they are not correlated or orthogonal with one another. When a data set has a significant variance, the first principal component (F1) absorbs and accounts for as much variance as feasible. The second component (F2) absorbs the remaining variation as feasible, and so on. The maximum number of PCs or principal components equals the total number of variables in a model unless otherwise specified. Because each standardized variable has one variance, the total variance accounted for by all of the Fi's will be equal to the number of variables. Only a few Fi numbers are maintained in the data processing process to facilitate comprehension. The Kaiser criterion determines the number of primary components preserved in the analysis. It is also possible to express the latent root as a proportion of the overall variance in the data set. The diagram showing methodology is given in Figure 2.

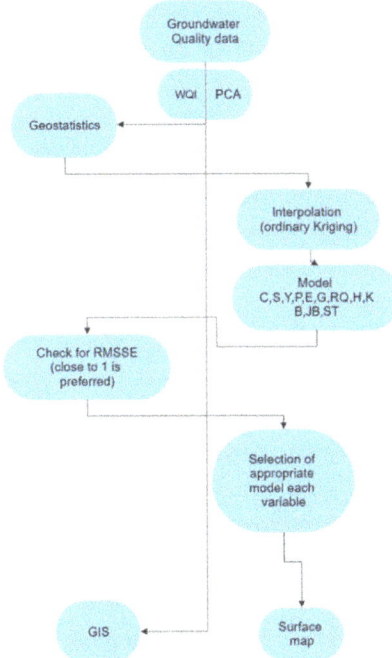

Figure 2. Flow chart of methodology used in the study.

4. Results and Discussion

4.1. Water Quality Index

The water quality index is represented in Table 1. Water quality classification based on WQI value. If the WQI value is between 0 and 25, it can be considered excellent, and it is good if it falls between 26 and 50. Some other ranges, including 51–75 (poor), 76–100 (very poor), and >100 (Unfit for consumption). The calculated WQI of this area is 900, and it is way beyond the recommended value.

Table 1. Water Quality Index.

Parameters	V_n	V_0	S_n	W_n	Q_n	WQI
Cr	0.015	0	0.05	20.000	29.523	29.523
Cu	0.002	0	2	0.500	0.079	0.079
K	8.964	0	20	0.050	44.820	44.820
Mn	0.002	0	0.4	2.500	0.589	0.589
Zn	0.005	0	3	0.333	0.156	0.156
Ba	0.166	0	0.7	1.429	23.704	23.704
As	0.022	0	0.01	100.000	220.996	220.996
TDS	863.049	0	500	0.002	172.610	172.610
EC	1478.488	0	400	0.003	369.622	369.622
NO_3	1.410	0	5	0.200	28.200	28.200
SO_4	23.570	0	250	0.004	9.428	9.428
pH	6.519	7	8.5	0.118	−32.065	−32.065
HCO_3	87.546	0	350	0.003	25.013	25.013
Total						900.52

The descriptive statistics presented in Table 2 reflect different mean values for each variable in this dataset. The cadmium concentration ranges from 0.17 to 0.183 ppm with a mean of 0.18, chromium concentration from 0.00048 to 0.023 ppm with a mean of 0.014 ppm, copper concentration from 0.000873 ppm to 0.004 ppm with a mean value of 0.001 ppm, potassium concentration observed to be from 2.704 to 17.202 ppm, with a mean value of 8.964 ppm. The mean value of the manganese concentration was found to be 0.0023 ppm. The Ni concentration ranges from 0.00049 to 0.0044 ppm with a mean value of 0.001 ppm. The Pb concentration ranges from 0.289 to 0.490 ppm with an observed mean value of 0.412 ppm, and the Zn concentration ranges from 0.000357 to 0.051 ppm with a mean value of 0.003 ppm. The mean concentration of Ba and Al is 0.166 and 0.990 ppm, respectively.

Table 2. Descriptive statistics.

Variable	Mean	Max	Min	SD
Al	0.990	1.450	0.339	0.233
As	0.022	0.029	0.008	0.004
Ba	0.166	0.457	−0.065	0.136
Cd	0.181	0.183	0.175	0.002
Cr	0.014	0.023	0.48×10^{-3}	0.006
Cu	0.001	0.004	0.873×10^{-3}	9.351×10^{-4}
EC	1478.488	3003	328	656.631
HCO_3	87.546	236.680	14.640	49.449
K	8.964	17.203	2.704	3.160
Mn	0.002	0.011	0.0002	0.003
Ni	0.001	0.004	0.0004	0.001
NO_3	1.410	2.486	0.426	0.557
Pb	0.412	0.490	0.289	0.048
pH	6.519	7.190	6.190	0.259
SO_4	23.570	45.794	4.129	9.255
TDS	863.049	1565	136	358.995
Temp	28.378	32.600	23.500	1.741
Zn	0.003	0.051	0.0357×10^{-2}	0.008

Hence, correlation analysis is presented in Figure 3. The correlation analysis was done to investigate the relationships between the parameters measured.

	Cr	Cu	K	Mn	Ni	Pb	Zn	Ba	Al	As	TDS	EC	NO$_3$	SO$_4$	pH	HCO$_3$	Temp
Cr	1.000																
Cu	0.151	1.000															
K	0.088	0.068	1.000														
Mn	-0.252	-0.008	0.127	1.000													
Ni	-0.010	-0.196	0.350	0.037	1.000												
Pb	0.031	0.036	0.842	0.078	0.291	1.000											
Zn	-0.086	0.145	-0.140	-0.019	-0.173	-0.180	1.000										
Ba	0.179	0.265	-0.230	-0.157	-0.060	-0.239	-0.141	1.000									
Al	-0.019	-0.047	0.507	0.210	0.291	0.640	0.355	-0.317	1.000								
As	-0.100	0.046	0.719	0.163	0.229	0.838	-0.053	-0.271	0.713	1.000							
TDS	0.045	0.045	0.836	0.165	0.216	0.738	-0.049	-0.198	0.597	0.719	1.000						
EC	0.138	0.305	0.713	-0.039	0.209	0.796	-0.177	0.054	0.521	0.599	0.565	1.000					
NO$_3$	-0.117	0.304	0.604	0.005	0.228	0.598	0.034	-0.094	0.538	0.680	0.660	0.549	1.000				
SO$_4$	0.002	-0.001	0.763	-0.017	0.167	0.649	0.080	-0.459	0.537	0.578	0.628	0.545	0.554	1.000			
pH	-0.012	0.028	-0.420	-0.022	-0.105	-0.513	0.249	0.446	-0.379	-0.604	-0.486	-0.287	-0.564	-0.418	1.000		
HCO$_3$	-0.141	-0.248	-0.063	-0.093	0.056	-0.029	-0.075	-0.372	-0.028	0.103	-0.144	-0.208	0.143	-0.043	-0.324	1.000	
Temp	0.139	-0.156	-0.355	-0.123	0.029	-0.242	-0.265	0.196	-0.254	-0.365	-0.397	-0.142	-0.195	-0.170	0.112	0.086	1.000

Figure 3. Correlation table.

The degree to which two variables are allied is weighed by a correlation coefficient, indicated by the letter r. This coefficient, named after its discoverer, Pearson, measures linear association used in statistics and education. It is necessary to utilize alternative, more though measures of the correlation if a curved line is required to describe the relationship. The correlation coefficient can be measured from + 1 to −1 and is expressed as a percentage. The degree of complete correlation between two variables is represented by the numbers + 1 or −1, respectively. Correlations are positive when one variable increases in response to another's increase and negative when one variable reduces in response to the other increases. The number zero represents the complete absence of association.

Cadmium is negatively correlated with Cr, K, Pb, TDS, and EC and positively correlated with Mn, Zn, As, and pH. This shows that the heavy metals with pH other than Cd are affecting the Cd concentrations and this can be due to anthropogenic causes. Cr is positively correlated with Cu, K, Ba, Temp, and EC and negatively correlated with Mn, As, NO$_3$, and HCO$_3$. This shows that the increase in Cr concentration is associated with the other heavy metals (Antropogenic) along with the EC and temperature (Natural Process). Cu is positively correlated with Zn, Ba, NO$_3$, Ni (Anthropogenic), EC, HCO$_3$, and Temp. K exhibits a positive correlation with Mn, Ni, Pb, Al, As, TDS, NO$_3$, SO$_4$, and EC and a negative correlation with Zn, Ba, Temp, and pH. Mn positively correlates with Al and As and negatively correlates with Ba and Temp. Ni is positively correlated with Pb, Al, As, TDS, NO$_3$, SO$_4$, and EC and negatively correlated with Zn and pH. Pb is positively correlated with Al, As, TDS, NO$_3$, SO$_4$, and EC and negatively correlated with Zn, Ba, Temp, and pH. Zn is positively correlated with Al and pH and negatively correlated with Temp and EC. Ba is positively correlated with Temp, pH, and EC and negatively correlated with Al, As, TDS, and HCO$_3$. Al is positively correlated with As, TDS, NO$_3$, SO$_4$, and EC and negatively correlated with Temp and pH. As is positively correlated with TDS, NO$_3$, SO$_4$, HCO$_3$, and EC and negatively correlated with Temp and pH. TDS is positively correlated with NO$_3$, SO$_4$, and EC and negatively correlated with HCO$_3$, Temp, and pH. NO$_3$ is positively correlated with SO$_4$, HCO$_3$, and EC and negatively correlated with Temp and pH. SO_4 exhibits a negative correlation with HCO_3, Temp, and pH and positively correlates with EC. HCO_3 is negatively correlated with pH and EC.

The observed cadmium value is 0.18 ppm in the study area. This water might not suit agriculture, irrigation, and drinking purposes. The increase in cadmium levels can be due to sewage sludge, fertilizers, battery alloys, and cigarette smoking. The increased levels of Cd can damage kidneys. Due to this, there will be disruption of the endocrine system and inhibition of sex hormones in humans. Long-time exposure to Cadmium may cause Itai-itai in humans. The recommended chromium standard in drinking water is 0.1 ppm (EPA) and 0.05 ppm (WHO). The maximum value of chromium in the water sample is 0.023, which is well within the permissible limits. Cu (0.004), Mn (0.01), Zn (0.051), Ba (0.457), NO_3 (2.4), SO_4 (45), Temperature (32.6), pH (7.19), and bicarbonates are within the normal range specified by WHO and US EPA. The EC values are at 3003, and it is problematic. Pb (0.49), As (0.028), and Al (1.45) are high in the water samples collected from the study area. The concentration of Cadmium in the observed samples is 0.184 (0.183984), and this is higher than the WHO recommended value of 0.003 mg/L. The concentration of lead observed in the groundwater samples is 0.4905 mg/L against 0.01 mg/L (WHO). The concentration of Aluminum in the samples analyzed is 1.4502 and is higher than the WHO recommended value of 0.9 mg/L.

4.2. Principal Component Analysis

Pearson correlation matrices and PCA construed the datasets. Principal components were generated using varimax rotation, and this yielded variables that contribute more and other variables that contribute less. Multi-variate analysis was used in the PCA to transform a significant set of correlated variables into a minor set of uncorrelated variables. The interrelationships among the variables can be highlighted using covariance by this tool, and it is also called a dimensionless reduction tool. We can use PCA to know the associated chemicals construed as variable loadings on certain groundwater quality factors. Two significant eigenvalues, i.e., PC1 and PC2, were observed in the 40 groundwater samples with 18 parameters, constituting 35 and 12% of the variance. PC3 exhibited 10% of variance but PC4 and PC5 exhibited variance less than 10%. The first five components exhibited eigenvalues that are greater than 0.5. It is assumed that the factor loading value near +/− 1 exhibits a strong correlation. If the value if greater than 0.5, it is significant. PC1 exhibits 35% of variance about significant loadings of K, Pb, Al, As, TDS, nitrate, sulfate, and pH, and is shown in Table 3.

Table 3. Principal Component Loadings.

Variable	PC1	PC2	PC3	PC4	PC5	Uniqueness
Cd			0.739			0.387
Cr			−0.680			0.489
Cu					0.841	0.203
K	0.908					0.156
Mn			0.709			0.412
Ni					−0.522	0.527
Pb	0.919					0.140
Zn				0.889		0.174
Ba		0.618				0.274
Al	0.708			0.405		0.288
As	0.868					0.156
Nitrate	0.867					0.231
Sulfate	0.768					0.246
HCO_3	0.738	−0.836				0.336
Temp						0.270
pH	−0.572	0.626		−0.524		0.586
EC	0.810					0.254
						0.204

PC2 showed 12% variance associated with significant loadings of Ba, HCO$_3$, and pH. The component characteristics and component loadings for this data are presented in Tables 3 and 4. PC2 is loaded with Ba, HCO$_3$, and pH, whereas PC3 is loaded with Cd, Cr, and Mn. PC4 is loaded with Zn, Al, and pH and PC5 is loaded with Cu and Ni. Temperature, Cr, Ni, and Mn exhibited higher uniqueness values, suggesting that these variables have limited commonality. As evidenced by systematic data analysis, PC1 exhibited significant cations and anions due to anthropogenic and natural sources.

Table 4. Component characteristics.

Variable	Model	Nugget Ratio (%)
Cd	S *	0
Cr	P	63.456
Cu	C	34.148
K	RQ	7.384
Mn	G	79.008
Mn	S *	79.008
Pb	J	64.935
Zn	H	100
Ba	H	40.879
Al	RQ	27.089
As	G	76.753
As	S *	76.753
Sulfate	T	1.817
HCO$_3$	J	56.035
Temp	H	93.559
PH	C	100
EC	J	27.262
Ni	C	100
NO$_3$	H	36.387
TDS	J	53

S * = stable model.

The cos values are employed to know the representation's quality, and the individuals closer to the center of the plot are assumed of limited or low importance for the reported first components. The low cos values are shown in blue and high cos values in red (Figure 4). The contribution of the variables with sample points is presented in Figure 5.

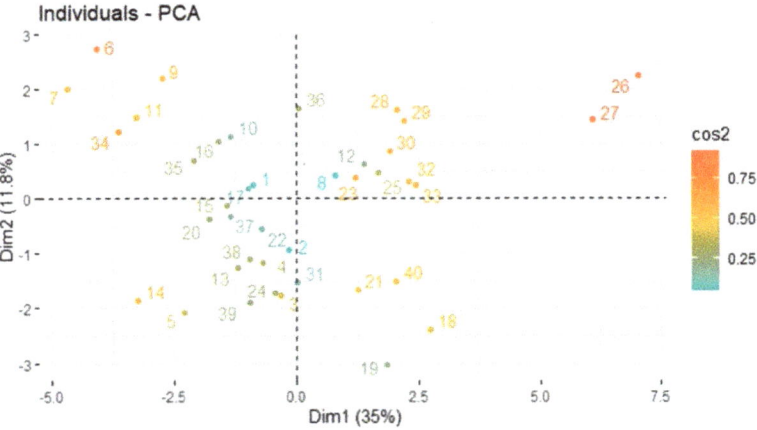

Figure 4. PCA for variables.

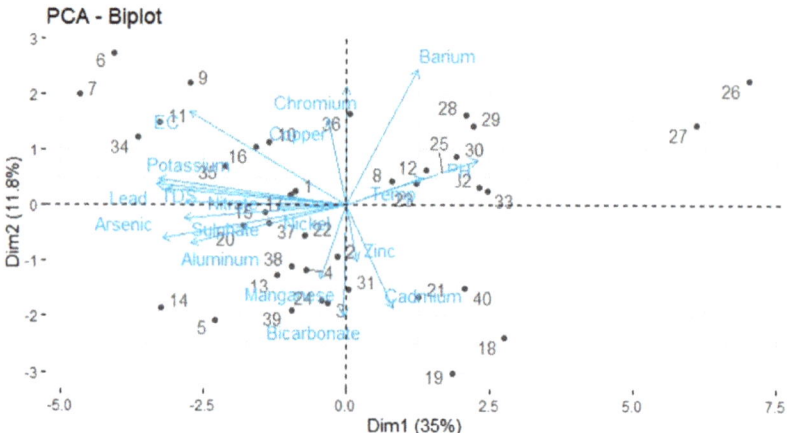

Figure 5. Biplot with sample points and variables.

4.3. Geostatistical Analysis of the Study Area (Spatial Distribution)

The statistical and geostatistical analysis was done using ArcGIS 10.2, R, R studio, and MS Excel 2019. The spatial interpolators can easily predict the values of a specific attribute at locations unknown to the observer utilizing the values of sample locations known previously. The deterministic interpolators can utilize mathematical formulae to know the predicted values. This can be interpreted as similarity among the neighboring points and smoothing extent. The geostatistical interpolation techniques use certain statistical properties of the points previously measured to estimate the value of the surface locations. Depending on the spatial structure of the datasets framed, we can investigate the spatial dependence among the variables. The ordinary kriging method was selected based on a comparative analysis of interpolation methods. Models such as Circular (C), Spherical (S), Tetraspherical (T), Pentaspherical (P), Exponential (E), Gaussian (G), Rational Quadratic (RQ), Hole effect (H), K-Bessel (K), J-Bessel (J), and Stable (S*) were used to arrive at appropriate semi variogram. The best-fitted models are RQ (Al and K), G (As), S* (As, Cd, and Mn), H (Ba, NO_3, Temp, Zn), P (Cr), C (Cu, Ni, and pH), and J (EC, HCO_3, Pb, and TDS). Suppose the nugget ratio is less than 25%. In that case, we can assume that there is strong spatial dependence; 25 to 75% can be reflected as moderate spatial dependence. If greater than 75%, we can expect least or weak spatial dependence. After nugget analysis, it is obvious that Cd (0), K (7.38%), and SO_4 (1.81%) variables exhibited strong spatial dependence. Al (27%), Ba (40.87%), Cr (63%), Cu (34%), EC (27%), HCO_3 (56%), NO_3(36%), Pb (64%), and TDS (53%) represented moderate spatial dependence. As (76%), Mn (79%), Ni (100%), pH (100%), Temp (93%), and Zn (100%) exhibited weak spatial dependence. Nugget analysis for selecting the appropriate model was presented in Table 5.

Table 5. The Spatial distribution and semi variogram element's observed sites.

S. No	Elements	Low	Moderate	High
1.	Aluminum (Al)	North west	Western end	Eastern end
2.	Arsenic	North west	Western end	South eastern
3.	Barium	Western end	South western and north eastern	South western
4.	Cadmium	Northern end	North-south (extended)	Western end
5.	Chromium	Northern end	North-south (extended)	Western end
6.	Copper	North-south (extended)	Western end	Eastern end
7.	Potassium	Western end	Partially spotted all over the area	North eastern
8.	Manganese	Western end	South east	North west

The spatial distribution maps and semi variogram plots of essential variables were presented in Figures 6–12. Table 5 shows the specific locations of elements distribution in the area.

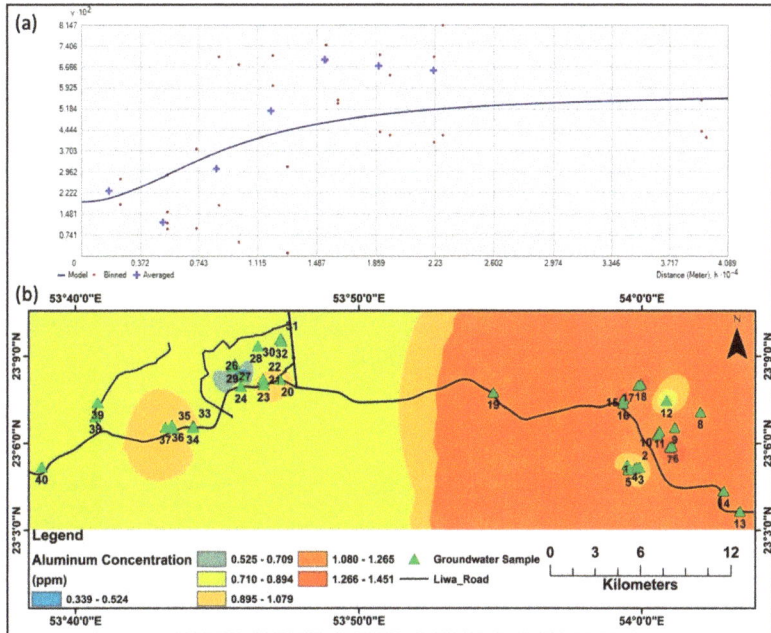

Figure 6. (**a**) Semi variogram, and (**b**) Spatial distribution mapping of Aluminum.

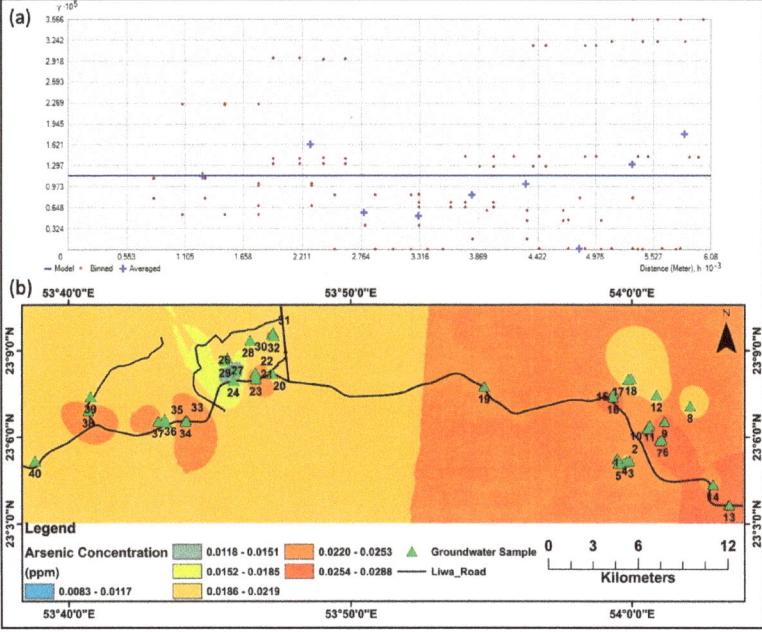

Figure 7. (**a**) Semi variogram, and (**b**) Spatial distribution mapping of Barium.

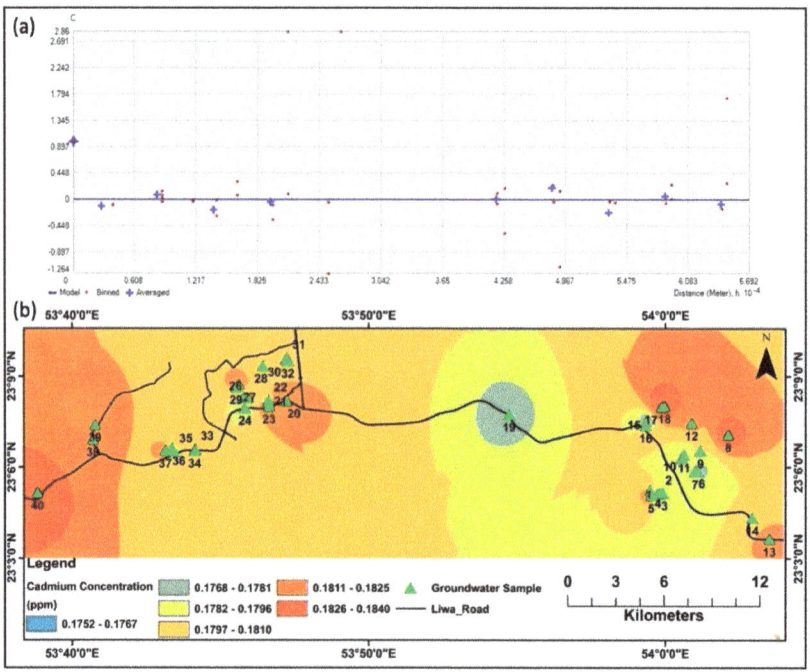

Figure 8. (**a**) Semi variogram, and (**b**) Spatial distribution mapping of Cadmium.

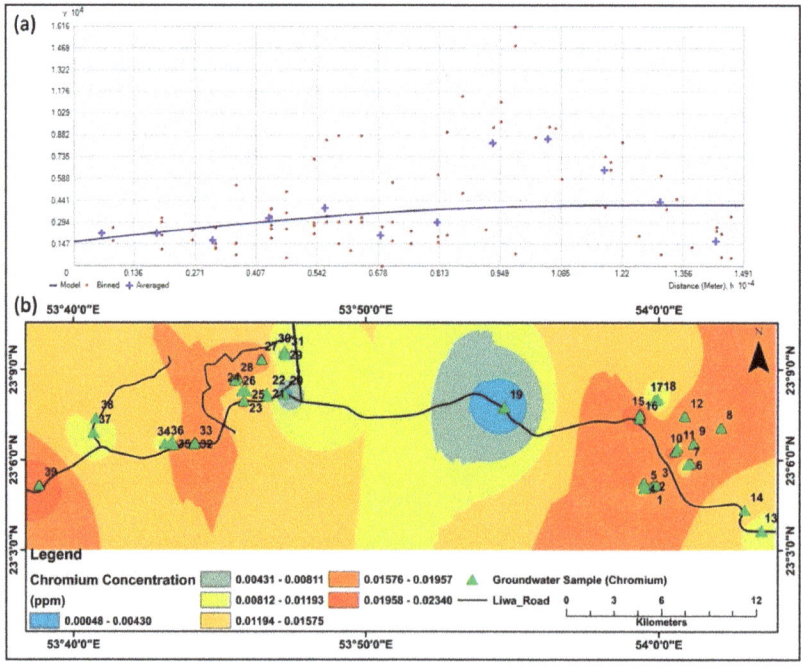

Figure 9. (**a**) Semi variogram, and (**b**) Spatial distribution mapping of Chromium.

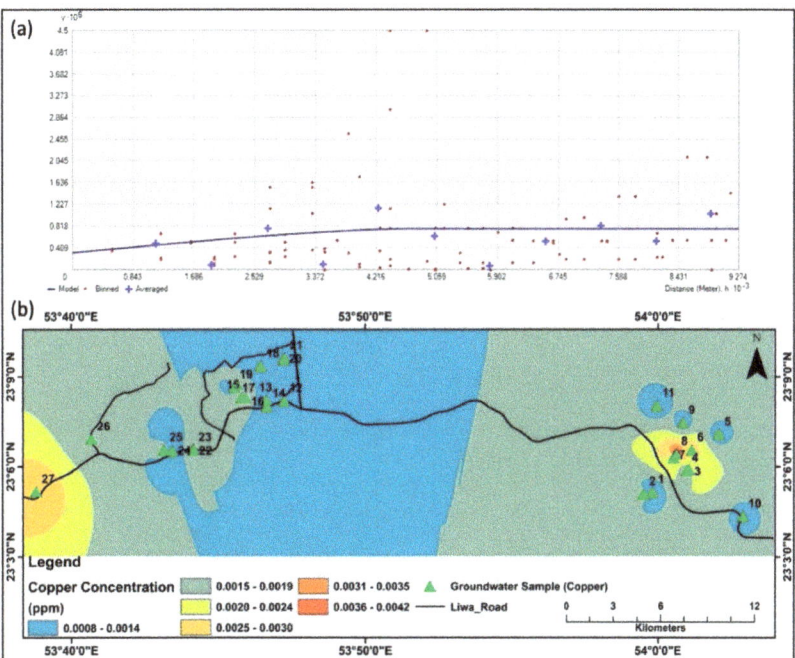

Figure 10. (a) Semi variogram, and (b) Spatial distribution mapping of Copper.

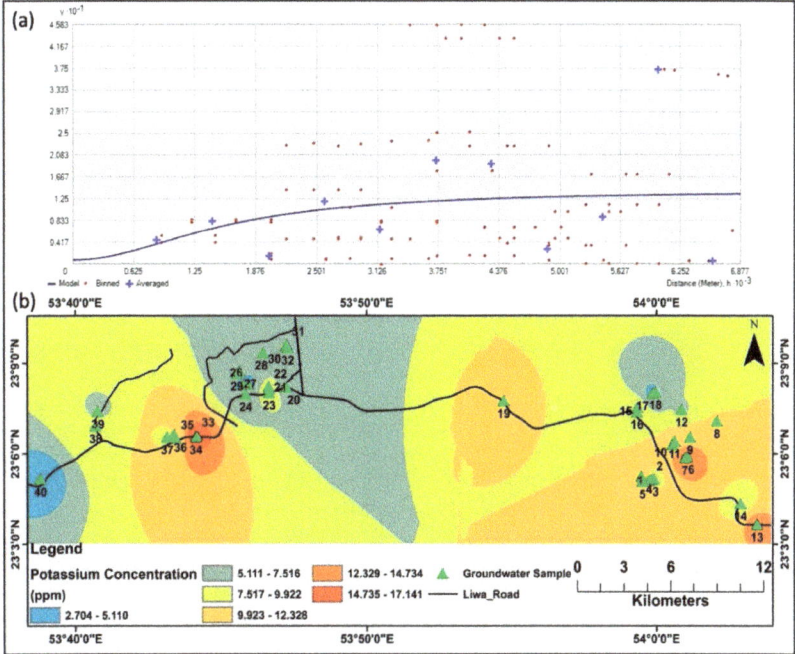

Figure 11. (a) Semi variogram, and (b) Spatial distribution mapping of Potassium.

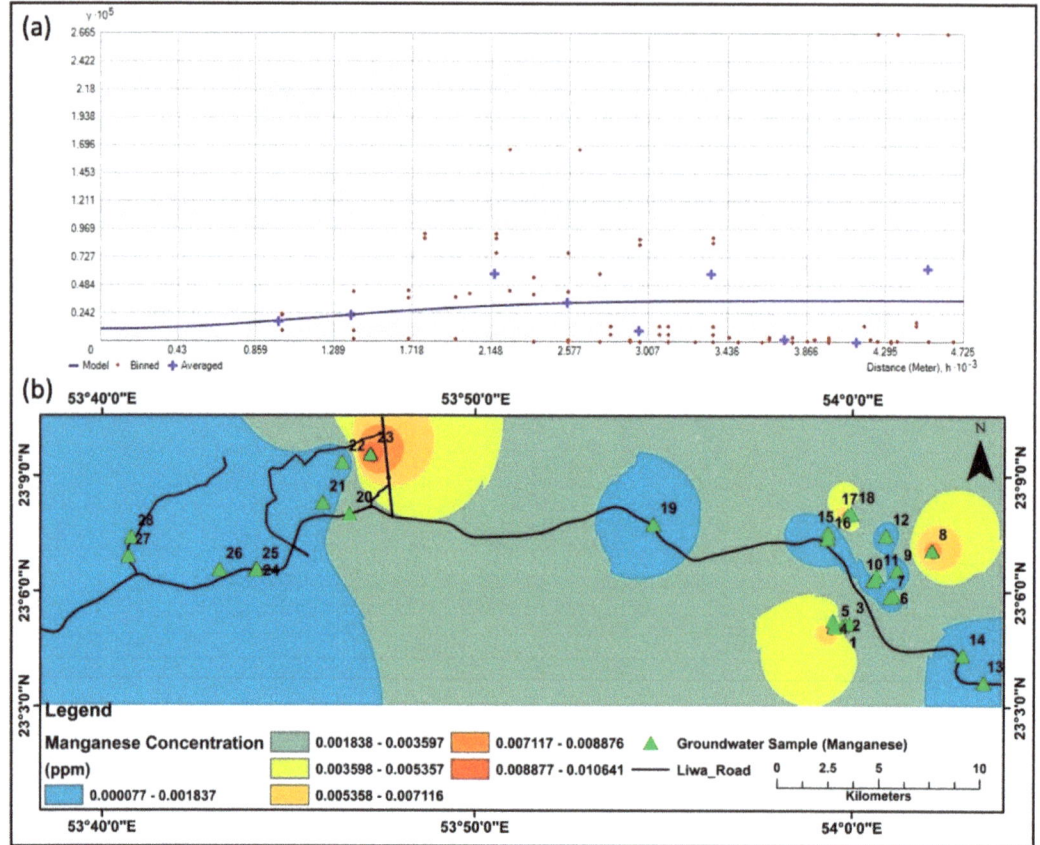

Figure 12. (a) Semi variogram, and (b) Spatial distribution mapping of Manganese.

The groundwater investigations in this area are expensive, and the sample locations were limited. Efforts are in progress to collect more samples with an increased temporal resolution to cross-validate the results obtained from the actual observations. Geostatistics in groundwater studies was never attempted for this study area. This study integrated physio-chemical, multivariate, and geostatistical analysis and water quality indices to analyze groundwater quality parameters. The parameters were correlated using correlation analysis, and both positive and negative correlations were found. To mention a few, Arsenic is positively correlated with nitrates, sulfates and EC and negatively correlated with pH and temperature. Lead is positively correlated with Al and As and negatively correlated with Ba and Zn. Ni positively correlated with As, Al, and Pb and negatively correlated with Zn. Chromium reflected a positive correlation with Ba and Cu and negatively correlated with As and Mn. Zinc is positively correlated with Al.

A PCA was performed, and the results showed that there were five main components, PC1 through PC5, which represented a variance of 35% (PC1) and 12% (PC2), 10% (PC3), <10% (PC4 and PC5), respectively. Significant loadings on PC1 are As, Al, Pb, K, pH, sulfate, and nitrate. PC2 accommodates significant loadings of Ba and bi-carbonates with pH.

The dataset was subjected to geostatistical analysis, and a suitable model was identified using standard kriging. The J-Bessel model was selected to represent Pb, TDS, HCO_3, and EC. Pentaspherical model is employed to represent Cr. The circular model was used to show the distribution of Ni, Cu, and pH. The hole effect model was used to describe the spatial distribution of Zn, Temp, NO_3, and Ba. A stable model was employed to reflect the distribution of Mn, Cd, and As. The rational quadratic model was used to represent K and Al. Based on nugget analysis, it was observed that Zn, As, Mn, Ni, pH, and TDS exhibited weak spatial dependence. Moderate spatial dependence was exhibited by Al, Ba, Cr, Cu, Ec, HCO_3, NO_3, Pb, and TDS. Strong spatial dependence was observed in Cr, K, and SO_4.

Groundwater quality deterioration has become a nightmare in this region, and this was due to limited surveillance and the never-ending injection of pollutants into this precious source. Most of the populace in this study area rely on this for industrial and drinking purposes after purification. Several studies were made to evaluate the potability of groundwater of Liwa aquifer mainly for agricultural needs; however, rigorous geostatistical methods were sparsely applied. This paper attempts to fill the void left in using geostatistics to represent groundwater quality. The variation among the sample clusters was studied previously using PCA. This paper uses optimal interpolation techniques and semivariogram analysis to produce statistically enriched results. Geological elements, and environmental and hydrological parameters were considered the core of these studies. Previous studies were made to emulate the subsurface hydrology characteristics with limited utilization of the geostatistics, and this work will add valuable inputs to the contemporary research on groundwater situation analysis and management. This paper presents a GIS-based approach with geostatistics in assessing groundwater quality at the Liwa region, UAE. The correlation matrix obtained supports PCA analysis. The results also shed light on groundwater quality deterioration due to anthropogenic activities. The exponential semivariogram model was systematically authenticated for each groundwater parameter. Analysis of groundwater samples reflects that cadmium, aluminum, and lead is in high proportions compared to other parameters like Cr, Cu, K, Mn, Ni, Zn, and Ba. The distribution maps are produced using the appropriate model of the kriging interpolation method for each variable.

5. Conclusions and Recommendation

In the present study, the detailed analyses show that this study offers background information on the groundwater parameters and factors that affect groundwater quality. This paper can aid water resource planners in coming up with management plans to safeguard the local population's health. The water quality index of this area is poor, and it is to be improved with the immediate inclusion of proposals for the rejuvenation of groundwater.

Liwa aquifer is exploited beyond the reasonable limit. It will permanently change the subsurface landscape in the coming decades. Given the observed water quality parameters, it is proposed that some of the stringent actions must be levied on the exploiters of this jewel of water in the Liwa basin. The water quality index of this region is alarmingly high, and the stakeholders in this region must mitigate the problem with immediate sustainable solutions.

The excess dumping of the wastes and improper groundwater extraction are assumed to be the main reasons behind this physicochemical variability observed in the water samples. The local geology might also affect this variability, and it is yet to be studied over all the study areas. This study combined the multivariate, geostatistical, and physicochemical analysis; however, complete hydrological analysis in the watershed, HRU, etc., could not be conducted. This is attributed to the local non-conducive conditions.

This area is devoid of the undulations and inundations of the terrain with almost indiscriminate relief. The data is collected once, and the second attempt to collect groundwater samples was not feasible due to budgetary and administrative constraints. If this had been materialized, there might have been a more intense comparative study with the dynamics of groundwater quality parameters. Due to the unavailability of the two-date data, there were no attempts to compare pre-monsoon, monsoon, and post-monsoon groundwater quality.

Author Contributions: Conceptualization, I.B.S. and M.S.; Methodology, M.S. and J.K.M.; Software, M.S. and J.K.M.; Validation, I.B.S., Y.N. and F.M.H.; Formal analysis, M.S., J.K.M. and C.M.X.; Investigation, M.S. and J.K.M.; Resources, Y.N. and F.M.H.; Data curation, M.S. and Y.N.; Writing—original draft preparation, I.B.S., M.S. and J.K.M.; Writing—review and editing, I.B.S., M.S., J.K.M. and C.M.X.; Visualization, M.S.; Supervision, I.B.S.; Funding acquisition, I.B.S. All authors have read and agreed to the published version of the manuscript.

Funding: This project was funded by the Research Office, Zayed University, United Arab Emirates (Project No. R 21005).

Institutional Review Board Statement: Not applicable.

Informed Consent Statement: Not applicable.

Data Availability Statement: The data used in this research work will be supplied for those interested upon request with no reservations.

Acknowledgments: The authors would like to thank Faculty and staff of Zayed University, Abu Dhabi, for their support in conducting and reporting this research.

Conflicts of Interest: The authors declare no conflict of interest. The funders had no role in the design of the study; in the collection, analyses, or interpretation of data; in the writing of the manuscript, or in the decision to publish the results.

References

1. Maliva, R.G. Aquifer Characterization and Properties. In *Aquifer Characterization Techniques: Schlumberger Methods in Water Resources Evaluation Series No. 4*; Maliva, R.G., Ed.; Springer Hydrogeology; Springer International Publishing: Cham, Switzerland, 2016; pp. 1–24. ISBN 978-3-319-32137-0.
2. Barbulescu, A.; Nazzal, Y.; Howari, F. Assessing the Groundwater Quality in the Liwa Area, the United Arab Emirates. *Water* **2020**, *12*, 2816. [CrossRef]
3. Park, B.; Kim, K.; Kwon, S.; Kim, C.; Bae, D.; Hartley, L.; Lee, H. Determination of the hydraulic conductivity components using a three-dimensional fracture network model in volcanic rock. *Eng. Geol.* **2002**, *66*, 127–141. [CrossRef]
4. Guo, Y.; Li, P.; He, X.; Wang, L. Groundwater quality in and around a landfill in northwest China: Characteristic pollutant identification, health risk assessment, and controlling factor analysis. *Expo. Health* **2022**. [CrossRef]
5. Nazzal, Y.; Orm, N.B.; Barbulescu, A.; Howari, F.; Sharma, M.; Badawi, A.E.; Al-Taani, A.A.; Iqbal, J.; Ktaibi, F.E.; Xavier, C.M.; et al. Study of Atmospheric Pollution and Health Risk Assessment: A Case Study for the Sharjah and Ajman Emirates (UAE). *Atmosphere* **2021**, *12*, 1442. [CrossRef]
6. Liu, L.; Wu, J.; He, S.; Wang, L. Occurrence and distribution of groundwater fluoride and manganese in the Weining Plain (China) and their probabilistic health risk quantification. *Expo. Health* **2021**. [CrossRef]
7. Wei, M.; Wu, J.; Li, W.; Zhang, Q.; Su, F.; Wang, Y. Groundwater geochemistry and its impacts on groundwater arsenic enrichment, variation, and health risks in Yongning County, Yinchuan Plain of northwest China. *Expo. Health* **2021**. [CrossRef]
8. Belkhiri, L.; Narany, T.S. Using Multivariate Statistical Analysis, Geostatistical Techniques and Structural Equation Modeling to Identify Spatial Variability of Groundwater Quality. *Water Resour. Manag.* **2015**, *29*, 2073–2089. [CrossRef]
9. Sankhla, M.S.; Kumar, R. Contaminant of Heavy Metals in Groundwater & Its Toxic Effects on Human Health & Environment. *IJESNR* **2019**, *18*, 1–5. [CrossRef]
10. Bodrud-Doza, M.; Bhuiyan, M.A.H.; Islam, S.M.D.-U.; Quraishi, S.B.; Muhib, M.I.; Rakib, M.A.; Rahman, M.S. Delineation of Trace Metals Contamination in Groundwater Using Geostatistical Techniques: A Study on Dhaka City of Bangladesh. *Groundw. Sustain. Dev.* **2019**, *9*, 100212. [CrossRef]
11. Verma, P.; Singh, P.K.; Sinha, R.R.; Tiwari, A.K. Assessment of Groundwater Quality Status by Using Water Quality Index (WQI) and Geographic Information System (GIS) Approaches: A Case Study of the Bokaro District, India. *Appl. Water Sci.* **2019**, *10*, 27. [CrossRef]
12. Tyler, S.W.; Muñoz, J.F.; Wood, W.W. The Response of Playa and Sabkha Hydraulics and Mineralogy to Climate Forcing. *Groundwater* **2006**, *44*, 329–338. [CrossRef] [PubMed]
13. Setianto, A.; Triandini, T. Comparison of Kriging and Inverse Distance Weighted (IDW) Interpolation Methodsin Lineament Extraction and Analysis. *J. Appl. Geol.* **2015**, *5*, 21–29. [CrossRef]
14. Alsharhan, A.S.; Rizk, Z.E. Conclusions. In *Water Resources and Integrated Management of the United Arab Emirates*; Alsharhan, A.S., Rizk, Z.E., Eds.; World Water Resources; Springer International Publishing: Cham, Switzerland, 2020; pp. 793–829. ISBN 978-3-030-31684-6.
15. Al-Taani, A.A.; Nazzal, Y.; Howari, F.M.; Iqbal, J.; Bou Orm, N.; Xavier, C.M.; Bărbulescu, A.; Sharma, M.; Dumitriu, C.-S. Contamination Assessment of Heavy Metals in Agricultural Soil, in the Liwa Area (UAE). *Toxics* **2021**, *9*, 53. [CrossRef] [PubMed]
16. Zheng, K.; Li, C.; Wang, F. Gaussian Radial Basis Function for Unsteady Groundwater Flow. *IOP Conf. Ser. Earth Environ. Sci.* **2019**, *304*, 022052. [CrossRef]

17. Elubid, B.A.; Huang, T.; Ahmed, E.H.; Zhao, J.; Elhag, M.; Abbass, W.; Babiker, M.M. Geospatial Distributions of Groundwater Quality in Gedaref State Using Geographic Information System (GIS) and Drinking Water Quality Index (DWQI). *Int. J. Environ. Res. Public Health* **2019**, *16*, 731. [CrossRef]
18. Venkatramanan, S.; Viswanathan, P.M.; Chung, S.Y. (Eds.) *GIS and Geostatistical Techniques for Groundwater Science*; Elsevier: Amsterdam, The Netherlands, 2019; ISBN 978-0-12-815413-7.
19. Żak, S. *Hydraulic Conductivity of Layered Anisotropic Media*; IntechOpen: Rijeka, Croatia, 2011; ISBN 978-953-307-470-2.
20. Iqbal, J.; Nazzal, Y.; Howari, F.; Xavier, C.; Yousef, A. Hydrochemical Processes Determining the Groundwater Quality for Irrigation Use in an Arid Environment: The Case of Liwa Aquifer, Abu Dhabi, United Arab Emirates. *Groundw. Sustain. Dev.* **2018**, *7*, 212–219. [CrossRef]
21. Al-Katheeri, E.S.; Howari, F.M.; Murad, A.A. Hydrogeochemistry and Pollution Assessment of Quaternary–Tertiary Aquifer in the Liwa Area, United Arab Emirates. *Environ. Earth Sci.* **2009**, *59*, 581. [CrossRef]
22. Eggleston, J.R.; Mack, T.J.; Imes, J.L.; Kress, W.; Woodward, D.W.; Bright, D.J. *Hydrogeologic Framework and Simulation of Predevelopment Groundwater Flow, Eastern Abu Dhabi Emirate, United Arab Emirates*; Scientific Investigations Report; U.S. Geological Survey: Reston, VA, USA, 2020; Volume 2018–5158, p. 60.
23. Dassargues, A. Hydrogeology: Groundwater Science and Engineering. Available online: https://www.routledge.com/Hydrogeology-Groundwater-Science-and-Engineering/Dassargues/p/book/9780367657147 (accessed on 18 August 2021).
24. Apollaro, C.; Di Curzio, D.; Fuoco, I.; Buccianti, A.; Dinelli, E.; Vespasiano, G.; Castrignanò, A.; Rusi, S.; Barca, D.; Figoli, A.; et al. A Multivariate Non-Parametric Approach for Estimating Probability of Exceeding the Local Natural Background Level of Arsenic in the Aquifers of Calabria Region (Southern Italy). *Sci. Total Environ.* **2022**, *806*, 150345. [CrossRef]
25. Figoli, A.; Fuoco, I.; Apollaro, C.; Chabane, M.; Mancuso, R.; Gabriele, B.; Rosa, R.D.; Vespasiano, G.; Barca, D.; Criscuoli, A. Arsenic-Contaminated Groundwaters Remediation by Nanofiltration. *Sep. Purif. Technol.* **2020**, *238*, 116461. [CrossRef]
26. Jiang, Y.; Guo, H.; Jia, Y.; Cao, Y.; Hu, C. Principal Component Analysis and Hierarchical Cluster Analyses of Arsenic Groundwater Geochemistry in the Hetao Basin, Inner Mongolia. *Geochemistry* **2015**, *75*, 197–205. [CrossRef]
27. Alexakis, D.E. Meta-Evaluation of Water Quality Indices. Application into Groundwater Resources. *Water* **2020**, *12*, 1890. [CrossRef]
28. Feng, J.; Sun, H.; He, M.; Gao, Z.; Liu, J.; Wu, X.; An, Y. Quality Assessments of Shallow Groundwaters for Drinking and Irrigation Purposes: Insights from a Case Study (Jinta Basin, Heihe Drainage Area, Northwest China). *Water* **2020**, *12*, 2704. [CrossRef]
29. Dirks, H.; Al Ajmi, H.; Kienast, P.; Rausch, R. Hydrogeology of the Umm Er Radhuma Aquifer (Arabian Peninsula). *Grundwasser* **2018**, *23*, 5–15. [CrossRef]
30. Sanford, W.E.; Wood, W.W. Hydrology of the Coastal Sabkhas of Abu Dhabi, United Arab Emirates. *Hydrogeol. J.* **2001**, *9*, 358–366. [CrossRef]
31. Boelens, R.; Hoogesteger, J.; Swyngedouw, E.; Vos, J.; Wester, P. Hydrosocial Territories: A Political Ecology Perspective. *Water Int.* **2016**, *41*, 1–14. [CrossRef]
32. SheikhyNarany, T.; Ramli, M.F.; Aris, A.Z.; Sulaiman, W.N.A.; Juahir, H.; Fakharian, K. Identification of the Hydrogeochemical Processes in Groundwater Using Classic Integrated Geochemical Methods and Geostatistical Techniques, in Amol-Babol Plain, Iran. *Sci. World J.* **2014**, *2014*, e419058. [CrossRef]
33. Nazzal, Y.; Howari, F.M.; Iqbal, J.; Ahmed, I.; Orm, N.B.; Yousef, A. Investigating Aquifer Vulnerability and Pollution Risk Employing Modified DRASTIC Model and GIS Techniques in Liwa Area, United Arab Emirates. *Groundw. Sustain. Dev.* **2019**, *8*, 567–578. [CrossRef]
34. Cariou, A. Liwa: The Mutation of an Agricultural Oasis into a Strategic Reserve Dedicated to a Secure Water Supply for Abu Dhabi. In *Oases and Globalization. Ruptures and Continuities*; Emilie Lavie, A.M., Ed.; Springer Geography; Springer International Publishing: Cham, Switzerland, 2017.
35. Hoummaidi, L.E.; Larabi, A.; Ahmad Al Shaikh, S. Mode Flow Map: An Innovative Enterprise Gis for Better Groundwater Management and Monitoring. *GIS Bus.* **2010**, *15*, 220–240. Available online: https://www.gisbusiness.org/index.php/gis/article/view/18255 (accessed on 18 August 2021). [CrossRef]
36. Ako, A.A.; Eyong, G.E.T.; Shimada, J.; Koike, K.; Hosono, T.; Ichiyanagi, K.; Richard, A.; Tandia, B.K.; Nkeng, G.E.; Roger, N.N. Nitrate Contamination of Groundwater in Two Areas of the Cameroon Volcanic Line (Banana Plain and Mount Cameroon Area). *Appl. Water Sci.* **2014**, *4*, 99–113. [CrossRef]
37. Kazemi, E.; Karyab, H.; Emamjome, M.-M. Optimization of Interpolation Method for Nitrate Pollution in Groundwater and Assessing Vulnerability with IPNOA and IPNOC Method in Qazvin Plain. *J. Environ. Health Sci. Eng.* **2017**, *15*, 23. [CrossRef]
38. Wackernagel, H. Ordinary Kriging. In *Multivariate Geostatistics: An Introduction with Applications*; Wackernagel, H., Ed.; Springer: Berlin/Heidelberg, Germany, 2003; pp. 79–88. ISBN 978-3-662-05294-5.
39. Joseph, V.R.; Kang, L. Regression-Based Inverse Distance Weighting with Applications to Computer Experiments. *Technometrics* **2011**, *53*, 254–265. [CrossRef]
40. Harahsheh, H.; Mashroom, M.; Marzouqi, Y.; Khatib, E.A.; Rao, B.R.M.; Fyzee, M.A. Soil Thematic Map and Land Capability Classification of Dubai Emirate. In *Developments in Soil Classification, Land Use Planning and Policy Implications: Innovative Thinking of Soil Inventory for Land Use Planning and Management of Land Resources*; Shahid, S.A., Taha, F.K., Abdelfattah, M.A., Eds.; Springer: Dordrecht, The Netherlands, 2013; pp. 133–146. ISBN 978-94-007-5332-7.

41. Arslan, H. Spatial and Temporal Mapping of Groundwater Salinity Using Ordinary Kriging and Indicator Kriging: The Case of Bafra Plain, Turkey. *Agric. Water Manag.* **2012**, *113*, 57–63. [CrossRef]
42. Dash, J.P.; Sarangi, A.; Singh, D.K. Spatial Variability of Groundwater Depth and Quality Parameters in the National Capital Territory of Delhi. *Environ. Manag.* **2010**, *45*, 640–650. [CrossRef] [PubMed]
43. Nazzal, Y.; Zaidi, F.K.; Ahmed, I.; Ghrefat, H.; Naeem, M.; Al-Arifi, N.S.N.; Al-Shaltoni, S.A.; Al-Kahtany, K.M. The Combination of Principal Component Analysis and Geostatistics as a Technique in Assessment of Groundwater Hydrochemistry in Arid Environment. *Curr. Sci.* **2015**, *108*, 1138–1145.
44. Nazzal, Y.; Bărbulescu, A.; Howari, F.; Al-Taani, A.A.; Iqbal, J.; Xavier, C.M.; Sharma, M.; Dumitriu, C.Ş. Assessment of metals concentrations in soils of Abu Dhabi Emirate using pollution indices and multivariate statistics. *Toxics* **2021**, *9*, 95. [CrossRef]
45. American Public Health Association (APHA). *Standard Methods for Examination of Water and Wastewater*; American Public Health Association: Washington, DC, USA, 2005.
46. Tian, R.; Wu, J. Groundwater quality appraisal by improved set pair analysis with game theory weightage and health risk estimation of contaminants for Xuecha drinking water source in a loess area in northwest China. *Hum. Ecol. Risk Assess.* **2019**, *25*, 132–157. [CrossRef]
47. Su, F.; Wu, J.; He, S. Set pair analysis-Markov chain model for groundwater quality assessment and prediction: A case study of Xi'an City, China. *Hum. Ecol. Risk Assess.* **2019**, *25*, 158–175. [CrossRef]
48. Su, F.; Li, P.; He, X.; Elumalai, V. Set pair analysis in earth and environmental sciences: Development, challenges, and future prospects. *Expo. Health* **2020**, *12*, 343–354. [CrossRef]
49. Li, P.; Wu, J.; Qian, H. Groundwater quality assessment based on rough sets attribute reduction and TOPSIS method in a semi-arid area, China. *Environ. Monit. Assess.* **2012**, *184*, 4841–4854. [CrossRef]
50. Li, P.; Qian, H.; Wu, J.; Chen, J. Sensitivity analysis of TOPSIS method in water quality assessment: I. Sensitivity to the parameter weights. *Environ. Monit. Assess.* **2013**, *185*, 2453–2461. [CrossRef]
51. Li, P.; Wu, J.; Qian, H.; Chen, J. Sensitivity analysis of TOPSIS method in water quality assessment II: Sensitivity to the index input data. *Environ. Monit. Assess.* **2013**, *185*, 2463–2474. [CrossRef] [PubMed]
52. Li, P.; He, S.; Yang, N.; Xiang, G. Groundwater quality assessment for domestic and agricultural purposes in Yan'an City, northwest China: Implications to sustainable groundwater quality management on the Loess Plateau. *Environ. Earth Sci.* **2018**, *77*, 775. [CrossRef]
53. Li, P.; Wu, J.; Tian, R.; He, S.; He, X.; Xue, C.; Zhang, K. Geochemistry, hydraulic connectivity and quality appraisal of multilayered groundwater in the Hongdunzi Coal Mine, northwest China. *Mine Water Environ.* **2018**, *37*, 222–237. [CrossRef]
54. Li, P.; He, X.; Guo, W. Spatial groundwater quality and potential health risks due to nitrate ingestion through drinking water: A case study in Yan'an City on the Loess Plateau of northwest China. *Hum. Ecol. Risk Assess.* **2019**, *25*, 11–31. [CrossRef]
55. Wu, J.; Zhou, H.; He, S.; Zhang, Y. Comprehensive understanding of groundwater quality for domestic and agricultural purposes in terms of health risks in a coal mine area of the Ordos basin, north of the Chinese Loess Plateau. *Environ. Earth Sci.* **2019**, *78*, 446. [CrossRef]
56. Wang, D.; Wu, J.; Wang, Y.; Ji, Y. Finding High-Quality Groundwater Resources to Reduce the Hydatidosis Incidence in the Shiqu County of Sichuan Province, China: Analysis, Assessment, and Management. *Expo. Health* **2020**, *12*, 307–322. [CrossRef]
57. Wang, Y.; Li, P. Appraisal of shallow groundwater quality with human health risk assessment in different seasons in rural areas of the Guanzhong Plain (China). *Environ. Res.* **2022**, *207*, 112210. [CrossRef]
58. Brown, R.M.; McClelland, N.I.; Deininger, R.A.; O'Connor, M.F. A Water Quality Index—Crashing the Psychological Barrier. In *Indicators of Environmental Quality*; Thomas, W.A., Ed.; Springer US: Boston, MA, USA, 1972; pp. 173–182.
59. Wu, J.; Li, P.; Wang, D.; Ren, X.; Wei, M. Statistical and multivariate statistical techniques to trace the sources and affecting factors of groundwater pollution in a rapidly growing city on the Chinese Loess Plateau. *Hum. Ecol. Risk Assess.* **2020**, *26*, 1603–1621. [CrossRef]
60. Li, P.; Tian, R.; Liu, R. Solute geochemistry and multivariate analysis of water quality in the Guohua Phosphorite Mine, Guizhou Province, China. *Expo. Health* **2019**, *11*, 81–94. [CrossRef]

Article

Hydrochemical Characteristics of Arsenic in Shallow Groundwater in Various Unconsolided Sediment Aquifers: A Case Study in Hetao Basin in Inner Mongolia, China

Zizhao Cai [1,2], Lingxia Liu [2,3,*], Wei Xu [4], Ping Wu [5] and Chuan Lu [2]

1. School of Water and Environment, Chang'an University, Yanta Road 126, Xi'an 710054, China; caizizhao@mail.cgs.gov.cn
2. The Institute of Hydrogeology and Environmental Geology, Chinese Academy of Geological Sciences, Shijiazhuang 050061, China; luchuancn@163.com
3. School of Environmental Studies, China University of Geosciences, Wuhan 430074, China
4. Hebei Prospecting Institute of Hydrogeology and Engineering Geology, Shijiazhuang 050021, China; xu30874959@163.com
5. Institute of Hydrogeology and Environmental Geology of Ningxia, Yinchuan 750011, China; pushmyluck@163.com
* Correspondence: llingxia2004@163.com

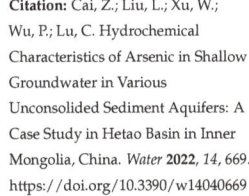

Citation: Cai, Z.; Liu, L.; Xu, W.; Wu, P.; Lu, C. Hydrochemical Characteristics of Arsenic in Shallow Groundwater in Various Unconsolided Sediment Aquifers: A Case Study in Hetao Basin in Inner Mongolia, China. *Water* 2022, *14*, 669. https://doi.org/10.3390/w14040669

Academic Editor: Daniel D. Snow

Received: 28 November 2021
Accepted: 17 February 2022
Published: 21 February 2022

Publisher's Note: MDPI stays neutral with regard to jurisdictional claims in published maps and institutional affiliations.

Copyright: © 2022 by the authors. Licensee MDPI, Basel, Switzerland. This article is an open access article distributed under the terms and conditions of the Creative Commons Attribution (CC BY) license (https://creativecommons.org/licenses/by/4.0/).

Abstract: This study focused on the entire Hetao Basin, which can fall into four hydrogeological units, the Houtao Plain, Sanhuhe Plain, Hubao Plain, and South Bank Plain of the Yellow River, all of which are under different geological and environmental conditions. To systematically investigate the hydrochemical characteristics and spatial distribution of high-As groundwater (As > 10 µg/L), 974 samples were collected from shallow groundwater. As indicated from the results, high-As groundwater had an extensive distribution, and its spatial distribution in the four hydrogeological units exhibited significant variability. Three concentrated distribution areas were reported with high-As groundwater, which were all in the discharge areas of groundwater, and the arsenic contents in the groundwater were found to exceed 50 µg/L. The hydrochemical types of high-As groundwater in the HT Plain and the SHH Plain consisted of $HCO_3 \cdot SO_4 \cdot Cl$ for anions and Na for cations, while those in the other two plains included HCO_3 for anions as well as $Na \cdot Mg \cdot Ca$ for cations. According to the pH values, the groundwater was weakly alkaline in the areas with high-As groundwater, and arsenic primarily existed as arsenite. Furthermore, high-As groundwater in the Hetao Basin was characterized by high contents of Fe (mean value of 2.77 mg/L) and HCO_3^- (mean value of 460 mg/L) and a low relative concentration of SO_4^{2-} (average value of 310 mg/L). This study did not identify any significant correlation between groundwater arsenic and other ions (e.g., Fe^{2+}, Fe^{3+}, HCO_3^-, SO_4^{2-}, NO_2^- and NO_3^-) in the entire Hetao Basin over a wide range of hydrogeological units. The results remained unchanged after the four hydrogeological units were analyzed. The special sedimentary environment evolution of the Hetao Basin was found as the prerequisite for the formation of high-arsenic groundwater. Furthermore, groundwater runoff conditions and hydrogeochemical processes in the basin were indicated as the factors controlling the formation of high-arsenic groundwater.

Keywords: high-As groundwater; shallow groundwater; geological environment; geochemistry; correlation; Hetao Basin

1. Introduction

High As groundwater has been reported as one of the most serious geological environmental issues facing the international community [1,2]. Drinking contaminated water such as high-As groundwater can directly jeopardize human health [3–8], which has aroused great interest from a wide range of organizations in the international community and numerous national government agencies [9,10]. Endemic arsenic poisoning attributed to drinking high-As groundwater has frequently occurred worldwide in places such as India,

Bangladesh, Vietnam and Thailand in Asia, Germany, Chile, Brazil, Argentina, Italy, and The United Kingdom in Europe, and the United States and Canada in America [11–20].

Several natural and anthropogenic sources are responsible for the contamination of As in groundwater. As occurs as a major constituent in more than 200 minerals, and the desorption and dissolution of naturally occurring As-bearing minerals are generally considered the principal source of As contamination in groundwater [21,22]. Arsenopyrite (FeAsS) is the most abundant As-containing mineral that commonly exists in anaerobic environments and can also be found in different concentrations in various rock-forming minerals like sulfide, oxide, phosphate, carbonate and silicate [23]. Moreover, other secondary As minerals (e.g., scorodite, $FeAsO_4\ 2H_2O$) can contain the pollutant as a main or trace component [24,25].

The Hetao Basin of Inner Mongolia in China has been found to be a highly typical area of high-As groundwater [26,27]. Due to drinking high-As groundwater, arsenic poisoning patients were reported in the early 1990s [28]. The As-affected area reached over 3000 km^2. The affected population exceeded 1 million, 400 thousand people drank high-As groundwater (As > 50 μg/L), and over 2000 residents of the 776 villages were confirmed as arsenic patients [29].

On the whole, groundwater As in the Hetao Basin has been considered to occur naturally in Later Pleistocene–Holocene alluvial–lacustrine aquifers (generally ranging from 10 to 50 m) [30–32]. In the typical arsenic poisoning area of the HT Plain, studies on the arsenic morphology and trace element concentration in groundwater have basically clarified the causes of arsenic enrichment in groundwater and its effect on arsenic poisoning [33,34]. Smedley et al. investigated the migration of As and other trace elements in the aquifer of the alluvial plain of Hohhot Basin in Inner Mongolia and highlighted that the high concentrations of As in groundwater were dependent on the strong reduction environment, as reflected in the high content of soluble Fe, Mn, NH_4–N and DOC in groundwater [23]. When initially studying the migration, enrichment and transformation of As in arsenic-affected areas of the Hetao Plain, Lin Nianfeng et al. highlighted that clay soil and humus soil could help enrich As and reduction environments could covert As^{5+} into As^{3+} [35]. Guo Huaming et al. performed an indoor microbial leaching experiment through the in situ collection of sediments and indigenous microorganisms from a high-As aquifer in the Hetao Plain, Inner Mongolia. As indicated by the above studies, the release and transformation processes of arsenic as impacted by indigenous microorganisms consisted of the release of As(V) in sediments, the reduction of As(V) in solution, as well as the release of As(III) in sediments. The reductive dissolution of Fe/Mn oxide minerals under the action of indigenous microorganisms was highlighted as a major cause of arsenic release in sediments [31]. Gao Cunrong et al. explored the distribution and hydrochemical characteristics exhibited by high-As groundwater in the riverfront area of the Hetao Plain and highlighted that the high-As groundwater was largely distributed in the depositional center of the Hetao Plain and in small patches locally. They also found that the arsenic content in groundwater a short distance inland varied significantly. The over-standard rate of iron content in groundwater in the high-As areas significantly exceeded the non-arsenic-rich area, and the formation of high-As groundwater in this area was considered to be highly dependent on the characteristics exhibited by the sedimentary environment and sediments [36,37].

Numerous scholars in and outside China have conducted considerable investigations and research work in the HT Plain and the HB Plain. Most of the relevant studies were local and did not comprehensively assess the regional spatial distribution and formation mechanisms of high-As groundwater. Besides, arsenic in natural water bodies shows significantly high spatial variability [23], and the conclusions of the local studies are generally difficult to verify in the region. However, the formation and evolution mechanisms of high-As groundwater should be explored from the time and space perspectives to select an appropriate typical area or a geological section.

This study focused on the entire Hetao Basin, which fell into four hydrogeological units under the different geological and environmental conditions (i.e., Houtao Plain (HT Plain), Sanhuhe Plain (SHH Plain), Hubao Plain (HB Plain), and South Bank Plain of Yellow River (SBYR Plain)). Considerable groundwater samples have been collected, especially samples from the SHH Plain and SBYR Plain, which have not yet been studied, and the distribution and hydrochemical characteristics exhibited by shallow high-As groundwater have been systematically studied.

2. Study Area

2.1. Geographical Conditions

The Hetao Basin is located in the central and western regions of the Inner Mongolia Autonomous Region (106°07′~112°15′ E, 40°10′~41°27′ N, Figure 1), stretching from the Ulanbuh Desert in the west to the west foot of the Manhan Mountain in the east, as well as from the Yinshan Mountains in the north to the Ordos Plateau in the south, which takes up a total area of nearly 3.2×10^4 km^2. It has been recognized as a vital grain production base in China and the most developed socio-economic area in the Inner Mongolia Autonomous Region, involving Hohhot (the capital city of the Inner Mongolia Autonomous Region) and Baotou (the largest industrial city in the Autonomous Region). Such an area is characterized by an average annual temperature of 6.5–7.8 °C, with an average monthly maximal temperature of 22.4 °C (July) and an average monthly minimum temperature of −12.0 °C (January). On average, the annual precipitation is 245.5 mm, and the annual precipitation of the eastern part exceeds the western part. The evaporation in the eastern part is 2185.2 mm, lower than that of the western part [38,39]. The Yellow River is known as the only perennial river in the study area, flowing from the west to the east via the whole area, and the other large seasonal rivers consist of the Dahei River and the Kundulun River [40].

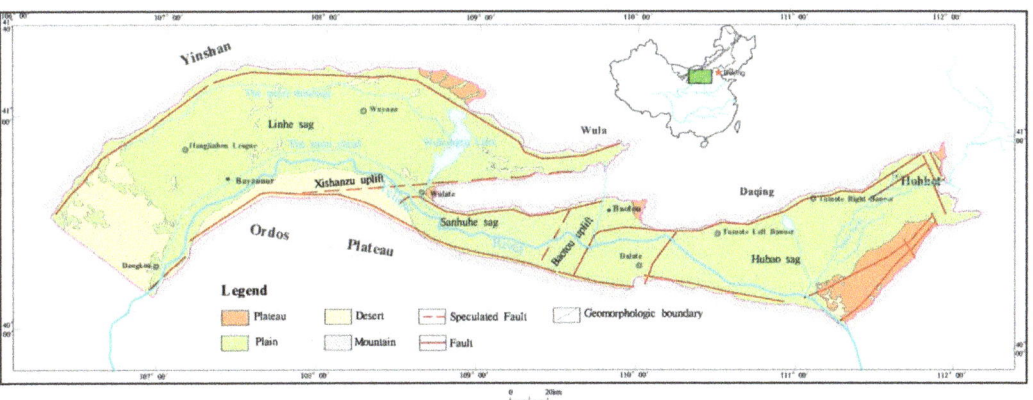

Figure 1. Geologic and geomorphologic map of the Hetao Basin.

2.2. Geological Backgrounds

The Hetao Basin can be divided into four geomorphic zones: mountains, plateaus, plains and desert. Spatially, the Hetao Basin exhibits a geologic structure of "three sags and two uplifts", and its tectonic pattern basically regulates the stratigraphic framework and sedimentary (phase) pattern of the Quaternary sedimentary sequence, as well as the spatial distribution of the sedimentary system [40]. It can fall into the Linhe sag area, Xishanzui uplift area, Sanhuhe sag area, Baotou uplift area, as well as the Hubao sag area (Figure 1).

The Quaternary strata can be split into four rock formations from top to bottom. The first rock formation (Q_{3-4}) is composed of grayish-yellow, light gray alluvial medium-fine sand, silty-fine sand mixed with medium-coarse sand, alluvial silty-fine sand, fine sand, silty clay, and other interbeds exhibiting different thicknesses with local muddy clay intercalation and thicknesses of 10–260 m. Overall, the second rock formation ($Q_2{}^2$) covers black-gray and grayish-brown limnetic facies mud, muddy clay, muddy silty-fine sand and silty clay, with local multilayers of mirabilite. Such a formation is extensively and continuously distributed in the basin and pertains to a landmark formation significantly correlated with the formation of high-As groundwater with thicknesses of 30–170 m [26,27]. The third rock formation ($Q_2{}^1$) comprises brown-yellow, light gray and gray-brown alluvial silty-fine sand, medium-fine sand, medium-coarse sand and thick layers of clay silt, as well as silty clay interbed, with local light grey muddy clay, which is the main confined aquifer with thicknesses of 20–319 m. The fourth rock formation (Q_1) covers the lacustrine facies-based gray, grayish-brown and grayish-green clay silt, silty clay and silty-fine sand, and medium-fine sand interbed. The bottom is sandy gravel, which is not reached [38–40].

2.3. Hydrogeological Conditions

On the whole, 48 hydrogeological profiles were prepared with the collected data of over 1300 boreholes, dynamic groundwater monitoring data of 20 years and the water level data in 4 periods from 2009 to 2010, illustrating the hydrogeological conditions of the study area and laying the conditions for analyzing high-As groundwater.

The shallow aquifer of the HT Plain in the piedmont alluvial-pluvial fan in the north comprises Holocene and Epipleistocene proluvium with gravel and medium-coarse sand containing gravel, 10–130 m thick [23]. The water level varies from 20–40 m at the top of the fan group to 3–5 m at the front edge. The alluvial lacustrine plain of the Yellow River in the south primarily covers Holocene-Epipleistocene alluvial lacustrine facies medium-fine sand, fine sand and silty-fine sand, which is dominated by semi-confined water with local phreatic water [31]. The thickness exhibited by the aquifer increases from 20–80 m in SE to 100–240 m in NW. The burial depth of groundwater level is shallow, generally ranging from 3–10 m. On the whole, the shallow groundwater is recharged by the lateral runoff, the underground flow in the valleys, the atmospheric precipitation, and the infiltration of irrigation with water from the Yellow River in the northern bedrock mountain area. In the northern piedmont fan zone, the groundwater is largely supplied by the north-south runoff, and the runoff conditions deteriorate in the foreland depression. In the alluvial plain of the Yellow River in the south, the groundwater flows as SW-NE runoff. The shallow groundwater forms a drainage zone by complying with the main drainage line and eventually flows as runoff eastward to the Wuliangsuhai Lake [38–40].

The shallow aquifer in the SHH Plain mostly comprises Holocene and Epipleistocene medium-fine sand and silty-fine sand [41,42]. The local river channel covers sandy gravel. From the north to the bank along the Yellow River, the thickness of the aquifer tends to decrease from 30–40 m to 20–30 m, and the groundwater depth declines from 5–10 m to 1–3 m. The groundwater is primarily recharged by the surface runoff, the subsurface flow, atmospheric precipitation and irrigation infiltration while flowing from NW to SE as runoff to the east of Xisanzui. It flows as runoff from the northeast to the southwest to the east of Baotou, which is mostly discharged by mining and lateral discharge to the Yellow River [41,42].

The shallow aquifers in the eastern and western regions in the HB Plain exhibit different characteristics attributed to a wide range of sedimentary environments. The aquifer in the eastern region is the phreatic water-micro-confined aquifer of the piedmont plain of the Dahei River. The compositions vary from proluvium and alluvium gravels, pebbles and sandy gravels gradually to alluvial gravels, medium-coarse sand and fine silty sand to the lake basin. From the west to the east, the groundwater level is altered from nearly 15 m to approximately 5 m, and the thickness of the aquifer is up-regulated from 10 m to 27 m [43]. The aquifer in the western region, composed of Holocene and Epipleistocene medium-fine sand and silty-fine sand, refers to the phreatic water-micro-confined aquifer of the alluvial lacustrine plain of the Yellow River [44]. The thickness of the aquifer varies from 80–100 m in the northwest to 20–40 m in the east. The burial depth of water level is 5–10 m close to the front of the mountain and 2–5 m generally following the

Yellow River and in the eastern region [45]. Laterally, the groundwater is recharged by the bedrock fissure water in some sections, subsurface flow of river valleys and surface runoff, as well as precipitation and irrigation infiltration in the mountainous area. Furthermore, the groundwater flows as runoff from the northeast to the southwest in the eastern region and flows as runoff from the northwest to the southeast and is discharged through mining, evaporation and lateral discharge to a canal for water release of Hasuhai Lake and the Yellow River.

The shallow aquifer of the SBYR Plain covers Holocene-Epipleistocene proluvium and alluvial lacustrine silty-fine sand, fine sand and silty sand [26,27]. The thickness of the aquifer rises from less than 50 m in the south to 50–100 m in the north, and it exceeds 100 m in local areas. In most areas, the burial depth of the groundwater level is less than 5 m, generally 5–15 m in the front of hills. The groundwater is largely recharged by atmospheric precipitation, lateral infiltration in the hilly southern area, infiltration of the valleys, and infiltration of the irrigation canals for water diversion from the Yellow River. The groundwater is mainly recharged from precipitation, runoffs from hills in the south, and water in irrigation channels from the Yellow River. It is discharged by evaporation, manual mining, and lateral seepage to the Yellow River [40–42].

3. Materials and Methods

3.1. Collection of Samples

The samples were collected from the HT Plain and the SHH Plain in September and October 2009 and from the HB Plain and the SBYR Plain from June to August 2010. A total of 974 sets of samples were collected in total from the shallow groundwater, including 450 from the HT Plain, 56 from the SHH Plain, 278 from the HB Plain, as well as 190 from the SBYR Plain. The sampling points were evenly distributed (Figure 2). 40 mL arsenic samples, 250 mL iron samples, 250 mL trace element samples and 1500 mL samples for full analysis were collected from the respective sampling points. After water temperature, electrical conductivity (EC), pH, and Eh were stable, groundwater samples were taken. All samples were membrane filtered (0.45 μm) in the field. The samples used for As speciation analysis were stored in new but pre-rinsed HDPE bottles (Nalgene) after adding 0.25 MEDTA. The filtered samples were then acidified to pH 1 by the addition of ultra-pure HCl for major and trace element analysis. Protective agents were also added to other samples, i.e., 2.5 mL of 1:1 sulfuric acid solution and 0.5 g of ammonium sulfate to iron samples, as well as 2.5 mL of 1:1 nitric acid solution to trace element samples. Except for the samples intended for full analyses, the other samples were stored in a freezer with the temperature regulated at 4 °C and analyzed within 5 days.

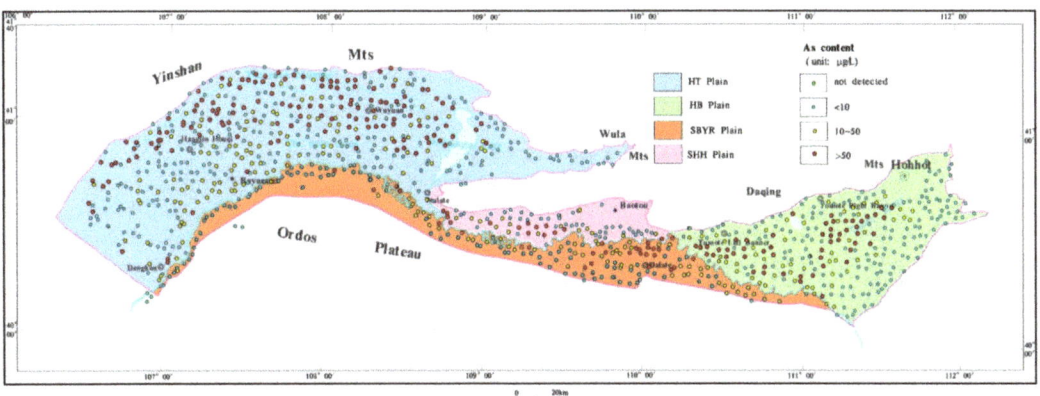

Figure 2. Arsenic content distribution of groundwater samples in the Hetao Basin.

3.2. Analytical Method

The hydrochemical indexes of high-As groundwater (pH value, conductivity, dissolved oxygen and oxidation-reduction potential) were determined in the field by employing American Hach sensION2 and Hach LDO TM HQ 10 portable testers. The NO_3^- and SiO_2 of groundwater were determined by adopting the UV8500 UV-vis spectrophotometer. Cations (e.g., Ca, Mg, Na, K, Fe, Mn, Sr, B and Ba) were determined with PHS-2C ICP-AES. Fe elements were examined in Fe^{2+} and Fe^{3+}. Se was determined under an XGY-1012 atomic fluorescence spectrometer. As^{3+} and As^{5+} were tested with XGY-1012 atomic fluorescence spectrometer, respectively. Other elements were analyzed by using the standard methods. To ensure the quality of the test data of groundwater samples, 5% repeated samples were added, and the error of all repeated samples was less than 5%. The samples were tested by the Key Laboratory of Groundwater Mineral Water and Environmental Supervising and Testing Center, Ministry of Land and Resources of the PRC.

4. Results

4.1. Content and Regional Distribution of Arsenic in Groundwater

The high-As samples of shallow groundwater took up 47.33% of the total samples, and the average concentration of arsenic in the groundwater reached 45.58 µg/L. Only 18 samples had lower contents than the detection limit (<0.1 µg/L), and the detection rate reached 98.15% (Table 1). According to the above table, the high-As groundwater in the Hetao Basin was widely distributed. The high-As groundwater was extensively distributed in the four hydrogeological units, and it achieved the widest distribution in the HT Plain. The high-As groundwater samples accounted for 53.78% of the 450 sample sets. The maximal arsenic content was 916.70 µg/L, and the arsenic content in groundwater was 64.13 µg/L on average, which was the highest in the whole region. Only three groundwater samples contained arsenic contents lower than the detection limit (<0.1 µg/L), and the detection rate took up 99.33%. The SHH Plain was the minimum. Though there were only 56 sets of samples, the detection rate was 100%. The average arsenic content reached 40.64 µg/L, only smaller than the HT Plain. The average arsenic content in the groundwater of the SBYR Plain was the minimum (22.51 µg/L) among the four hydrogeological units. Furthermore, the number of high-As groundwater samples in the HB Plain accounted for 38.49% of the total number of samples, which was the minimum in the whole region.

Table 1. Statistical table of arsenic content in shallow groundwater of the Hetao Basin.

	Area (km^2)	Number	0.1 µg/L (%)	>10 µg/L (%)	Mean (µg/L)	Medium (µg/L)	Max (µg/L)
HT Plain	13,880	450	99.33	53.78	64.13	14.44	916.7
SHH Plain	1623	56	100	51.79	40.64	12.48	615.4
HB Plain	11,382	278	95.32	38.49	32.31	3.76	398.9
SBYR Plain	4995	190	99.47	43.68	22.51	6.31	238.3
Hetao Baisn	31,880	974	98.15	47.33	45.58	7.95	916.7

As revealed by the distribution of high-As groundwater points (Figure 1), high-As groundwater in the Hetao Basin was distributed throughout the region with an extremely uneven spatial distribution. The variability of arsenic contents in groundwater was analyzed by using SPSS (IBM, Armonk, NY, USA). The spatial variation coefficient of arsenic content was 2.07, and the variation coefficient of arsenic content in groundwater of the four hydrogeological units ranged from 1.57 to 2.18, which demonstrated that the spatial variability of arsenic content in groundwater was significantly large in the whole basin and the four hydrogeological units. There were three relatively concentrated distribution areas of high-As groundwater in the whole basin (Figure 3), with arsenic contents of underground water usually larger than 50 µg/L. The mentioned areas consisted of (1) an intersection zone of the front edge of the piedmont alluvial-pluvial fan of the HT Plain and the alluvial lacustrine plain of the Yellow River, (2) both sides of the modern river channel of the Yellow

River from the HT Plain to the SHH Plain and the HB Plain, as well as (3) the central region of the HB Plain, located in the middle east of Tumd Left Banner.

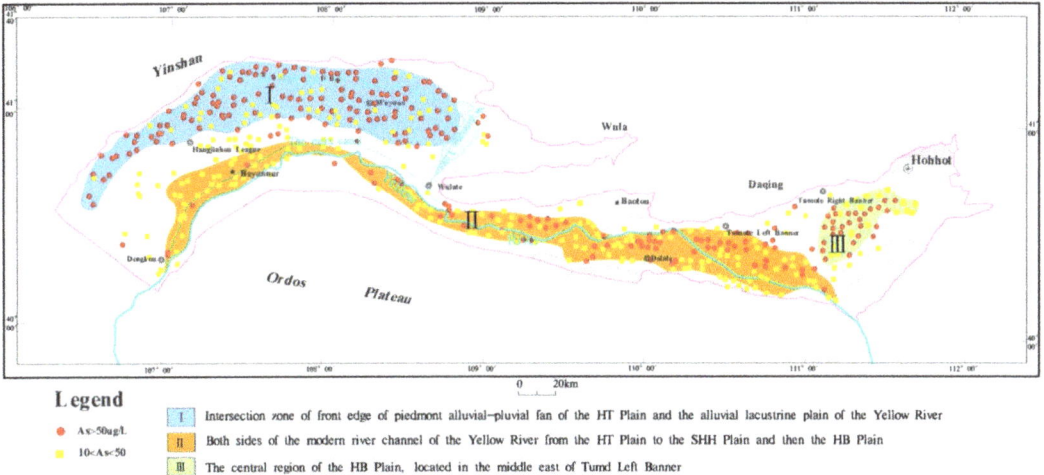

Figure 3. Contour map of high-arsenic shallow groundwater points and groundwater level elevation in Hetao Basin.

4.2. Hydrochemical Characteristics of Shallow High As Groundwater

In this part, 462 sets of shallow high-As groundwater samples are discussed, including 242 from the HT Plain, 29 from the SHH plain, 107 from the HB Plain, and 87 from the SBYR Plain. The groundwater was weakly alkaline, achieving a pH value of 7.15–9.26. The pH value of the HT plain approached that of the SBYR Plain, with levels of 7.15–9.26 and 7.37–9.25, respectively. The pH value of the SHH Plain approached that of the HB Plain, with levels of 7.44–8.65 and 7.26–8.76, respectively.

4.2.1. Hydrochemical Types

According to the Piper classification method [40], there were 13 hydrochemical types in the SHH Plain and 19 types in each of the other three hydrogeological units (Figure 4). However, they showed different hydrochemical characteristics. The major hydrochemical type of high-As groundwater in the HT Plain and the SHH Plain referred to the $HCO_3 \cdot SO_4 \cdot Cl$ type, followed by the $HCO_3 \cdot Cl$ type for anions, and mainly Na·Mg·Ca and Na·Mg for cations. The hydrochemical type of high-As groundwater in the HB Plain was largely the HCO_3 type, followed by the $HCO_3 \cdot SO_4 \cdot Cl$ type and $HCO_3 \cdot Cl$ for anions, and mainly Na and Na·Mg·Ca for cations. The hydrochemical type of high-As groundwater in the SBYR Plain is mainly the HCO_3 type, the $HCO_3 \cdot SO_4 \cdot Cl$ type, the $HCO_3 \cdot SO_4$ type, and the $HCO_3 \cdot Cl$ type, all of which are approximately equally for anions, as well as Na, Na·Mg, and Na·Mg·Ca for cations.

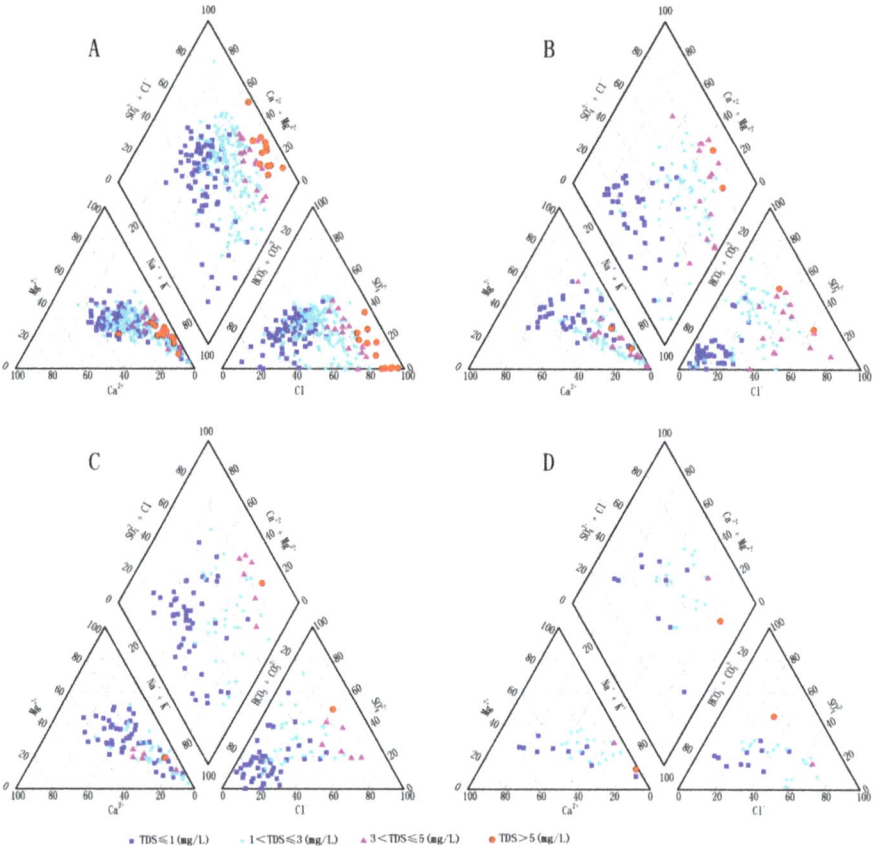

Figure 4. Piper diagram of high-arsenic shallow groundwater quality in four hydrogeological units. (**A**) HT Plain, (**B**) HB Plain, (**C**) SBYR Plain, and (**D**) SHH Plain.

4.2.2. Main Ionic Components

Given the statistical table of characteristic values of As, As(III), As(V), Fe, Fe^{2+}, Fe^{3+}, HCO_3^- and SO_4^{2-} in shallow high-As groundwater in the Hetao Basin (Table 2), the average content of As in the shallow groundwater with high-As in the Hetao Basin reached 93.46 μg/L, and the maximal content in the HT Plain was reported to be 116.93 μg/L, nearly 2.5 times of that in the SBYR Plain; these values for the SHH Plain and the HB Plain were similar, reaching 76.20 μg/L and 80.65 μg/L, respectively. In addition, As(III) was reported to be the major form of As, and the average concentration of As took up 85.2% of the total As. The minimal value was 76.1% in the SHH Plain, and the values in the other three hydrogeological units were nearly unchanged, ranging from 84.0% to 86.1%, with the maximal value in the HT Plain. The ratios of As(III) to As(V) in the 4 plains are all very large, reaching 19.27 in the highest As area in the SHH Plain (Table 3), indicating that the groundwater in the entire Hetao Basin is basically in a reducing environment. Impacted by the toxicity of As(III) and such a high content of As(III) in the shallow high-As groundwater in the Hetao Basin, it is not difficult to understand that there are large-scale arsenic poisoning areas in the Hetao Basin and the safety of the drinking water of over one million people is affected [28,29].

Table 2. Statistical table of characteristic values of the main ions in high-As shallow groundwater in the Hetao Basin.

Item	As (µg/L)			As(III) (µg/L)			As(V) (µg/L)			HCO_3^- (mg/L)		
	Mean	Median	Max	Mean	Median	Max	Mean	Median	Max	Mean	Median	Max
HT Plain	116.93	58.21	916.70	100.69	50.99	719.40	16.24	6.95	224.60	521.34	480.70	1327.00
SHH Plain	76.20	30.17	615.40	58.01	26.81	376.80	18.19	6.66	238.60	548.77	524.70	1347.00
HB Plain	80.65	48.16	398.90	68.74	41.02	383.70	11.92	7.03	120.10	704.32	620.40	2123.00
SBYR Plain	47.60	35.41	238.30	40.02	32.29	227.10	7.57	4.98	39.10	547.50	504.10	2220.00
Hetao Baisn	93.46	49.64	916.70	79.67	41.02	719.40	13.80	6.52	238.60	570.25	505.00	2220.00

Item	Fe (mg/L)			Fe^{2+} (mg/L)			Fe^{3+} (mg/L)			SO_4^{2-} (mg/L)		
	Mean	Median	Max	Mean	Median	Max	Mean	Median	Max	Mean	Median	Max
HT Plain	2.63	1.60	17.00	1.26	0.75	11.80	1.37	0.53	16.00	306.27	235.35	2494.00
SHH Plain	5.84	2.40	40.00	2.59	1.20	30.00	3.25	0.80	24.00	291.90	227.90	2004.00
HB Plain	2.60	1.32	25.00	1.17	0.48	17.00	1.43	0.52	22.00	360.65	263.20	2419.00
SBYR Plain	2.33	0.88	34.00	1.02	0.28	17.00	1.32	0.52	17.00	264.77	172.20	2544.00
Hetao Baisn	2.77	1.50	40.00	1.28	0.60	30.00	1.49	0.54	24.00	310.51	225.60	2544.00

Table 3. As(III)/As(IV) ratios in shallow groundwater from Hetao Basin.

Item	As < 10 µg/L			As ≥ 10 µg/L		
	As(III) (µg/L)	As(V) (µg/L)	As(III)/As(V)	As(III) (µg/L)	As(V) (µg/L)	As(III)/As(V)
HT Plain	1.74	0.94	1.85	100.69	16.24	6.20
SHH Plain	1.61	0.84	1.91	60.93	19.27	3.16
HB Plain	1.2	0.56	2.14	71.89	12.71	5.66
SBYR plain	2.12	0.97	2.19	39.35	7.38	5.33

The content of total Fe in the Hetao Basin was higher, with an average concentration of 2.77 mg/L, and the maximum value of 5.84 mg/L was identified in the SHH Plain. The average values of the other 3 hydrogeological units between were basically the same, at 2.33–2.63 mg/L (Figure 5). The average concentration of Fe^{2+} was 1.28 mg/L, accounting for 46% of the total Fe. The average concentration of Fe^{3+} was 1.49 mg/L, and the ratio of the average concentration of Fe^{2+} approached 1:1. The ratio of the average concentrations of Fe^{2+}/Fe^{3+} in the 4 hydrogeological units approached 1:1 across the entire Hetao Basin.

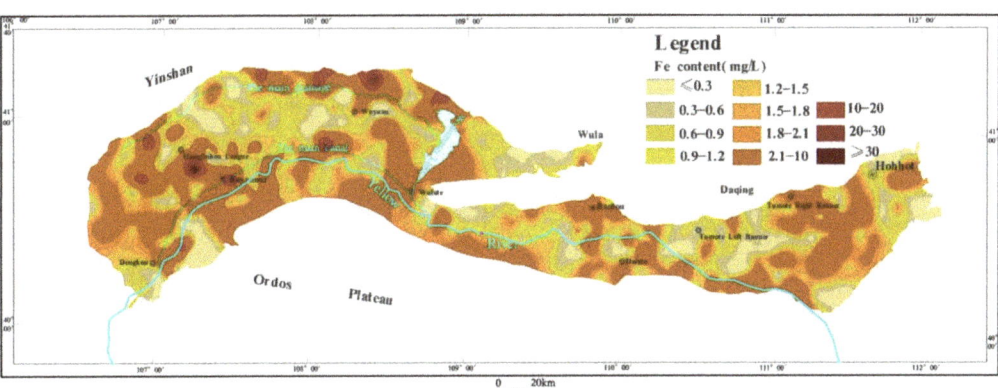

Figure 5. Contour map of total Fe content in Hetao Basin.

The average concentration of SO_4^{2-} in the Hetao Basin reached 310 mg/L, and the relative concentration was relatively low. The maximal value was 2455 mg/L in the SBYR Plain.

The average concentration of SO_4^{2-} in the 4 hydrogeological units was 264~360 mg/L. Given the SO_4^{2-} ion concentration partition of shallow high-As groundwater (Figure 6), the distribution area with concentrations of less than 100 mg/L was 4953 km^2, and the distribution area with concentrations of 100~200 mg/L was 4493 km^2. There was almost no SO_4^{2-} in the groundwater in the north of the HT Plain and in the eastern part of the HB Plain, which demonstrated the effect of desulfurization.

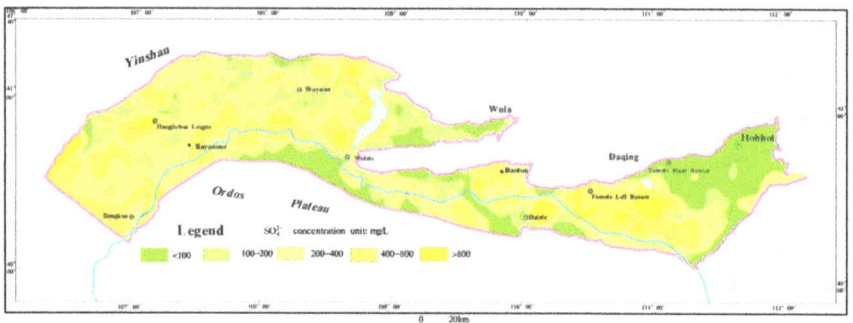

Figure 6. Zoning map of SO_4^{2-} concentration in Hetao Basin.

The average concentration of HCO_3^- of the Hetao Basin was 570 mg/L, and the relative concentration was relatively high. In four hydrogeological units, the average concentration of HCO_3^- of the HB Plain reached 704 mg/L, and the values of the other 3 hydrogeological units were relatively close, at 521~547 mg/L. The HCO_3^- concentration in the HB Plain tended to rise from the pre-piedmont recharge area to the central part of the plain (Figure 7). The highest concentration reached 2123 mg/L, which demonstrated the effect of organic carbon oxidation.

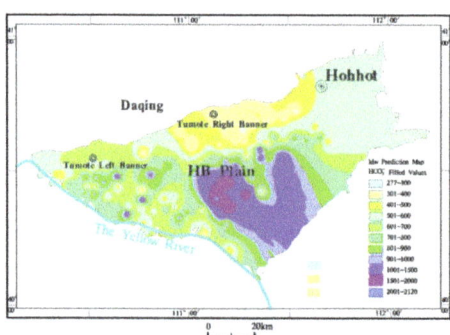

Figure 7. Contour map of HB Plain HCO_3^- concentration.

5. Discussion

5.1. The Hydrogeochemical Process of As Mobilization in Aquifers

Most of the potential sources of arsenic in the shallow groundwater of the Hetao Plain are believed to be from arsenic-bearing Quaternary strata derived from local aquifer sediments [31,32,46]. Under natural conditions, where the water-rock interaction is strong and the geochemical environment of certain aquifers is suitable for arsenic migration and accumulation, the aquifers often have higher arsenic concentrations. The migration of arsenic in the groundwater of the Hetao Basin occurs in a strong reducing environment with rich organic matter. The anaerobic environment where the surface lacustrine clay deposits is located is particularly conducive to the formation of As(III), causing As(III) to be the dominant valence state in the high-As groundwater in the study area. pH is an important factor affecting arsenic migration in aqueous systems. In the area of high-As groundwater in the Hetao Plain, the pH is also relatively high. Arsenic exists in the groundwater mainly in the form of two valence states as As(V) or As(III) anions, so it is more likely to be absorbed by positively charged substances in an aqueous medium, such as iron and manganese oxides, goethite, gibbsite and ferrihydrite. The increase of pH will reduce the adsorption of colloid and clay minerals to arsenate or arsenite in the form of anions and enhance their migration performance [47].

Numerous studies show that the finer the soil particles are, the greater amounts the arsenic adsorption are, and the higher the arsenic content is [48,49]. The silty-fine sand layer or the silty-fine sand interbeds with clay and silty clay are widely spread in the research area as identified by drilling data, and the organic matter content of the sediment is relatively high. Moreover, as the depth increases, the sediment particles become finer, and the arsenic content tends to increase [31]. The lacustrine silty clay and sandy clay in the sediment have a strong adsorption capacity for arsenic. When the arsenic element enters a depression with an alluvial or groundwater flow system, part of the arsenic is adsorbed by the clay and deposited directly, and part of the arsenic interacts with the ferric hydroxide colloid in the river or lake water to form insoluble precipitates. At the same time, the study area is rich in Fe^{2+} and Fe^{3+}, and iron ions have a strong ability to fix arsenic. Once groundwater rich in iron and arsenic ions flow to the central area of the plain, the slow flow of groundwater, caused by the low hydraulic gradient, poor aquifer permeability and small recharge, will limit aquifer flushing, solute (As) transport and its removal from the system. In areas with slow groundwater movement, aqueous As concentrations are very sensitive to releases of small amounts of arsenic from the various hydrogeochemical processes described above. These reasons also explain the high-As distribution in the three main regions.

5.2. Correlation Analysis between Arsenic and Main Oxidation-Reducing Ions

Pearson correlation analysis was applied to reveal the relationship between arsenic and the main oxidation-reducing ions in this study [50–54]. As revealed by the results of the Pearson correlation analysis of arsenic with major ions (e.g., As, Fe, Fe^{2+}, Fe^{3+}, HCO_3^-, SO_4^{2-}, NO_2^- and NO_3^-) for all shallow groundwater samples ($n = 974$) (significance level at $p < 0.01$, Table 4), no significant correlation was reported between arsenic and the ions, and the correlation coefficient was primarily less than 0.4. As indicated by the results of the separate correlation analysis regarding the groundwater samples of the four hydrogeological units (Tables 5–8), an insignificant correlation was reported between arsenic and other ions; only in the SHH Plain was a correlation coefficient of 0.6 found between arsenic and Fe, and the correlation coefficient between arsenic and NO_2^- was 0.5, which was more significant. No significant correlation was identified between the arsenic and the iron contents in the groundwater, probably due to the formation of insoluble iron sulfides by Fe^{2+} from the reduction of iron oxides and by S^{2-} from the reduction of SO_4^{2-}. Iron sulfides have been extensively distributed in many high-As aquifers worldwide [55]. No significant correlation was identified between arsenic and HCO_3^- concentration in the groundwater, whereas high-As groundwater samples generally contained HCO_3^- in high concentrations. In this study area, the average concentration of HCO_3^- was relatively high, up to 570 mg/L, which might also be related to the process of microbial activity. Microorganisms can reduce iron's release of arsenic and oxidize the organic matter in the aquifer to produce a large amount of HCO_3^-. In addition, it might be related to high concentrations of HCO_3^- and the competitive adsorption of arsenate and arsenite on the surface of iron oxide.

Table 4. Pearson correlation analysis between As and the main redox ions in the Hetao Plain ($n = 974$).

	As	Fe	As(III)	As(V)	Fe^{2+}	Fe^{3+}	HCO_3^-	NO_2^-	NO_3^-	SO_4^{2-}
As	1	0.242 **	0.992 **	0.797 **	0.170 **	0.231 **	0.177 **	−0.007	−0.093 **	−0.044
Fe	0.242 **	1	0.213 **	0.315 **	0.812 **	0.843 **	0.139 **	−0.013	−0.080 *	0.171 **
As(III)	0.992 **	0.213 **	1	0.712 **	0.167 **	0.188 **	0.171 **	−0.010	−0.095 **	−0.050
As(V)	0.797 **	0.315 **	0.712 **	1	0.140 **	0.375 **	0.159 **	0.007	−0.062	−0.004
Fe **	0.170 **	0.812 **	0.167 **	0.140 **	1	0.378 **	0.054	−0.011	−0.093 **	0.105 **
Fe **	0.231 **	0.843 **	0.188 **	0.375 **	0.378 **	1	0.175 **	−0.010	−0.042	0.178 **
HCO_3 *	0.177 **	0.139 **	0.171 **	0.159 **	0.054	0.175 **	1	0.068 *	0.081 *	0.382 **
NO_2 *	−0.007	−0.013	−0.010	0.007	−0.011	−0.010	0.068 *	1	0.270 **	0.011
NO_3 *	−0.093 **	−0.080 *	−0.095 **	−0.062	−0.093 **	−0.042	0.081 *	0.270 **	1	0.155 **
SO_4 **	−0.044	0.171 **	−0.050	−0.004	0.105 **	0.178 **	0.382 **	0.011	0.155 **	1

** Significantly correlated at the level of 0.01 (bilateral); * Significantly correlated at the level of 0.05 (bilateral).

Table 5. Pearson correlation analysis between As and the main redox ions in the Houtao Plain ($n = 190$).

	As	Fe	As(III)	As(V)	Fe **	Fe **	HCO_3 *	NO_2 *	NO_3 *	SO_4 **
As	1	0.182 **	0.993 **	0.797 **	0.115 *	0.181 **	0.212 **	0.137 **	0.056	−0.124 **
Fe	0.182 **	1	0.161 **	0.243 **	0.821 **	0.796 **	0.166 **	0.025	−0.020	0.240 **
As(III)	0.993 **	0.161 **	1	0.720 **	0.121 *	0.141 **	0.210 **	0.144 **	0.057	−0.137 **
As(V)	0.797 **	0.243 **	0.720 **	1	0.057	0.345 **	0.174 **	0.066	0.040	−0.027
Fe **	0.115 *	0.821 **	0.121 *	0.057	1	0.308 **	0.030	0.045	−0.024	0.191 **
Fe **	0.181 **	0.796 **	0.141 **	0.345 **	0.308 **	1	0.246 **	−0.007	−0.008	0.198 **
HCO_3 *	0.212 **	0.166 **	0.210 **	0.174 **	0.030	0.246 **	1	0.125 **	−0.041	0.283 **
NO_2 *	0.137 **	0.025	0.144 **	0.066	0.045	−0.007	0.125 **	1	0.042	0.092
NO_3 *	0.056	−0.020	0.057	0.040	−0.024	−0.008	−0.041	0.042	1	0.189 **
SO_4 **	−0.124 **	0.240 **	−0.137 **	−0.027	0.191 **	0.198 **	0.283 **	0.092	0.189 **	1

** Significantly correlated at the level of 0.01 (bilateral); * Significantly correlated at the level of 0.05 (bilateral).

Table 6. Pearson correlation analysis between As and the main redox ions in the Sanhuhe Plain ($n = 190$).

	As	Fe	As(III)	As(V)	Fe **	Fe **	HCO_3 *	NO_2 *	NO_3 *	SO_4 **
As	1	0.574 **	0.988 **	0.958 **	0.348 **	0.629 **	0.368 **	0.476 **	−0.167	−0.043
Fe	0.574 **	1	0.560 **	0.563 **	0.826 **	0.849 **	0.117	0.174	−0.218	−0.073
As(III)	0.988 **	0.560 **	1	0.901 **	0.359 **	0.597 **	0.378 **	0.495 **	−0.203	−0.040
As(V)	0.958 **	0.563 **	0.901 **	1	0.305 *	0.648 **	0.327 *	0.411 **	−0.091	−0.044
Fe **	0.348 **	0.826 **	0.359 **	0.305 *	1	0.433 **	0.096	0.131	−0.182	−0.062
Fe **	0.629 **	0.849 **	0.597 **	0.648 **	0.433 **	1	0.131	0.180	−0.170	−0.040
HCO_3 *	0.368 **	0.117	0.378 **	0.327 *	0.096	0.131	1	0.489 **	0.020	0.687 **
NO_2 *	0.476 **	0.174	0.495 **	0.411 **	0.131	0.180	0.489 **	1	−0.141	0.210
NO_3 *	−0.167	−0.218	−0.203	−0.091	−0.182	−0.170	0.020	−0.141	1	0.406 **
SO_4 **	−0.043	−0.073	−0.040	−0.044	−0.062	−0.040	0.687 **	0.210	0.406 **	1

** Significantly correlated at the level of 0.01 (bilateral); * Significantly correlated at the level of 0.05 (bilateral).

Table 7. Pearson correlation analysis between As and the main redox ions in the Hubao Plain ($n = 278$).

	As	Fe	As(III)	As(V)	Fe **	Fe **	HCO$_3$ *	NO$_2$ *	NO$_3$ *	SO$_4$ **
As	1	0.288 **	0.993 **	0.810 **	0.237 **	0.229 **	0.217 **	−0.045	−0.175 **	0.020
Fe	0.288 **	1	0.285 **	0.238 **	0.704 **	0.876 **	0.182 **	−0.044	−0.087	0.277 **
As(III)	0.993 **	0.285 **	1	0.733 **	0.242 **	0.221 **	0.208 **	−0.049	−0.181 **	0.018
As(V)	0.810 **	0.238 **	0.733 **	1	0.157 **	0.216 **	0.214 **	−0.018	−0.107	0.027
Fe **	0.237 **	0.704 **	0.242 **	0.157 **	1	0.275 **	0.119 *	−0.043	−0.145 *	0.156 **
Fe **	0.229 **	0.876 **	0.221 **	0.216 **	0.275 **	1	0.167 **	−0.030	−0.019	0.269 **
HCO$_3$ *	0.217 **	0.182 **	0.208 **	0.214 **	0.119 *	0.167 **	1	0.056	0.101	0.417 **
NO$_2$ *	−0.045	−0.044	−0.049	−0.018	−0.043	−0.030	0.056	1	0.298 **	0.013
NO$_3$ *	−0.175 **	−0.087	−0.181 **	−0.107	−0.145 *	−0.019	0.101	0.298 **	1	0.278 **
SO$_4$ **	0.020	0.277 **	0.018	0.027	0.156 **	0.269 **	0.417 **	0.013	0.278 **	1

** Significantly correlated at the level of 0.01 (bilateral); * Significantly correlated at the level of 0.05 (bilateral).

Table 8. Pearson correlation analysis between As and the main redox ions in the Southern Plain of the Yellow River ($n = 190$).

	As	Fe	As(III)	As(V)	Fe **	Fe **	HCO$_3$ *	NO$_2$ *	NO$_3$ *	SO$_4$ **
As	1	0.320 **	0.993 **	0.823 **	0.315 **	0.283 **	0.231 **	−0.074	−0.175 *	0.006
Fe	0.320 **	1	0.303 **	0.334 **	0.903 **	0.945 **	0.186 *	−0.046	−0.096	0.195 **
As(III)	0.993 **	0.303 **	1	0.750 **	0.321 **	0.251 **	0.227 **	−0.073	−0.176 *	−0.004
As(V)	0.823 **	0.334 **	0.750 **	1	0.223 **	0.376 **	0.201 **	−0.063	−0.133	0.050
Fe **	0.315 **	0.903 **	0.321 **	0.223 **	1	0.713 **	0.118	−0.034	−0.118	0.099
Fe **	0.283 **	0.945 **	0.251 **	0.376 **	0.713 **	1	0.214 **	−0.049	−0.067	0.243 **
HCO$_3$ *	0.231 **	0.186 *	0.227 **	0.201 **	0.118	0.214 **	1	0.043	−0.011	0.491 **
NO$_2$ *	−0.074	−0.046	−0.073	−0.063	−0.034	−0.049	0.043	1	0.075	0.036
NO$_3$ *	−0.175 *	−0.096	−0.176 *	−0.133	−0.118	−0.067	−0.011	0.075	1	0.072
SO$_4$ **	0.006	0.195 **	−0.004	0.050	0.099	0.243 **	0.491 **	0.036	0.072	1

** Significantly correlated at the level of 0.01 (bilateral); * Significantly correlated at the level of 0.05 (bilateral).

5.3. Correlation between Formation of High As Groundwater and Geological Environment

Large-scale high-As groundwater deposits tend to be found in two types of environments, i.e., inland or closed basins in arid or semi-arid regions and aquifers derived from alluvium under strong reducing conditions [23]. Both environments tend to contain geologically young sediments and to be in flat, low-lying areas where groundwater flow is sluggish. The Hetao Basin, which is located in the west of Inner Mongolia, pertains to the typical temperate continental arid and semi-arid climate. The annual precipitation is 245.5 mm, and the average water evaporation is 2100 mm [38,39], which is 9 times the average rainfall.

The Hetao Basin is a Mesozoic-Cenozoic rifted basin formed in the late Jurassic, located between the Yinshan uplift and the Ordos platform. The tectonic movement of the Hetao Basin is very active, and tectonic faults have developed since the Cenozoic. The tectonic systems are dominated by high-angle normal faults and fault-bending belts. The deep east-west fault formed in the piedmont of Yinshan Mountain made the Ordos block continue to squeeze to the northwest, and the Hetao Basin between these two began to sink into depression. Since the Cenozoic, the Ordos block has been pushed from the northwest direction, and the squeeze turned into a southeastward pulling, which caused the Hetao Basin to sink significantly, forming a tectonic pattern termed as "three sags and two uplifts", in which the basin and mountains intersect each other and the depression and uplift are adjacent to each other (Figure 1). The fault depressions are favorable places for arsenic enrichment. Strong evaporation and weathering accelerate the decomposition rate and biogeochemical cycle of minerals in the bedrock weathering zone and promote the migration and enrichment of arsenic in water. Moreover, in the depression, the underground flow is stagnant, forming a structural water storage area and a gathering place for various elements. The enclosed tectonic conditions and long-term inheritance and subsidence of the Hetao Basin has resulted in the formation of fine-grained clastic sediments dominated by inland lacustrine facies. The thickness of the Quaternary sedimentary reached 200–1500 m. Center areas of the low-lying lake basins located in piedmonts contain fine-grained sediments and high amounts of organic matter, and these reducing environments create favorable conditions for the release of arsenic from the sediments and the accumulation of arsenic in the water. The coverage of fine-grained sediments isolates the air exchange of oxygen between the aquifer and the surface, forming a closed reducing environment, prompting the release and migration of arsenic.

Mineral-water interactions can make arsenic enter groundwater from the sediment adsorption phase. However, arsenic in groundwater must accumulate to a certain concentration before it can form high-As groundwater [23]. Obviously, in the middle of a low-lying basin, especially in areas with a large thickness of lacustrine sediments, slow groundwater runoff and evaporation as the main form of discharge, it is particularly easy to develop high-As groundwater [56]. There are three relatively concentrated areas of high-As groundwater points in the Hetao Basin, located in the discharge area of groundwater according to the contours of the groundwater table (Figure 8), where there is a small groundwater depth and a hydraulic gradient of less than 0.8‰, and the groundwater flow is slow [55]. As the mentioned three areas are situated in the pluvial and alluvial-lacustrine interlace lowland and interfluvial lowland, characterized by poor water flow, strong evaporation and widespread saline-alkali land, the sediments are saturated in long periods. These factors enhance the release of arsenic from sediments into the water, causing its high concentration.

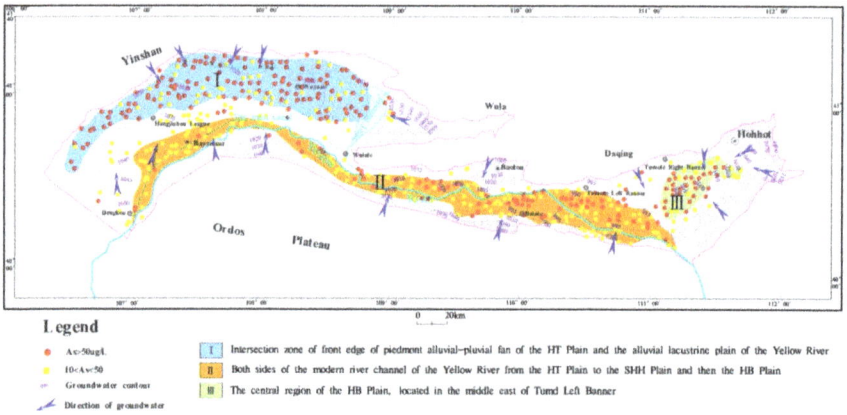

Figure 8. Contour map of groundwater table in Hetao Basin.

Due to the low rainfall and large evaporation, agricultural production in the Hetao Basin is primarily supported by the water from the Yellow River, and the introduction of considerable Yellow River water up-regulates the groundwater water level and forms a large area of soil salinization. Furthermore, the rising water level makes air unable to enter the reduction environment formed by the stratum. Besides, the pH value of the surface water in the arid and semi-arid climate environment is basically high, thereby creating favorable conditions for the dissolution of arsenic from the stratum.

6. Conclusions

Arsenic has been extensively distributed in the shallow groundwater of the Hetao Basin, and high-As groundwater was identified in the four hydrogeological units. In particular, the average concentration of arsenic in the shallow groundwater in the Hetao Basin reached 45.58 μg/L. The spatial distribution of arsenic content in the shallow groundwater was significantly uneven in the whole basin, and the variation coefficient ranged from 2.07 to 1.57–2.18, with high variability.

All high-As groundwater in the HT Basin was weakly alkaline, achieving a pH value of 7.15–9.26. As(III) was found to be the major form of As, and the average concentration of As took up 85% of the total As. The content of total Fe was higher, with a concentration of 2.77 mg/L on average, and the average concentration of Fe^{2+} took up 46% of the total Fe. The average concentration of SO_4^{2-} was 310 mg, and the relative concentration was lower. It is noteworthy that there was nearly no SO_4^{2-} in the groundwater in the north of the HT Plain and in the eastern part of the HB Plain, which demonstrated the effect of desulfurization. The concentration of HCO_3^- was 460 mg/L on average, and the relative concentration was higher. The HCO_3^- concentration in the HB Plain tended to rise from the pre-piedmont recharge area to the central part of the plain, and the maximal concentration was 2123 mg/L, which demonstrated the effect of organic carbon oxidation.

There were three concentrated areas with high-As groundwater in the Hetao Basin where the arsenic contents of groundwater were more than 50 µg/L. According to the analysis of hydrogeological conditions, three concentrated distribution areas with high-As groundwater were all in the discharge areas of groundwater with a shallow groundwater level, a water gradient of less than 0.8‰ and slow groundwater flow.

The distribution of high-As groundwater was found to be dependent on specific geological and geographical backgrounds. The special sedimentary environment evolution of the Hetao Basin lays the prerequisite for creating high-As groundwater, and the groundwater runoff condition and hydrogeochemical process of the basin can control the formation of high-As groundwater. The high pH and organic-rich reducing environment in the Hetao Basin, the dissolution, reduction, and precipitation of various minerals and the mutual transformation of different forms of arsenic contribute to the origin of the high-arsenic groundwater. The long-term lacustrine-dominated paleogeographic environment and closed tectonic conditions, slow refreshment and stagnant groundwater, arid-semi-arid climatic conditions, as well as sediments with high arsenic content and regional arsenic-rich environments constitute the feasible conditions for high-arsenic groundwater.

Author Contributions: L.L. and Z.C. designed the idea and structure of the article, W.X. analyzed the data, P.W. drew the graph, and Z.C., L.L. and C.L. wrote the paper. All authors have read and agreed to the published version of the manuscript.

Funding: This research was funded by the National Natural Science Foundation of China (No.41672241) and the Geological Survey Project of China Geological Survey (NO.DD20190433).

Institutional Review Board Statement: Not applicable.

Informed Consent Statement: Not applicable.

Data Availability Statement: Not applicable.

Conflicts of Interest: The authors declare no conflict of interest.

References

1. He, X.; Li, P.; Ji, Y.; Wang, Y.; Su, Z.; Elumalai, V. Groundwater arsenic and fluoride and associated arsenicosis and fluorosis in China: Occurrence, distribution and management. *Expo. Health* **2020**, *12*, 355–368. [CrossRef]
2. He, X.; Li, P.; Wu, J.; Wei, M.; Ren, X.; Wang, D. Poor groundwater quality and high potential health risks in the Datong Basin, northern China: Research from published data. *Environ. Geochem. Health* **2021**, *43*, 791–812. [CrossRef] [PubMed]
3. Nsabimana, A.; Li, P.; He, S.; He, X.; Alam, S.M.K.; Fida, M. Health risk of the shallow groundwater and its suitability for drinking purpose in Tongchuan, China. *Water* **2021**, *13*, 3256. [CrossRef]
4. Wang, Y.; Li, P. Appraisal of shallow groundwater quality with human health risk assessment in different seasons in rural areas of the Guanzhong Plain (China). *Environ. Res.* **2022**, *207*, 112210. [CrossRef]
5. Zhang, Q.; Li, P.; Lyu, Q.; Ren, X.; He, S. Groundwater contamination risk assessment using a modified DRATICL model and pollution loading: A case study in the Guanzhong Basin of China. *Chemosphere* **2022**, *291*, 132695. [CrossRef] [PubMed]
6. Li, Y.; Li, P.; Cui, X.; He, S. Groundwater quality, health risk, and major influencing factors in the lower Beiluo River watershed of Northwest China. *Hum. Ecol. Risk Assess.* **2021**, *27*, 1987–2013. [CrossRef]

7. Wu, J.; Zhou, H.; He, S.; Zhang, Y. Comprehensive understanding of groundwater quality for domestic and agricultural purposes in terms of health risks in a coal mine area of the Ordos basin, north of the Chinese Loess Plateau. *Environ. Earth Sci.* **2019**, *78*, 446. [CrossRef]
8. Wu, J.; Zhang, Y.; Zhou, H. Groundwater chemistry and groundwater quality index incorporating health risk weighting in Dingbian County, Ordos basin of Northwest China. *Geochemistry* **2020**, *80*, 125607. [CrossRef]
9. Hu, G.; Mian, H.R.; Dyck, R.; Mohseni, M.; Jasim, S.; Hewage, K.; Sadiq, R. Drinking water treatments for arsenic and manganese removal and health risk assessment in white rock, Canada. *Expo. Health* **2020**, *12*, 793–807. [CrossRef]
10. Wei, M.; Wu, J.; Li, W.; Zhang, Q.; Su, F.; Wang, Y. Groundwater geochemistry and its impacts on groundwater arsenic enrichment, variation, and health risks in Yongning County, Yinchuan Plain of Northwest China. *Expo. Health* **2021**, 1–20. [CrossRef]
11. Alam, M.O.; Shaikh, W.A.; Chakraborty, S.; Avishek, K.; Bhattacharya, T. Groundwater arsenic contamination and potential health risk assessment of gangetic plains of Jharkhand, India. *Expo. Health* **2016**, *8*, 125–142. [CrossRef]
12. Li, Y.; Ma, L.; Abuduwaili, J.; Li, Y.; Uulu, S.A. Spatiotemporal distributions of fluoride and arsenic in rivers with the role of mining industry and related human health risk assessments in Kyrgyzstan. *Expo. Health* **2021**, 1–14. [CrossRef]
13. Fano, D.; Vásquez-Velásquez, C.; Aguilar, J.; Gribble, M.O.; Wickliffe, J.K.; Lichtveld, M.Y.; Steenland, K.; Gonzales, G.F. Arsenic concentrations in household drinking water: A cross-sectional survey of pregnant women in Tacna, Peru, 2019. *Expo. Health* **2020**, *12*, 555–560. [CrossRef] [PubMed]
14. Sathe, S.S.; Mahanta, C.; Subbiah, S. Hydrogeochemical evaluation of intermittent alluvial aquifers controlling arsenic and fluoride contamination and corresponding health risk assessment. *Expo. Health* **2021**, *13*, 661–680. [CrossRef]
15. Owusu, C.; Silverman, G.S.; Vinson, D.S.; Bobyarchick, A.; Paul, R.; Delmelle, E. A spatial autologistic model to predict the presence of arsenic in private wells across Gaston County, North Carolina using geology, well depth, and pH. *Expo. Health* **2021**, *13*, 195–206. [CrossRef]
16. Joardar, M.; Das, A.; Mridha, D.; De, A.; Chowdhury, N.R.; Roychowdhury, T. Evaluation of acute and chronic arsenic exposure on school children from exposed and apparently control areas of West Bengal, India. *Expo. Health* **2021**, *13*, 33–50. [CrossRef]
17. Rehman, U.; Khan, S.; Muhammad, S. Ingestion of arsenic-contaminated drinking water leads to health risk and traces in human biomarkers (hair, nails, blood, and urine), Pakistan. *Expo. Health* **2020**, *12*, 243–254. [CrossRef]
18. Wu, C.; Fang, C.; Wu, X.; Zhu, G. Health-risk assessment of arsenic and groundwater quality classification using random forest in the yanchi region of Northwest China. *Expo. Health* **2020**, *12*, 761–774. [CrossRef]
19. Apollaro, C.; Di Curzio, D.; Fuoco, I.; Buccianti, A.; Dinelli, E.; Vespasiano, G.; De Rosa, R. A multivariate non-parametric approach for estimating probability of exceeding the local natural background level of arsenic in the aquifers of Calabria region (Southern Italy). *Sci. Total Environ.* **2022**, *806*, 150345. [CrossRef]
20. Figoli, A.; Fuoco, I.; Apollaro, C.; Chabane, M.; Criscuoli, A. Arsenic-contaminated groundwaters remediation by nanofiltration. *Sep. Purif. Technol.* **2019**, *238*, 116461. [CrossRef]
21. Matschullat, J. Arsenic in the geosphere—A review. *Sci. Total Environ.* **2000**, *249*, 297–312. [CrossRef]
22. Polizzotto, M.L.; Harvey, C.F.; Li, G.; Badruzzman, B.; Ali, A.; Newville, M.; Sutton, S.; Fendorf, S. Solid-phases and desorption processes of arsenic within bangladesh sediments. *Chem. Geol.* **2006**, *228*, 97–111. [CrossRef]
23. Smedley, P.L.; Kinniburgh, D.G. A review of the source, behaviour and distribution of arsenic in natural waters. *Appl. Geochem.* **2002**, *17*, 517–568. [CrossRef]
24. Drahota, P.; Filippi, M. Secondary arsenic minerals in the environment: A review. *Environ. Int.* **2009**, *35*, 1243–1255. [CrossRef] [PubMed]
25. Fuoco, I.; De Rosa, R.; Barca, D.; Figoli, A.; Gabriele, B.; Apollaro, C. Arsenic polluted waters: Application of geochemical modelling as a tool to understand the release and fate of the pollutant in crystalline aquifers. *J. Environ. Manag.* **2022**, *301*, 113796. [CrossRef]
26. Cao, W.; Guo, H.; Zhang, Y.; Rong, M.; Zhao, R. Controls of paleochannels on groundwater arsenic distribution in shallow aquifers of alluvial plain in the Hetao Basin, China. *Sci. Total Environ.* **2017**, *613–614*, 958. [CrossRef]
27. Guo, H.M.; Wen, D.G.; Liu, Z.Y.; Jia, Y.F.; Guo, Q. A review of arsenic-rich groundwater in Mainland and Taiwan, China: Distribution, characteristics and geochemical processes. *Appl. Geochem.* **2014**, *41*, 196–217. [CrossRef]
28. Shen, Y.F.; Sun, D.J.; Zhao, X.H.; Yu, G.Q. Screening report in areas of endemic arsenism and high content of arsenic in China. *Chin. J. Endem.* **2005**, *24*, 172–175. Available online: http://en.cnki.com.cn/Article_en/CJFDTOTAL-ZDFB20050200N.htm (accessed on 22 October 2020).
29. Jin, Y.L.; Liang, C.K.; He, G.L.; Cao, J.X.; Ji, R.C. Study on distribution of endemic arsenism in China. *J. Hyg. Res.* **2003**, *32*, 519–540. [CrossRef]
30. Smedley, P.L.; Zhang, M.; Zhang, G.; Luo, Z. Mobilisation of arsenic and other trace elements in fluviolacustrine aquifers of the Huhhot Basin, Inner Mongolia. *Appl. Geochem.* **2003**, *18*, 1453–1477. [CrossRef]
31. Guo, H.; Yang, S.; Tang, X.; Li, Y.; Shen, Z. Groundwater geochemistry and its implications for arsenic mobilization in shallow aquifers of the Hetao Basin, Inner Mongolia. *Sci. Total Environ.* **2008**, *393*, 131–144. [CrossRef] [PubMed]
32. Deng, Y.; Wang, Y.; Teng, M. Isotope and minor element geochemistry of arsenic-rich groundwater from Hangjinhouqi, the Hetao Plain, Inner Mongolia. *Appl. Geochem.* **2009**, *24*, 587–599. [CrossRef]
33. Gong, Z.; Lu, X.; Watt, C.; Bei, W.; He, B.; Mumford, J.; Ning, Z.; Xia, Y.; Le, X.C. Speciation analysis of arsenic in groundwater from Inner Mongolia with an emphasis on acid-leachable particulate arsenic. *Anal. Chim. Acta* **2006**, *555*, 181–187. [CrossRef]

34. Jun, H.E.; Teng, M.A.; Deng, Y.; Yang, H.; Wang, Y. Environmental geochemistry of arsenic-rich groundwater at western Hetao plain, Inner Mongolia. *Front. Earth Sci. China* **2009**, *3*, 63. [CrossRef]
35. Lin, N.F.; Tang, J.; Bian, J.M. The study on environmental geo-chemical characteristics in Arseniasis Area in the Inner Mongolia. *World Geol.* **1999**, *18*, 83–88. Available online: http://en.cnki.com.cn/Article_en/CJFDTOTAL-SJDZ902.011.htm (accessed on 22 October 2020).
36. Gao, C.R.; Chao-Xin, L.I.; Zhou, X.H.; Liu, B.; Liu, W.B.; Cai, L.I.; Feng, D.Y. Occurrence and hydrochemical characteristics of As-rich groundwater in the Linhe district of the Hetao Plain. *Hydrogeol. Eng. Geol.* **2008**, *35*, 22–28. [CrossRef]
37. Gao, C.; Liu, W.; Feng, C.E.; Chen, Y.; Zhang, G.; Song, J. Research on the formation mechanism of arsenic-rich groundwater in arid and semi-arid regions: A case study of Hetao Plain in Inner Mongolia, China. *Earth Sci. Front.* **2014**, *21*, 13–29. [CrossRef]
38. Jia, L.; Zhang, X.; Ye, P.; Zhao, X.; He, Z.; He, X.; Zhou, Q.; Li, J.; Ye, M.; Wang, Z.; et al. Development of the alluvial and lacustrine terraces on the northern margin of the Hetao Basin, Inner Mongolia, China: Implications for the evolution of the Yellow River in the Hetao area since the Late Pleistocene. *Geomorphology* **2016**, *263*, 87–98. [CrossRef]
39. Guo, H.; Zhang, Y.; Xing, L.; Jia, Y. Spatial variation in arsenic and fluoride concentrations of shallow groundwater from the town of Shahai in the Hetao basin, Inner Mon-golia. *Appl. Geochem.* **2012**, *27*, 2187–2196. [CrossRef]
40. Guo, H.; Liu, Z.; Ding, S.; Hao, C.; Xiu, W.; Hou, W. Arsenate reduction and mobilization in the presence of indigenous aerobic bacteria obtained from high arsenic aquifers of the hetao basin, inner mongolia. *Environ. Pollut.* **2015**, *203*, 50–59. [CrossRef]
41. Deng, Y.; Wang, Y.; Ma, T.; Gan, Y. Speciation and enrichment of arsenic in strongly reducing shallow aquifers at western hetao plain, northern china. *Environ. Geol.* **2009**, *56*, 1467–1477. [CrossRef]
42. Ni, P.; Guo, H.; Cao, Y.; Jia, Y.; Jiang, Y.; Zhang, D. Aqueous geochemistry and its influence on the partitioning of arsenic between aquifer sediments and groundwater: A case study in the northwest of the Hetao Basin. *Environ. Earth Sci.* **2016**, *75*, 356.1–356.13. [CrossRef]
43. Wang, M. Quaternary Sedimentary and Structure Features of the Hohholt-Baotou Basin, Inner Mongolia. Master's Thesis, China University of Geosciences, Beijing, China, 2006. (In Chinese with English Abstract)
44. Xiang, H.; Deng, H.; Zheng, C.; Cao, G. Hydrogeochemical signatures and evolution of groundwater impacted by the bayan obo tailing pond in northwest china. *Sci. Total Environ.* **2016**, *543*, 357–372.
45. Fang, L. *Groundwater Monitoring Comprehensive Report from 2001 to 2005 in Baotou City*; Inner Mongolia Institute of Geological Environment Monitoring: Hohhot, China, 2006.
46. Guo, H.; Zhang, B.; Li, Y.; Berner, Z.; Tang, X.; Norra, S.; Stüben, D. Hydrogeological and biogeochemical constrains of arsenic mobilization in shallow aquifers from the Hetao Basin, Inner Mongolia. *Environ. Pollut.* **2011**, *159*, 876–883. [CrossRef]
47. Wang, Y.; Guo, H.; Yan, S. *Geochemical Evolution of Shallow Groundwater Systems and Their Vulnerability to Contaminants: A Case Study at Datong Basin, Shanxi Province*; Science Press: Beijing, China, 2004; p. 62. (In Chinese)
48. Goldberg, S. Chemistry of the solid-water interface: Processes at the mineral-water and particle-water interface in natural systems. *Geochim. Cosmochim. Acta* **1992**, *57*, 205. [CrossRef]
49. Smedley, P.L.; Kinniburgh, D.G.; Macdonald, D.; Nicolli, H.B.; Barros, A.J.; Tullio, J.O. Arsenic associations in sediments from the loess aquifer of la pampa, argentina. *Appl. Geochem.* **2005**, *20*, 989–1016. [CrossRef]
50. Li, P.; Zhang, Y.; Yang, N.; Jing, L.; Yu, P. Major ion chemistry and quality assessment of groundwater in and around a mountainous tourist town of China. *Expo. Health* **2016**, *8*, 239–252. [CrossRef]
51. Li, P.; Tian, R.; Liu, R. Solute geochemistry and multivariate analysis of water quality in the Guohua Phosphorite Mine, Guizhou Province, China. *Expo. Health* **2019**, *11*, 81–94. [CrossRef]
52. Ren, X.; Li, P.; He, X.; Su, F.; Elumalai, V. Hydrogeochemical processes affecting groundwater chemistry in the central part of the Guanzhong Basin, China. *Arch. Environ. Contam. Toxicol.* **2021**, *80*, 74–91. [CrossRef] [PubMed]
53. Wu, J.; Li, P.; Wang, D.; Ren, X.; Wei, M. Statistical and multivariate statistical techniques to trace the sources and affecting factors of groundwater pollution in a rapidly growing city on the Chinese Loess Plateau. *Hum. Ecol. Risk Assess.* **2020**, *26*, 1603–1621. [CrossRef]
54. Wu, J.; Li, P.; Qian, H.; Duan, Z.; Zhang, X. Using correlation and multivariate statistical analysis to identify hydrogeochemical processes affecting the major ion chemistry of waters: Case study in Laoheba phosphorite mine in Sichuan, China. *Arab. J. Geosci.* **2014**, *7*, 3973–3982. [CrossRef]
55. Zhang, Y.; Cao, W.; Wang, W.; Dong, Q. Distribution of groundwater arsenic and hydraulic gradient along the shallow groundwater flow-path in hetao plain, northern china. *J. Geochem. Explor.* **2013**, *135*, 31–39. [CrossRef]
56. Wang, Y.X.; Su, C.L.; Xie, X.J.; Xie, Z.M. The genesis of arsenic-rich groundwater:a case study in Datong basin. *Geol. China* **2010**, *37*, 771–780. [CrossRef]

Article

Potential Toxic Impacts of Hg Migration in the Disjointed Hyporheic Zone in the Gold Mining Area Experiencing River Water Level Changes

Ruiping Liu [1,2,3,4], Fei Liu [5], Jiangang Jiao [6,7,*], Youning Xu [1,2,3,4], Ying Dong [1,3,4], El-Wardany R.M. [7,8], Xinshe Zhang [1,*] and Huaqing Chen [1,2]

1. Xi'an Center, China Geological Survey, Ministry of Natural Resources, Xi'an 710054, China
2. Field Base of Scientific Observation of Shannxi Tongguan, Ministry of Natural Resources, Xi'an 710054, China
3. Shaanxi Province Water Resources and Environment Engineering Technology Research Center, Xi'an 710054, China
4. Key Laboratory for Geo-hazards in Loess Area, Ministry of Natural Resources, Xi'an 710054, China
5. Dalian Ocean and Fishery Comprehensive Administration Supervision Lochus, Dalian 116000, China
6. Key Laboratory of Western China's Mineral Resource and Geological Engineering, Ministry of Education, Xi'an 710054, China
7. School of Earth Science and Land Resources of Chang'an University, Xi'an 710054, China
8. Faculty of Science, Al-Azhar University, Assiut Branch 71524, Egypt
* Correspondence: jiangang@chd.edu.cn (J.J.); zxinshe@mail.cgs.gov.cn (X.Z.); Tel.: +86-130-9696-4973 (J.J.); +86-153-9902-9451 (X.Z.)

Abstract: In order to study the occurrence form, vertical migration and transformation and the potential ecological risk of Hg in the disjointed hyporheic zone in the gold mining area is investigated. Through field investigation, in-situ test, and test analysis, the results show that: (1) the form of mercury in the original stratum where the river water-groundwater hydraulic connection is disconnected is mainly in the residual state, accounting for 77.78% of the total mercury; (2) after the water content increases or the water level changes, the various forms of occurrence in the soil surface layer decrease, and the residual state is still the main form; the main forms of mercury in the sand and pebble layer are diversified, including the residual state, strong organic state and humic acid state; (3) the mercury content in the subsurface zone in winter is higher than that in summer; (4) although the mercury content in groundwater has not been detected, the potential ecological risk of mercury in the disjointed Hyporheic zone near the river in the study area is much higher than the extreme ecological hazard threshold, which has a value of 320. The risk of groundwater pollution caused by mercury during the long-term runoff of the river is higher than that during the flood period. Therefore, relevant departments need to rectify the river as soon as possible, from the source to reduce the ecological risk of heavy metals to groundwater. The results will provide a scientific basis for groundwater control.

Keywords: disjointed type; hyporheic zone; mercury; migration and transformation; gold mining area; ecological risks

1. Introduction

The hyporheic zone is an important transitional area for the dynamic interaction of surface water and groundwater. Its physical properties and biogeochemical environment jointly carry the changing process of water flow, material, and energy exchange.

The hyporheic zone (HZ) is an essential member of a river ecosystem which can be saturated with interstitial sediment beneath the streambed and close to the riverbed (Figure 1a,b [1–3]). It is the zone of fraternization, exchange, and interaction between groundwater and surface water. The surface water and groundwater interacted and combine to create a new interstitial or transitional zone of water with different properties.

The HZ can be divided into two zones, the "surface hyporheic" zone, and the "interactive hyporheic" zone. The new characteristics of the hyporheic zone are dynamic zone with hydraulic pressure at numerous scales (catchment scale and smaller scale) [4–8]. It has a higher concentrations of chemicals with low oxygen content, special hydrological processes, and ecological systems. Many hydrogeologists consider the hyporheic zone as part of the groundwater system, since it contains subsurface water in the saturated zone. The physiochemical characteristics of water flow in the subsurface are the main key to material and energy exchange between surface water and groundwater [9]. Due to the differences in physiochemical characteristics such as redox, pH, and temperature between river water and groundwater, drives variation in biological, chemical, and even temperature gradients between surface and groundwater.

Heavy metals are the extremely distributed deposits in the soil sediment and water nearby gold mining [10–12]. Heavy metals are classified into essential and nonessential metals. Heavy metals sources are categorized as natural sources such as ore deposits, bedrock weathering, geological weathering, atmospheric precipitation, storms, and wind bioturbation, whereas anthropogenic sources such as agrochemicals activities, mining processing, shipping, industrial products, manufacturing development and process (e.g., batteries and pharmaceutical products), smelting, fuel generation, electroplating, sludge discharge, energy transmission, dense urban areas, wastewater irrigation and [13–15]. Globally, the water discharges from the mining of ores, heaps leaching, milling and ore processes to liberate the metal and remove the waste are predominantly sources of metals in soil, surface, and groundwater pollution around these mine areas. Mine waters are typically rich in metals and elements such as iron, copper, manganese, arsenic, mercury, lead, and zinc [16–19]. By water-rock interaction, the heavy metals move to water in order to sediment and can accumulate in water or soil by bioaccumulation processes, which may enter the food chain and create a health risk. Consequently, detecting the processes of surface and groundwater interface is highly significant to evaluate and assume the behavior and consequences of mine pollution. The main four attributes which controlled the element behavior in certain heavy metals are the valence state, compound state, binding state, and structural state [20]. In addition to the stability of the element in the soil organized the occurrence form of this element.

Mercury is one of the most critical pollutions and ranked the third most toxic elements by the US Government Agency for Toxic Substances and Disease Registry [21–23]. As well as this, it is one of the main pollutant structures in the gold mining areas [24,25]. It can reach the human body through food and water which can act naturally by ore rock interaction or by human activities (by using dental amalgam, batteries, thermometers, barometers, and medical waste) in modern life. Mercury (inorganic or as organic) can bioaccumulate in soil and water and cause bad effects on human health [26–28].

When mercury enters the soil, various forms will be redistributed among the soil solid phases. MA et al. [29] revealed that there are significant differences in the mass concentrations of dissolved mercury (DHg) and total methyl mercury (TMeHg) in the stratum water samples in the water storage period (September to October), submerged period (November to December), water withdrawal period (February to March) and drying period (May to June). Changes in the surface water level will cause the content of heavy metals and the occurrence of environmental changes [30,31]. Wang [32] studied the change in mercury under laboratory conditions, he concluded that the content in different forms under dry and wet environments after flooding, shows the mercury Migration ability is: acid-soluble mercury > inert mercury > Alkali-soluble mercury > water-soluble mercury; and under the condition of drying, inert mercury > alkali-soluble mercury > acid-soluble mercury > water-soluble mercury. On the other hand, Tang [33] studied the speciation distribution of mercury in wetland sediments with different water level gradients, and he noticed that the change in groundwater level is directly affected by the river water level during high and low water seasons, and the fluctuation of surface water level causes different changes in the content of water and polluting elements in the soil layer [34]. The river water and groundwater level in the hyporheic zone in the wet season and dry season directly affect the risk of groundwater pollution [35,36], and the disconnected hyporheic zone is the bottom line for an early warning of regional pollution risks, so the development of disconnected river reaches. It is very important to study the vertical migration of the mercury information profile and its ecological risk.

Analysis of water, sediments, and members of indigenous biota can be used to determine the relative pollution of aquatic habitats by mercury and other heavy and trace metals. This analysis includes different monitoring techniques, and scientific problem usually include soil and water quality, hydrological processes composition of vegetation, and animal population. Previous studies mainly focus on reports on the potential ecological hazard assessment of heavy metal pollution in static water and soil environments [37,38]. The migration and transformation rules and ecological risks of mercury occurrence forms in the subsurface under the conditions of river water level changes in wet and dry periods are not well studied. The research topic provides a scientific basis for the prevention and control of groundwater pollution along the river.

Xiaoqinling gold-mining area is considered the third largest gold-producing region in China, and gold production in the area began about 900 years ago. In Mayu alone, there are 72 abandoned ancient mines, and production from 1980 to 2003 reached 562,500 t [39]. Furthermore, the large-scale exploitation of gold and mining activities began in 1975 [40]. Currently, more than 29 large gold mining companies have many gold mills, which are mainly in the Shuangqiao river. Based on the geochemical characteristics landscape of the region, five areas are included (Figure 1a,b): the prockbase mountain area, piedmont alluvial-pluvial inclined tableland, loess ravine tableland, Yellow river, Weihe and Shuangqiao river alluvial terrace [40]. Both agricultural activities and gold-mining activities were commonly affected in the study area. The more complicated structures include faults and folds in the north and south, the main regional faults are the Taiyao, Guanyintang, Xiaohe, and Huanchiyu. The structure system influences the heavy mineral distribution in the study area. Several gold mills are scattered along numerous streams which are tributaries of the Yellow river. Since mining pactivities during the 1980s to 1990s, artisanal gold in which elemental Hg has been used widely for gold processing [39]. Up to the present time, wastewater is still discharged directly into rivers and streams due to more than eight million tons of tailings having been produced with high-risk pollution of heavy metals for the soil and water. The main sources of heavy and trace elements especially Hg in the polluted soil were identified through the correlation analysis and comparing the maps of these heavy metal concentrations distribution and land-use types [41]

The Shuangqiao River in the Xiaoqinling gold mining area is a typical area of mercury pollution [42,43]. Xu et al. [44] studied mercury pollution in the sediment. The soil content is significantly higher than that of the three geochemical landscapes: the piedmont alluvial-proluvial slope, the loess gully residual plateau, and the Huangwei River alluvial plain; the accumulation or pollution of heavy metals caused by mining activities becomes the soil Hg, Pb, Cd, Cu, and Zn in the study area. Owing to the chemical factor affecting the enrichment of heavy metals in soil [45], the gold mining activities are closely related to Hg in the Hyporheic zone of the study area, the authors studied the disjointed continuous type in dry season. The migration law and influencing factors of Hg morphological characteristics in the subsurface zone in the vertical soil layer [42].

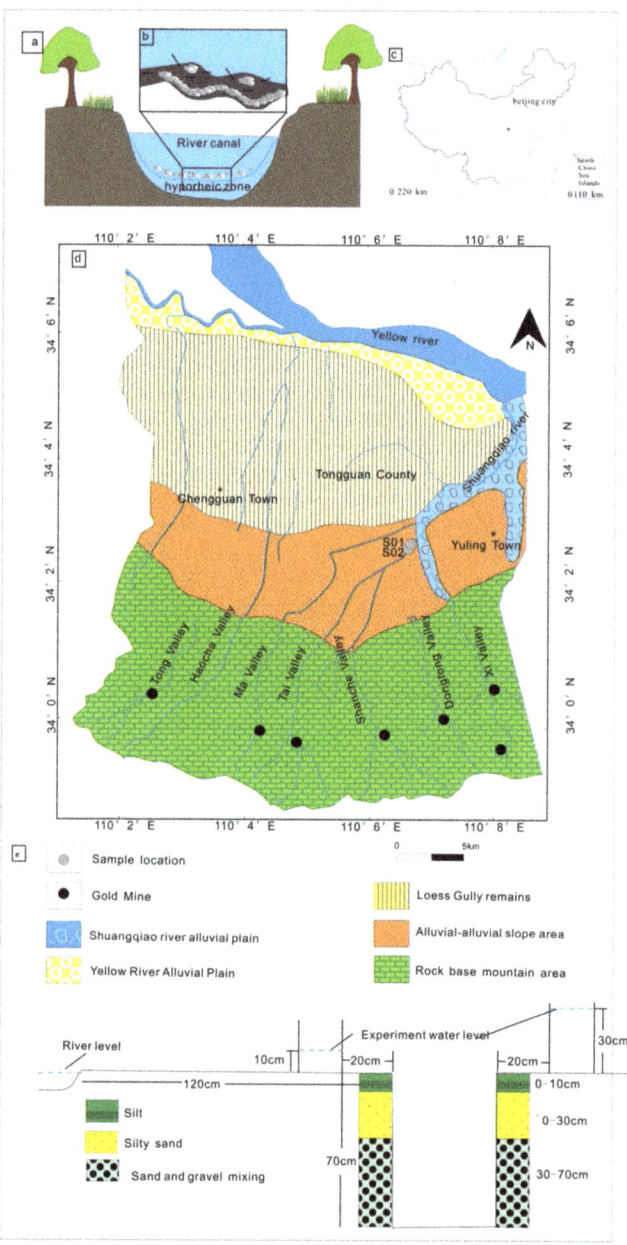

Figure 1. (**a**) view of a river shows the position of the hyporheic zone. (**b**) the directional flow of water through the hyporheic zone. (after [2,3]), (**c**) location map of the study area. (**d**) The map of the sample and its landscape distribution after [9]. (**e**) Distribution of water seepage test.

2. Background of the Study Area

The Shuangqiao River originates from the northern foot of the Xiaoqinling Mountains and is formed by the confluence of Mayu, Taiyu, Shancheyu, Dongtongyu, and Xiyu. region (Figure 1c,d [46]). Its flow direction is from south to north, with a length of 14.8 km. It belongs to the warm temperate continental monsoon semi-humid-humid climate zone, with dry winters and little rain, and humid and rainy summers. The annual precipitation is 587.4 mm and the evaporation is 1193.6 mm. At present, it is mainly used for industrial water. Shuangqiao river channel has not been fully regulated by 2020. The groundwater types in this area are mainly Archean bedrock fissure water and quaternary pore water. There are complex hydraulic relationships among different geological and geomorphological types (alluvial fan, loess plateau, and river alluvial terrace) in the study area. The field investigation found that there was a phenomenon of water-saturated zone-aeration zone-upper stagnant zone-relative impermeable zone-aeration zone-saturated zone in a river in the Little Qin Mountains region. According to the shallow water level in the middle and lower reaches of the river in the mining area, the groundwater is not polluted in the area with good hydraulic alternation between the river and groundwater in the alluvial-alluvial slope.

3. Survey Analysis and Evaluation Methods

3.1. In-Situ Test and Sample Analysis

The principal three sampling techniques with a variety of methods that can be carried out for the HZ are in situ, coring, and tracing. The methods will be chosen according to the purpose of research.

In this paper, in-situ seepage tests were carried out. After the seepage tests, we collected soil samples and tested and analyzed the contents of soil (according to Figure 1e), groundwater, and surface water in HZ. Mercury is volatile and therefore has been tested in winter and summer. Based on the collecting data of moisture content, temperature and content, the law of migration and transformation was analyzed. Among them, the monitoring data of soil profiles were analyzed, and 7 forms were analyzed. The stability of each form in natural soil was different, and the residue state was the most stable, the effect of leaching on morphology has not been studied. The ecological risk was analyzed based on the soil monitoring data.

During the wet season (June) and the dry season (December) in 2021, the survey and sampling were conducted by the Shuangqiao River in the alluvial-proluvial landform area. The sampling point is Sidi Village, where the groundwater and surface water are disconnected hydraulically. In winter and summer, 24 soil samples, 1 groundwater sample and 1 surface water sample were collected, respectively. The test lasted 30 min. Regarding different river water level settings, when the flood occurs, the flood level is 30 cm (the difference between the floodplain and the river bed), and the low water level is the average depth of the current river water level of 10 cm. The test includes occurrence form, total amount, water content, pH, and hydrogeological conditions of mercury and sees the sampling, analysis and testing methods for details [42].

The samples of water collected during the wet season (June) and the dry season (December) in 2021 were tested and analyzed using an atomic fluorescence spectrometer (AFS) in the test center of the Xi'an Institute of Geology and Mineral Resources. Instead, the mercury distinguishes into 48 sets of soil samples which were determined using AFS, as well as the ion-selective electrode analysis in the Hefei Mineral Resources Supervision and Testing Center, Ministry of Land and Resources. All of the tests were analyzed according to the standards DZ/T0279-2016, DD2005-03, and DZ/T0130.2-2006, specifying the detection limit and determination range of the method, which are 0.005 mg/kg and 0.02–6 mg/kg, respectively. The standard DD2005-03 divides the elements in soil into seven forms, namely the water-soluble form extracted by the extractant of deionized water, the ion-exchange form extracted by the extractant of magnesium chloride, the carbonate bound form extracted by the extractant of acetic acid-sodium acetate, the weak organic (humic acid) bound form extracted by the extractant of sodium pyrophosphate, the Fe-Mn bound form extracted by the extractant of hydroxylamine hydrochloride, the strong organic bound form extracted by the extractant of hydrogen peroxide, and the residual form extracted by hydrofluoric acid.

3.2. Potential Ecological Risks

According to the single-factor ecological hazard coefficient calculation formula, combined with the toxicity response coefficient of heavy metal elements to calculate the Ei value, the single-factor potential ecological hazard coefficient formula is as follows:

$$E_{Hg} = (T_{Hg} \times C)/C_0 \quad (1)$$

where E_{Hg} is the potential ecological risk factor of Hg, T_{Hg} is the toxic response factor of Hg, C is the concentration of Hg, C_0 is the concentration of Hg in the background. In this study, the T of Hg was 40 [47]. Hakanson proposed that the highest value of heavy metals in sediments before industrialization was used as a reference value [48]. In the process of practical application, the potential ecological risk assessment of sedimentary river bottoms is used as a reference value, but when the risk assessment object is extended to the soil, The selection of the reference value is not uniform. Considering the characteristics of the study area and the collected soil samples, this work selects the background value of heavy metals in uncontaminated soil in the area [49,50] as the reference value for evaluation. See Tables 1 and 2.

Table 1. Toxicity coefficients of Hg and its reference values.

Depth	T	C_0 (mg/kg)
0–20	40	0.070
20–30	40	0.036
30–50	40	0.032
50–70	40	0.018

Table 2. Potential ecological risk coefficient and classification of ecological risk degree.

Potential Ecological Hazard Coefficient (E_{Hg})	Potential Ecological Risk Degree Grading
<40	Mild ecological hazard
40–80	Moderate ecological hazard
80–160	Intensity Ecological Hazard
160–320	Very high ecological hazard
>320	Extremely ecologically hazardous

4. Results

4.1. The Influence of the Occurrence Form and Content of Soil Vertical Hg in the Hyporheic Zone of the Alluvial-Proluvial Area in the High and Low Water Season

It can be seen from Figure 2 that the stable environment of mercury in the undercurrent zone with pH between 8.39–8.87. The mercury forms in the wet and dry periods are mainly in the residual state, indicating that the source of Hg in the undercurrent zone of the gold mining area mainly is the mineral lattice [51], which may be implicated in mining wastewater into the undercurrent zone. The total mercury content (25.22 mg/kg) in the main enriched soil layers in the subsurface in winter was higher than that in summer (11.61 mg/kg). Mercury content displays a negative correlation with pH values. The diagrammatic cross-section in Figure 1e shows that the total amount of mercury in the undisturbed soil and the characteristics of the residue state in the vertical direction of the silt-sand-cobble-tailings-silt-sand-cobble alluvial layer, whether it was the wet season (summer) or the dry season (winter). Furthermore, the high value appears in both 0–10 cm silt containing tailings slag and 30–40 cm silt, however, the highest value is in 30–40 cm silt. In summer, the total and residual contents of mercury in 0–10 cm silt containing tailings were 1.48 mg/kg and 1.45 mg/kg, respectively; the total and residual mercury contents in 30–40 cm silt were 11.64 mg/kg and 10.00 mg/kg, respectively. Kg. Nevertheless, in winter, the total and residual mercury contents in 0–10 cm silt containing tailings are 1.447 mg/kg and 1.445 mg/kg, respectively; the total mercury and residual mercury contents in 30–40 cm sand and pebbles are 19.49 mg/kg and 20.63 mg/kg, to the 40–80 cm sand and pebble layer, this means the total mercury content decreased.

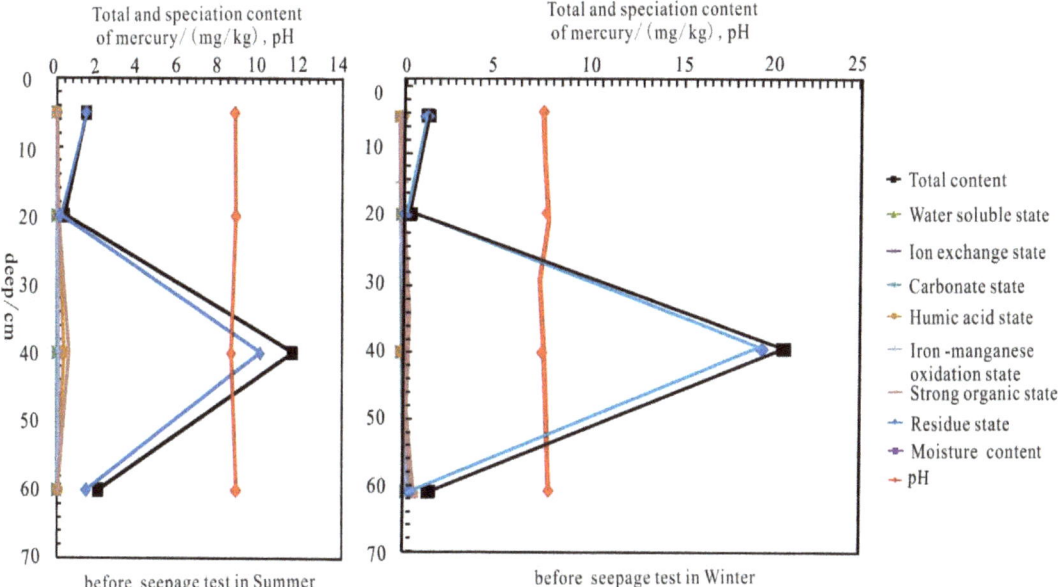

Figure 2. Occurrence forms of mercury in the soil layer of the vadose zone before the water seepage test was performed in the dry and wet periods.

4.2. The Influence of the Occurrence Form and Content of Vertical Hg in the Hyporheic Zone under Different Hydrodynamic Excitation Conditions during High and Low Water Seasons

The morphological characteristics of mercury occurrence of the vertical Hg in the subsurface under high and low water level seepage conditions in wet and dry periods (As shown in Figure 3): the mercury content is still mainly in the residual state, and the highest value is in 30–40 cm silt. Furthermore, the mercury content in the stratum in winter is still higher than that in summer, as well as the total and residual mercury content under high-water seepage conditions is higher than the corresponding mercury content. Whereas the total and residual mercury content in wet 30–40 cm silt sand under high water level seepage conditions is 10.21 mg/kg and 9.91 mg/kg, respectively, while the total amount of mercury in 30–40 cm silt sand under low water level seepage conditions in the wet season and residual state were 1.65 mg/kg and 1.14 mg/kg, respectively. On the other hand, the total amount and residual state of mercury in 30–40 cm silt were 13.02 mg/kg and 12.00 mg/kg, respectively, however, the total amount and residual state of mercury under the condition of high and low water level seepage in the dry season in the 30–40 cm silt under the conditions were 3.01 mg/kg and 2.01 mg/kg (as shown in Figure 4).

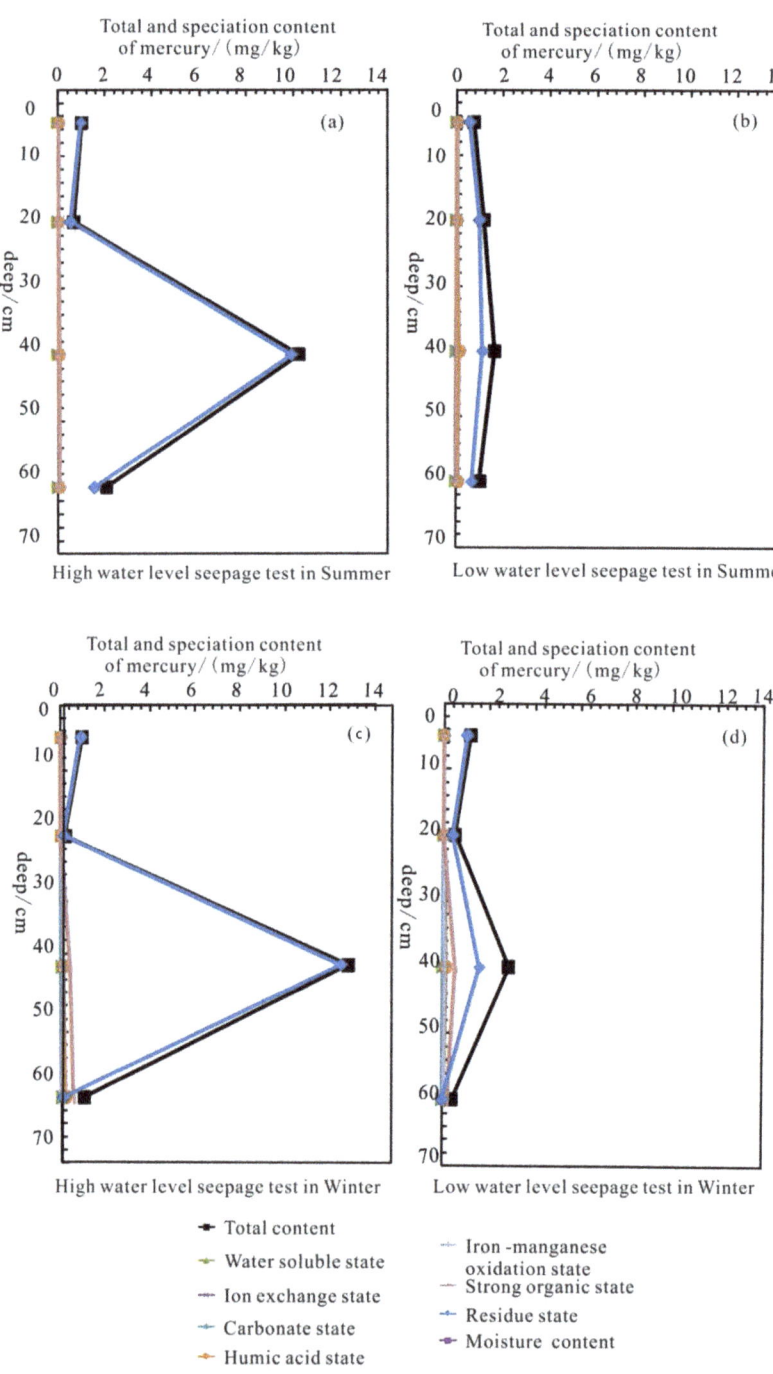

Figure 3. Characteristics of mercury transport in the subsurface under different hydrodynamic excitation conditions in wet and dry periods.

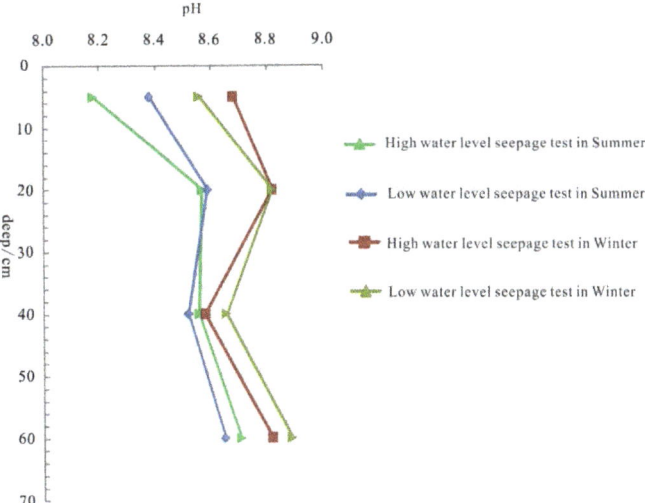

Figure 4. Characteristics of pH transport in the subsurface under different hydrodynamic excitation conditions in wet and dry periods.

5. Discussion

5.1. Factors Affecting Mercury Migration and Transformation

(1) The form of mercury in wet and dry seasons is mainly in the residual state, and the mercury content in the main enriched soil layer in the Hyporheic zone in winter (20.64 mg/kg) and higher than in summer (11.61 mg/kg). The temperature in the wet season is 29–36.9 °C, and the temperature in the dry season is 2–11.5 °C. Moreover, affected by temperature, soil mercury release flux shows that it is restricted by the atmospheric temperature with a positive correlation. The higher the atmospheric temperature, the greater the soil mercury release flux. Even air temperature affects mercury volatilization, but other factors also control soil volatilization, such as humidity, topographical conditions, and meteorology [52].

(2) Under the condition of water seepage, the mercury form changes regardless of the dry or the wet season, the various occurrence forms of 0–30 cm mercury in the vertical direction are decreasing, and still the main one. Nonetheless, in the sand and pebble layer, the main forms of mercury are diversified, including residue state, strong organic state, and humic acid state (see Tables 3 and 4). Currently, the pH of the layer with the largest change in morphological content also tends to be acidified (as shown in Figure 5). Acidic conditions may increase the content of available mercury in the soil, which leads to an increase in its bioavailability and enhanced migration ability. Adsorption-desorption has a great influence on the migration and transformation process. Occasionally, the influence mechanism of soil pH on the complexation-chelation, oxidation-reduction and methylation reactions in the process of mercury migration and transformation is reflected in ion competitive adsorption, valence state change, and methylation reaction, respectively. and promote the synthesis of factors [53]. The change in humic acid will promote the continuous decrease in water-soluble mercury, exchangeable mercury, and acid-soluble mercury in soil with the increase in alkali-soluble mercury, organically bound mercury and residual mercury [54]. But to the sand pebble layer, this is because in the disturbed system, almost all forms of mercury have mercury release [42].

Table 3. List of mercury speciation content and its percentage in subsurface flow zone in the dry season.

Different Hydrodynamic Excitation	Sample No.	pH	Moisture Content	Total	The Hg Speciation Content							The Percentage of Hg Speciation Content in the Total Amount						
					Water Soluble State	Ion Exchange State	Carbonate State Humic Acid State	Humic Acid State	Iron-Manganese Oxidation State	Strong Organic State	Residue State	Water Soluble State	Ion Exchange State	Carbonate State Humic Acid State	Humic Acid State	Iron Manganese Oxidation State	Strong Organic State	Residue State
			%	mg/kg	mg/kg	mg/kg	mg/kg	mg/kg	mg/kg	mg/kg	mg/kg	%	%	%	%	%	%	%
Low water level infiltration test	dd-1	8.56	8.67	1.038	0.001	0.002	0.005	0.010	0.008	0.003	1.009	0.09	0.19	0.50	0.92	0.78	0.29	97.24
	dd-2	8.82	16.52	0.423	0.001	0.001	0.002	0.005	0.004	0.003	0.407	0.20	0.25	0.56	1.24	0.98	0.60	96.17
	dd-3	8.49	11.39	4.106	0.008	0.005	0.008	0.245	0.086	0.562	3.194	0.19	0.11	0.18	5.96	2.08	13.69	77.78
	dd-4	8.89	3.24	0.355	0.010	0.001	0.001	0.054	0.045	0.226	0.018	2.81	0.32	0.37	15.09	12.63	63.82	4.95
High water level infiltration test	gd-1	8.68	8.3	0.923	0.001	0.001	0.006	0.003	0.003	0.004	0.904	0.13	0.14	0.67	0.34	0.28	0.44	97.99
	gd-2	8.82	12.14	0.152	0.001	0.001	0.002	0.002	0.002	0.003	0.141	0.39	0.67	1.18	1.23	1.28	2.22	93.02
	gd-3	8.35	17.26	25.220	0.014	0.020	0.018	0.024	0.143	0.376	24.626	0.05	0.08	0.07	0.10	0.57	1.49	97.64
	gd-4	8.82	4.82	0.853	0.028	0.002	0.002	0.184	0.010	0.547	0.081	3.31	0.18	0.18	21.59	1.21	64.08	9.44
Befurer infiltration test	yd-1	8.75	9.1	1.473	0.002	0.002	0.002	0.004	0.004	0.011	1.449	0.11	0.16	0.11	0.26	0.27	0.74	98.34
	yd-2	8.8	15.48	0.367	0.000	0.001	0.002	0.083	0.006	0.114	0.162	0.11	0.28	0.44	22.45	1.63	30.91	44.18
	yd-3	8.39	12.55	20.004	0.011	0.026	0.012	0.037	0.139	0.285	19.494	0.05	0.13	0.06	0.18	0.70	1.43	97.45
	yd-4	8.78	5.57	0.970	0.056	0.002	0.002	0.217	0.015	0.521	0.158	5.71	0.21	0.16	22.39	1.54	53.70	16.29

Table 4. List of mercury speciation content and its percentage in subsurface flow zone in the wet season.

Different Hydrodynamic Excitation	Sample No.	pH	Moisture Content (%)	The Hg Speciation Content (mg/kg)	Ion Exchange State (mg/kg)	Carbonate State / Humic Acid State (mg/kg)	Humic Acid State (mg/kg)	Iron-Manganese Oxidation State (mg/kg)	The Hg Speciation Content (mg/kg)	Total (mg/kg)	Water Soluble State (%)	Ion Exchange State (%)	Carbonate State / Humic Acid State (%)	Humic Acid State (%)	The Hg Speciation Content (mg/kg)	Total (mg/kg)	Residue State (%)
Low water level infiltration test	dd-1	8.38	9.98	0.625	0.001	0.001	0.054	0.007	0.005	0.555	0.21	0.19	0.21	8.70	1.17	0.75	88.77
	dd-2	8.59	10.04	1.042	0.001	0.001	0.041	0.006	0.003	0.987	0.14	0.11	0.13	3.97	0.59	0.31	94.75
	dd-3	8.52	12.91	1.325	0.002	0.001	0.160	0.004	0.012	1.145	0.14	0.08	0.09	12.10	0.32	0.87	86.39
	dd-4	8.65	10.60	0.868	0.002	0.001	0.099	0.012	0.044	0.708	0.28	0.14	0.13	11.38	1.43	5.05	81.59
High water level infiltration test	gd-1	8.18	12.58	1.029	0.001	0.001	0.033	0.005	0.004	0.984	0.14	0.12	0.12	3.20	0.45	0.40	95.58
	gd-2	8.57	13.16	0.577	0.002	0.001	0.033	0.003	0.009	0.528	0.30	0.21	0.21	5.70	0.57	1.52	91.48
	gd-3	8.56	12.77	9.986	0.002	0.001	0.057	0.007	0.003	9.914	0.02	0.01	0.01	0.57	0.07	0.03	99.27
	gd-4	8.71	12.58	1.682	0.002	0.001	0.060	0.007	0.048	1.563	0.13	0.07	0.08	3.57	0.42	2.84	92.90
Befurer infiltration test	yd-1	8.75	9.10	1.474	0.002	0.002	0.004	0.004	0.011	1.449	0.11	0.16	0.11	0.26	0.27	0.74	98.87
	yd-2	8.80	10.48	0.368	0.001	0.002	0.083	0.006	0.114	0.162	0.11	0.28	0.44	22.45	1.63	30.91	44.18
	yd-3	8.59	12.15	11.109	0.015	0.007	0.336	0.110	0.596	10.003	0.38	0.14	0.06	3.03	0.99	5.37	90.04
	yd-4	8.82	4.24	1.650	0.004	0.004	0.052	0.037	0.064	1.466	1.39	0.25	0.22	3.12	2.23	3.91	88.88

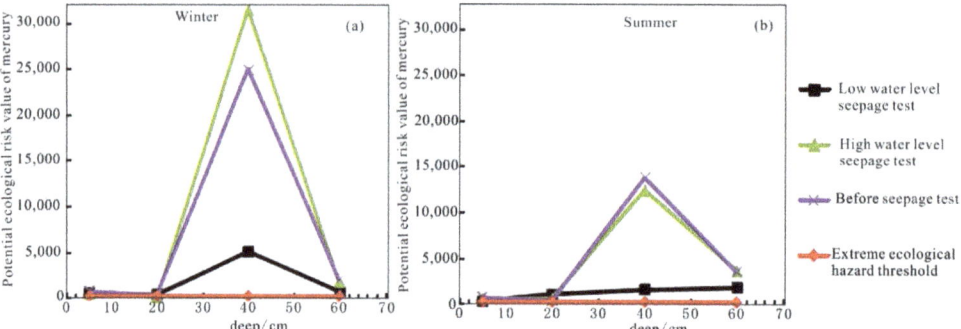

Figure 5. Line chart of potential ecological risks of mercury in the subsurface under changing river water levels. (**a**)—The line chart of the ecological risk of mercury in the subsurface under the condition of changing river water level in winter; (**b**)—The line chart of the ecological risk of mercury in the subsurface under the condition of changing the river water level in winter.

(3) The convective exchange law of mercury content caused by the pressure gradient in the same layer is reflected in the fact that 0–40 cm does not change significantly after leaching due to the low original content, while the leaching amount of mercury (the amount of mercury leaching under the condition of high-water level seepage at 40 cm) is not obvious. The original value (the mercury content after the water seepage test) is lower than the leaching amount of mercury under low water seepage conditions. The authors believe that the reason is the structure and hydrogeological characteristics of the flow sediments strongly influence the surface and groundwater exchange. Nevertheless, the silt and tailings sand are deposited on the riverbed; in addition, the silts bed has very low permeability and porosity, therefore it also reduces the exchange of undercurrent, and the low permeability rate promotes leaching and water-rocking, and a huge amount of mercury ions and their complexed forms are dissolved out. The inhomogeneous structure of the riverbed also creates effective anisotropy, and it causes restriction of the penetration of solutes, especially in the vertical direction [55]. Salehin et al. [56] conducted a series of experiments on the river-groundwater exchange, and constructed a two-dimensional inhomogeneous riverbed structure by artificial methods. Consequently, he discovered that the water flow prefers to pass through the high permeability area, and the interface flow increases in this area.

5.2. Potential Ecological Risks

During the history of gold mining areas, there was disorderly and random mining. Consequently, mercury rollers which is a beneficiation process liable to cause mercury pollution are used for illegal gold extraction in the mining valleys in the mountains from the 1990s to around 2010, thus the wastewater was directly discharged into the Shuangqiao River and other rivers with the mine water. Indisputably, it leads to serious ecological problems in rivers. Xu et. al., [57], comparatively investigated four types of soil and water pollution in the study area: atmospheric deposition, river irrigation, mining sewage irrigation, and tailings slag leaching. They summarized that the largest risk is river irrigation type, followed by tailings leaching type [58], and heavy metal elements in river sediments are homologous to those in tailings. As of Table 5 and Figure 5, it can be realized that the potential ecological risk of mercury in the disjointed Hyporheic zone is much higher than the extreme ecological hazard with a threshold (320), regardless of whether the hydraulic connection is in winter or summer, and the maximum risk of mercury in the original soil layer in winter and summer reaches 25,005.5 and 13,886.48, respectively, 78.14 and 43.40 times of the extreme ecological hazard threshold. Likewise, the pollution risk of 30–50 cm sand and pebble are the greatest, while the ecological risk of 0–30 cm silt layer and waste residue layer is relatively small. After leaching, the potential ecological risk of mercury in most soil layers in the Hyporheic zone has been reduced, especially after leaching at a constant water level. On the other hand, the risk value of the sand and pebble layer in winter has dropped from the original 25,005.5 to 5133, and the risk value of the sand and pebble layer in winter has been reduced from the original. Moreover, 13,886.48 dropped to 1656.07, which means that mercury entered the groundwater body. When a flood occurs (within 30 min), the mercury content in the surface layer decreases rapidly, attributable to many influencing factors described above, which produce an increase in the sand and pebble layer below (changed to 31,525); in addition, the potential ecological risk of mercury in the soil layer under normal water level leaching decreased, increasing the ecological risk of groundwater.

Table 5. The potential ecological risk value of mercury in the subsurface zone under the condition of changing river water level.

Season	Depth (cm)	The Potential Ecological Risk Value of Mercury after Seepage Test		The Potential Ecological Risk Value of Mercury before Water Seepage Test
		Low Water Level	High Water Level	Original Stratigraphic Section
Winter	0–20	592.91	527.2	841.94
	20–30	470.22	168.56	408.11
	30–50	5133	31525	25005.5
	50–70	788.67	1896.44	2156.22
Summer	0–20	357.26	588.13	842.19
	20–30	1157.3	641.64	408.59
	30–50	1656.07	12,482.71	13,886.48
	50–70	1927.79	3737.81	3666.46

6. Conclusions

The form of mercury in the undercurrent zone where the river water-groundwater hydraulic connection is disconnected from the gold mining area is mainly in the residual state. Nonetheless, after the water content increases or the water level changes, creates variation in the occurrence forms of mercury are all decreasing, and the residual form is still the main form, whereas the main forms of mercury in the sand and pebble layer are diversified, but also including residual form, strong organic form, and humic acid form. This facilitates the entry of mercury into water bodies. The mercury content in the subsurface in winter is higher than that in summer. Relevant departments should monitor heavy metal pollution in groundwater and Hyporheic zones in different seasons.

Although the mercury content in groundwater has not been detected, the potential ecological risk of mercury in the disjointed Hyporheic zone near the river in the study area has reached an extreme ecological hazard. In a short 30-min constant water level leaching test, the potential risk of mercury to groundwater is higher. Furthermore, during the flood period, the long-term runoff of rivers at a constant water level is bound to pose a greater risk of groundwater pollution. Consequently, the following measures can be taken: (1) control the sources of heavy metals at source, (2) reduce the ecological risks of heavy metals to groundwater by physical chemistry, bioremediation and pesticide control in soil as soon as possible on the basis of river regulation.

Author Contributions: R.L., F.L. and J.J. wrote this manuscript. R.L. conceives the original idea of this study. E.-W.R.M. polished this manuscript. All authors (R.L., F.L., J.J., Y.X., Y.D., E.-W.R.M., X.Z. and H.C.) modified the manuscript. Authorship must be limited to those who have contributed substantially to the work reported. All authors have read and agreed to the published version of the manuscript.

Funding: This research was funded by the survey projects initiated by the Ministry of Natural and Resources of the People's Republic of China (DD20211317, DD20189220, 1212010741003,1212011220224, and 121201011000150022), the public welfare scientific research project launched by the Ministry of Natural and Resources of the People's Republic of China (201111020), the project of 2015 Natural Science Basic Research Program of Shaanxi (2015JM4129), and the project of 2016 Fundamental Research Funds for the Central Universities (open fund) (310829161128). Formation of extensional crenulation cleavage in the ductile of Xiaoqinlin Metamorphic Core Complex (2021KJXX-73), A research on intelligent sensing, coupling evolution and driving mechanism of mine geological environment The authors 'thanks also go to the experiment and testing lab of Xi'an Center, China Geological Survey for the statistical analysis of the data conducted.

Conflicts of Interest: The authors declare no conflict of interest.

References

1. White, D.S. Perspectives on defining and delineating hyporheic zones. *J. N. Am. Benthol. Soc.* **1993**, *12*, 61–69. [CrossRef]
2. Stanford, J.A.; Ward, J.V. The hyporheic habitat of river ecosystems. *Nature* **1988**, *335*, 64–66. [CrossRef]
3. Biddulph, M. Chapter: 3.11.1: Hyporheic Zone. In *Situ Sampling*; Cook, S.J., Clarke, L.E., Nield, J.M., Eds.; Geomorphological Techniques (Online Edition); British Society for Geomorphology: London, UK, 2015; ISSN 2047-0371.
4. Triska, F.J.; Kennedy, V.C.; Avanzino, R.J. Retention and transport of nutrients in a third-order stream in northwestern California: Hyporheic processes. *Ecology* **1989**, *70*, 1893–1905. [CrossRef]
5. Elliott, A.H.; Brooks, N.H. Transfer of nonsorbing solutes to a streambed with forms: Theory. *Water Resour. Res.* **1997**, *33*, 123–136. [CrossRef]
6. Kasahara, T.; Wondzell, S.M. Geomorphic controls on hyporheic exchange flow in mountain streams. *Water Resour. Res.* **2003**, *39*, 1–14. [CrossRef]
7. Wondzell, S.M. Effect of morphology and discharge on hyporheic exchange flows in two small streams in the Cascade Mountains of Oregon, USA. *Hydrol. Processes* **2006**, *20*, 267–287. [CrossRef]
8. Boulton, A.J. Hyporheic rehabilitation in rivers: Restoring vertical connectivity. *Freshw. Biol.* **2007**, *52*, 632–650. [CrossRef]
9. Zhang, L.; Li, P.; He, X. Interactions between surface water and groundwater in selected tributaries of the Wei River (China) revealed by hydrochemistry and stable isotopes. *Hum. Ecol. Risk Assess.* **2022**, *28*, 79–99. [CrossRef]
10. Singh, M.; Ansari, A.A.; Muller, G.; Singh, I.B. Heavy metals in freshly deposited sediments of the Gomati River (a tributary of the Ganga River) effects of human activities. *Environ. Geol.* **1997**, *29*, 246–252. [CrossRef]
11. Calmano, W.; Ahlf, W.; Forstner, U. Exchange of Heavy Metals Between Sediment Components and Water. In *Metal Speciation in the Environment*; Springer: Berlin/Heidelberg, Germany, 1990; pp. 503–522.
12. Batley, G.E. *Trace Element Specification: Analytical Methods and Problems*; CRC Press: Boca Raton, FL, USA, 2000; ISBN 9780849347122.
13. Altın, A.; Filiz, Z.; Iscen, C.F. Assessment of seasonal variations of surface water quality characteristics for Porsuk Stream. *Environ. Monit. Assess.* **2009**, *158*, 51–65. [CrossRef]
14. Muhammad, S.; Shah, M.T.; Khan, S. Heavy metal concentrations in 845 soil and wild plants growing around Pb-Zn sulfide terrain in the Kohistan region, northern Pakistan. *Microchem. J.* **2011**, *99*, 67–75. [CrossRef]
15. Sun, Z.; Mou, X.; Tong, C.; Wang, C.; Xie, Z.; Song, J.; Sun, W.; Lv, Y. Spatial variations and bioaccumulation of heavy metals in intertidal zone of the Yellow River estuary, China. *Catena* **2015**, *126*, 43–52. [CrossRef]
16. Younger, P.L.; Banwart, S.A.; Hedin, R.S. *Mine Water: Hydrology, Pollution, Remediation*; Kluwer Academic Publishers: London, UK, 2002; 442p.
17. Stamatis, G.; Voudouris, K.; Karefilakis, F. Groundwater pollution by heavy metals in historical mining area of Lavrio, Attica, Greece. *Water Air Soil Pollut* **2001**, *128*, 61–83. [CrossRef]
18. Bird, G.; Macklin, M.G.; Brewer, P.A.; Zaharia, S.; Balteanu, D.; Driga, B.; Serban, M. Heavy metals in potable groundwater of mining-affected river catchments, northwestern Romania. *Env. Geochem. Health* **2009**, *31*, 741–758. [CrossRef] [PubMed]
19. Al-Hobaib, A.S.; Al-Jaseem QKh Baioumy, H.M.; Ahmed, A.H. Heavy metals concentrations and usability of groundwater at Mahd Adh Dhahab gold mine, Saudi Arabia. *Arab. J. Geosci.* **2013**, *6*, 259–270. [CrossRef]
20. Liu, Q.J.; Huang, Y.; Hu, S.Y.; Deng, K.W.; Huang, D.; Cao, L. Chemical Forms of Hg, Pb, Cd, Cu and Zn in Beijing Urban Soil and Its Environmental Effects. *Urban Geol.* **2015**, *10*, 11–15.
21. Clifton, J.C., 2nd. Mercury exposure and public health. *Pediatr. Clin. N. Am.* **2007**, *54*, 237–269. [CrossRef]
22. US Department of Health and Human Services, Public Health Service. *Toxicological Profile for Mercury*; US Department of Health and Human Services: Atlanta, GA, USA, 1999; pp. 1–600.
23. Rice, G.E.; Ambrose, R.B., Jr.; Bullock, O.R., Jr.; Smawtout, J. *Mercury Study Report to Congress*; US Environmental Protection Agency: Durham, NC, USA, 1997; pp. 1.1–6.30.
24. Marcello, M.; Veiga, O.F. A review of the failed attempts to curb mercury use at artisanal gold mines and a proposed solution. *Extr. Ind. Soc.* **2020**, *7*, 1135–1146.
25. Mantey, J.; Nyarko, K.B.; Owusu-Nimo, F.; Awua, K.A.; Bempah, C.K.; Amankwah, R.K.; Akatu, W.E.; Appiah-Effah, E. Mercury contamination of soil and water media from different illegal artisanal small-scale gold mining operations (galamsey). *Heliyon* **2020**, *6*, 1–13. [CrossRef]

26. Goldman, L.R.; Shannon, M.W. American Academy of Pediatrics: Committee on Environmental Health. Technical report: Mercury in the environment: Implications for pediatricians. *Pediatrics* **2001**, *108*, 197–205. [CrossRef]
27. Harada, M.; Nakachi, S.; Cheu, T.; Hamada, H.; Ono, Y.; Tsuda, T.; Yanagida, K.; Kizaki, T.; Ohno, H. Monitoring of mercury pollution in Tanzania: Relation between head hair mercury and health. *Sci. Total Environ.* **1999**, *227*, 249–256. [CrossRef]
28. Guzzi, G.; La Porta, C.A. Molecular mechanisms triggered by mercury. *Toxicology* **2008**, *244*, 1–12. [CrossRef] [PubMed]
29. Ma, W.B.; Chen, Q.Y.; Yin, D.L.; Sun, T.; Wang, Y.M.; Wang, D.Y. Temporal and Spatial Variation of Mercury in the Water of the Ruxi River Estuary, a Typical Tributary of the Three Gorges Reservoir Area. *Environ. Sci.* **2019**, *40*, 2211–2218.
30. Zhang, Y.F. Study on Migration Mechanism of Lead in Undulating Zone of Groundwater. Master's Thesis, North China University of Water Resources and Hydropower, Zhengzhou, China, 2021.
31. Xiong, J.Q. Study on the Effect of DOM on the Bioavailability of Heavy Metals under Poyang Lake Water Level. Master's Thesis, Nanchang University, Nanchang, China, 2021.
32. Wang, P.A. Dynamic Variation of Mercury Species in Soil in Wet-Dry Rotation Environment. Master's Thesis, Southwest University, Chongqing, China, 2007.
33. Tang, X.C.; Lin, T.; Xia, P.H.; Huang, X.F.; Ma, L.; Yang, Y. Speciation distribution and risk assessment of Hg and As in sediment of Lake Caohai wetlands under different water level gradients, Guizhou Province. *J. Lake Sci.* **2020**, *32*, 100–110.
34. Zhang, C.; Chen, H.; Sun, R.G.; Zhang, X.; Wang, D.Y. Risk Assessment of Mercury in Soil at Different Water-level Altitudes in Drawdown Areas of Three Gorges Reservoir. *J. Soil Water Conserv.* **2014**, *28*, 0242–0246.
35. Wu, Z.R. Experimental Study on the Vertical Migration of Typical Heavy Metals from the Saturated Zone to Hyporheic zone. Master's Thesis, Hefei University of Technology, Hefei, China, 2020.
36. Tian, Y.; Yang, W.C. Analysis on Migration of Heavy Metals in Groundwater Fluctuation Zone. *Environ. Sci. Manag.* **2020**, *44*, 40–44.
37. Li, Y.; Li, P.; Liu, L. Source Identification and Potential Ecological Risk Assessment of Heavy Metals in the Topsoil of the Weining Plain (Northwest China). *Expo. Health* **2022**, *14*, 281–294. [CrossRef]
38. Liu, Y.; Zhao, J.E.; Zhang, F.S. Pollution status and potential ecological risk assessment of heavy metals in the surface sediments of Qingdao intertidal zone. *Environ. Pollut. Control* **2021**, *43*, 492–501.
39. Dai, Q. Environmental geochemistry study of mercury contamination from gold mining areas by amalgamation technique in China—a case study from Tongguan county of Shaanxi Province. Master's Thesis, Institute of Geochemistry, Chinese Academy of Sciences, Beijing, China, 2004.
40. Xu, Y.N.; Ke, H.L.; Zhao, A.I. Assessment of heavy metals contamination of farmland soils in some gold mining area of Xiao Qinling. *Chin. J. Soil* **2007**, *4*, 24.
41. Jiang, N.; Xu, J.; Song, M. Fluid inclusion characteristics of mesothermal gold deposits in the Xiaoqinling district, Shaanxi and Henan Provinces, People's Republic of China. *Min. Depos.* **1999**, *34*, 150–162.
42. Liu, R.P.; Xu, Y.N.; Rui, H.C.; El-Wardany, R.M.; Ying, D. Migration and speciation transformation mechanisms of mercury in undercurrent zones of the Tongguan gold mining area, Shaanxi Loess Plateau and impact on the environment. *China Geol.* **2021**, *4*, 311–328.
43. Xiao, R.; Wang, S.; Li, R.H.; Wang, J.J.; Zhang, Z.Q. Soil heavy metal contamination and health risks associated with artisanal gold mining in Tongguan, Shaanxi, China. *Ecotoxicol. Environ. Saf.* **2017**, *141*, 17–24. [CrossRef] [PubMed]
44. Xu, Y.N.; Zhang, J.H. Contents of heavy metals in bottom sediments of the Taiyu River in the Tongguan gold mining ar ea, Shaaxi, China, and contamination assessments. *Geol. Bull. China* **2008**, *27*, 1263–1671.
45. Li, G.; Lu, N.; Wei, Y.; Zhu, D.W. Relationship between Heavy Metal Content in Polluted Soil and Soil Organic Matter and pH in Mining Areas. *IOP Conf. Ser. Mater. Sci. Eng.* **2018**, *394*, 052081. [CrossRef]
46. Wu, Y.; Xu, Y.; Zhang, J.; Hu, S.; Liu, K. Heavy metals pollution and the identification of their sources in soil over Xiaoqinling gold-mining region, Shaanxi, China. *Environ. Earth Sci.* **2011**, *64*, 1585–1592. [CrossRef]
47. Li, Z.; Liu, G.J.; Shen, M.C.; Liu, Y. Potential ecological and health risks of heavy metals for indoor and corresponding outdoor dust in Hefei. *Cent. China. Chemosphere* **2022**, *302*, 134864.
48. Hakanson, L. An Ecological risk index for quality pollution control:a sedimentological approach. *Water Res.* **1980**, *14*, 975–1001. [CrossRef]
49. Zhao, A.N. The Assessment of Contamination And Corerlation on Heavy Metal between Farmland and Crops in Tongguan Gold Mining Area in Shaanxi. Master's Thesis, Chang'an University, Xi'an, China, 2006; pp. 1–73.
50. Wang, X.W. The Research about the Relationship between Heavy metals contamination and Crops on farmland soi1 in the Goldfield. Master's Thesis, Chang'an University, Xi'an, China, 2010; pp. 1–84.
51. Zhou, Y.; Wang, X.M.; Jiang, Y.Z.; Zhao, Y.F.; Ji, H.B. Speciation and ecological risk assessment of arsenic and mercury in soil around a gold mining area in pinggu district, beijing. *Environ. Eng.* **2021**, *39*, 164+204–210.
52. Liu, R.P.; Xu, Y.N.; Zhang, J.H. The Relationship and Influencing Factors of Water-Gas Interface Mercury Emission Flux and Water Suspended Mercury of a Gold Mining Area River. In *Sustainable Development of Water Resources and Hydraulic Engineering in China*; Springer Cham: Cham, Switzerland, 2018.
53. Dong, W.; Lian, Y.Q.; Zhang, Y. (Eds.) *Environmental Earth Sciences*; Springer Science+Business Media: Berlin, Germany, 2018; Volume 35, pp. 407–415.

54. DOU, W.Q.; Yi, A.N.; Li, Q.; Zeng, Q.N. Xia, Q. Research progress on effects of soil pH on migration and transformation of mercury. *J. Agric. Resour. Environ.* **2019**, *36*, 1–8.
55. Ran, M.X.; Yang, H.Y.; Huang, J.Y.; Li, Z.; Chen, Y.; Feng, Q.Z.; Li, Y.; Liu, L.Y. Effect of humic acids on the mercury speciation in soils of Wanshan area in Guizhou Province. *J. Anhui Agric. Univ.* **2020**, *47*, 95–102.
56. Guang, Q.; Li, L. Advancement in the hyporheic exchange in rivers. *Adv. Water Sci.* **2008**, *19*, 285–293.
57. Salehin, M.; Packman, A.; Paradis, M. Hyporheic exchange with heterogeneous streambeds: Laboratory experiment sand modeling. *Water Resour.* **2004**, *40*, W11504.
58. Xu, Y.N.; Zhang, J.H.; Chen, S.B.; Ke, H.L. Distribution of Contents of Heavy Metals in Soil Profile Contaminated in Different Ways in Xiaoqinling Gold Mining Area. *J. Agro-Environ. Sci.* **2008**, *27*, 0200–0206.

Article

Hydrochemical Characteristics and Formation Mechanism of Strontium-Rich Groundwater in Tianjiazhai, Fugu, China

Chengcheng Liang [1,2], Wei Wang [1,2,*], Xianmin Ke [1,2], Anfeng Ou [1,2] and Dahao Wang [1,2]

1. School of Water and Environment, Chang'an University, No. 126 Yanta Road, Xi'an 710054, China; 2020129022@chd.edu.cn (C.L.); 2020029009@chd.edu.cn (X.K.); 2020129020@chd.edu.cn (A.O.); 2018129018@chd.edu.cn (D.W.)
2. Key Laboratory of Subsurface Hydrology and Ecological Effects in Arid Region, Chang'an University, Ministry of Education, No. 126 Yanta Road, Xi'an 710054, China
* Correspondence: wangweichd@chd.edu.cn

Abstract: Strontium-rich groundwater exists in the underlying carbonate rocks of the Tianjiazhai Shimachuan River basin, Fugu, China. In this study, the hydrochemical characteristics and formation mechanisms of Sr-rich groundwater were assessed using mathematical statistics and traditional water chemistry, combining geological and hydrogeological conditions, as well as hydrogeochemical theory. The results showed that the Sr^{2+} content range in Sr-rich groundwater was 0.85~2.99 mg·L^{-1}, which is weakly alkaline fresh water. HCO_3^- Ca·Mg·Na was the main facies type of Sr-rich groundwater. Sr-rich groundwater has relatively stable contents of chemical elements. The water–rock interaction was the main factor controlling the hydrochemical characteristics of Sr-rich groundwater, particularly carbonate dissolution, influenced by some degree of cation exchange. The Sr element in groundwater mainly comes from the dissolution of the sandstone of the Yanchang Formation. The higher the degree of weathering and the longer the water–rock reaction time, the more favorable the dissolution and enrichment of Sr in groundwater. Moreover, the large weathering thickness and fracture development of the rocks in the Tianjiazhai area provide favorable conditions for the formation of Sr-rich groundwater. The results of this study provide a scientific basis for developing effective policies to protect Sr-rich groundwater resources.

Keywords: strontium-rich groundwater; hydrochemical characteristics; formation mechanisms; hydrogeochemistry

1. Introduction

As an important source of water supply for human beings and an important component of the hydrological cycle, groundwater is of great importance for the sustainable development of the socio-economy and the protection of ecological stability [1,2]. Groundwater hydrochemical characteristics are controlled by natural and anthropogenic factors and are the result of long-term interactions between groundwater and the environment. Therefore, hydrochemical analysis of groundwater can reveal the hydrochemical formation mechanisms [3,4]. In natural conditions, the hydrochemical characteristics of groundwater are mainly controlled by dissolution, concentration, desulfation/carbonation, cation exchange, and mixing processes [5,6]. However, various changes in the hydrochemical characteristics of groundwater may occur under the influence of anthropogenic activities [7,8]. Under natural conditions, elements in rock layers are enriched into groundwater by hydrogeochemical processes, which can lead to mineralization and contamination of groundwater [9]. For example, under natural conditions, large amounts of dissolution of soluble rocks (carbonate rocks) can lead to mineralization of groundwater. In addition, trace elements (arsenic and fluorine) may be enriched to levels threatening human health under natural hydrogeochemistry [10,11]. A series of groundwater pollution problems have emerged under the influence of human activities, such as nitrate pollution of groundwater.

Multivariate statistical analysis and hydrogeochemistry are often used to investigate the hydrochemical characteristics of groundwater and their determining factors [12]. Recent studies on groundwater hydrochemistry have used a combination of several methods (mathematical statistics, water chemistry, hydrogeochemical simulation, etc.) to assess the hydrochemical characteristics of groundwater [13–15].

Strontium (Sr) is one of the essential trace elements for human health, which maintains human physiological functions [16,17]. Strontium in groundwater is derived from Sr in the surrounding rock and enters the groundwater through a leaching process via infiltration of atmospheric precipitation [18,19]. Suitable hydrogeological conditions are the basic conditions for the formation of strontium-rich groundwater. In fact, the extent of Sr-enrichment in the groundwater depends on the hydrogeochemical processes occurring in aquifers [20]. The amount of Sr in the surrounding rock entering the groundwater system is positively correlated with the degree of rock weathering. In addition, CO_2 content, temperature, and redox conditions are the important factors affecting the Sr enrichment in groundwater [21]. The Sr abundance in the surrounding rocks determines the Sr content in groundwater. Long-distance runoff and long-path circulation of groundwater increase the water–rock reaction time, thus promoting strontium enrichment [22]. The formation of Sr-rich groundwater is controlled by tectonic magmatic activity, Sr content in surrounding rocks, and hydrogeochemical conditions [23]. In recent years, numerous researchers have devoted considerable attention to the formation mechanism and enrichment factors of Sr in groundwater [24–26].

Fugu County is located in the Loess Plateau area in Northwest China, where surface water resources are limited. Indeed, groundwater has long been the source of local water supply in the area. However, in recent years, with the exploitation of coal resources and the development of the industrial economy, groundwater resource in Fugu County has become overexploited and inefficient. In the Shimachuan River basin in the Tianjiazhai area of Fugu County, Sr-rich groundwater exists in the underlying carbonate strata. The population of Shimachuan River Basin is 10,647 people, and the main water source for the production and living of local residents is groundwater. However, there are few studies on the hydrochemical characteristics and formation mechanism of Sr-rich groundwater in this area. Therefore, in this study, the hydrochemical characteristics of Sr-rich groundwater and its formation mechanism were assessed using descriptive analyses, Piper trilinear diagrams, Gibbs plots, ionic ratios, and correlation analysis. In addition, the formation conditions of Sr in groundwater were analyzed by combining geological and hydrogeological conditions as well as hydrogeochemical theory to reveal the formation mechanism of Sr in groundwater.

2. Materials and Methods

2.1. Study Area

The study area is located in the Shimachuan River basin of Fugu County in the eastern part of the Loess Plateau in North Shaanxi in China, belonging to the typical semi-arid continental monsoon climate (Figure 1). The terrain of the study area is generally high in the west and low in the east, with elevations ranging from 940 m to 1100 m. The geomorphology of the study area mainly consists of Loess mountains and river valley terraces. Loess mountains are mainly distributed on both sides of Shimachuan valley, with an elevation of 1000~1100 m and a width of 100~250 m, and incline to both sides of the valley at a slope angle of 10~20°. The Loess peak is narrow, along the watershed there are large ups and downs, and the ground is very broken, forming a unique Loess landform. The Loess Mountain area has uniform lithology, good water permeability, small catchment area and serious soil erosion. Valley terraces are mainly distributed along the Shimachuan river, with an elevation of 960~1100 m. The valley terraces are long, sporadic and intermittent. The area of the valley terraces is small. The front edge of the valley terraces is 2~3 m higher than the river bed, 15~40 m wide, and most of the terraces tilt to the river bed at a slope angle of 3~5°. The overall surface water network is poorly developed in the study area. The Shimachuan River in the study area is a first-class tributary of the west bank of the Yellow

River. In addition, groundwater is the main water supply source in the study area, stored at the bottom of the Quaternary formation and in bedrock fissures. Sr-rich groundwater in the study area is mainly stored in the weathered and fractured aquifer (unconfined) of the Yanchang Formation. Its recharge mainly occurs from atmospheric precipitation and the unconfined water of the loose layer of the Quaternary formation and surface water, which eventually discharges at the downstream outlet of the Shimachuan River. The stratigraphy of the Yanchang formation is covered by loose sediments of the Quaternary formation, with outcrops in the river valley; the thickness of strata is 200~411 m, the maximum exposed thickness is 65 m, and the thickness of the weathering layer is 20~43.26 m. In addition, the lithology of the Yanchang formation consists of a thickly laminated light gray-green medium to fine-grained feldspathic sandstone with medium sorting and roundness. It has large inclined beddings, wedge-shaped beddings and massive beddings. The study area is located in the northeast of the secondary tectonic units of Ordos Basin, with stable geological structures and oscillatory rise caused by the neotectonic movement. There is no trace of magmatic activity in the area. The overall structure is monoclinal with a dip direction of 292° and a dip angle of 1~4°.

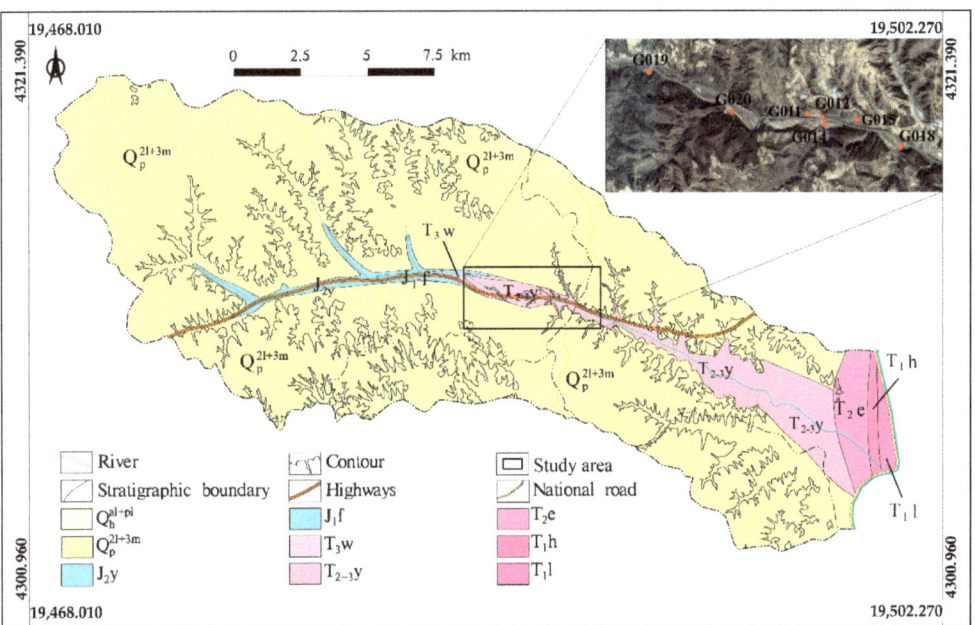

Figure 1. Regional geological map of Tianjiazhai East Ditch, Fugu County. The geographical location of the study area belongs to the 19th projection band of the Gaussian projection six-degree band, and the coordinates used in the figure are the Gaussian projection six-degree band coordinate system.

2.2. Sampling

In this study, 30 groundwater samples were first collected from hydrogeological boreholes in the study area during the March 2018 to September 2019 period, according to the Technical Requirements for Water Sample Collection and Delivery issued by the China Geological Survey. These were then analyzed for groundwater quality, the test indexes including pH, total dissolved solids (TDS), total hardness, cation (K^+, Na^+, Ca^{2+} and Mg^{2+}), anion (Cl^-, HCO_3^-, SO_4^{2-} and NO_3^-) and strontium elements. The rock type of the sampling point was sandstone of Yanchang Formation. In addition, 25 and 24 rock samples were collected at different depths from drilling G019 and G020 in the study area, respectively, while 22 rock samples were collected from a selected profile between drilling G020 and the Temple River Gully. All the rock samples were analyzed for Sr content.

2.3. Methods

The pH value was determined by water quality tester (SD150), the total dissolved solids (TDS) was determined by gravimetric method (GR), the total hardness was determined by ethylene diamine tetraacetic acid disodium titration, and the contents of cations K^+, Na^+, Ca^{2+} and Mg^{2+} were determined by atomic absorption spectrometer (ICP-AES) [27–29]. The contents of anions Cl^-, SO_4^{2-} and NO_3^- were determined by ion chromatograph (IC-90), and HCO_{3-} was determined by hydrochloric acid titration [30,31]. Sr^{2+} was determined by plasma mass spectrometry (ICP-MS). The content of strontium (Sr) in rock was determined by EDTA volumetric method [32,33]. The free carbon dioxide in groundwater was determined by the volumetric method of standard alkali solution [34].

In this study, the descriptive statistics of physicochemical parameters of groundwater were performed using SPSS 22.0 software. Mineral saturation indexes (SI) were calculated by hydrochemical simulation software Phreeqc, and the database was the phreeqc.dat thermodynamic database [35]. Pearson correlation analysis was performed to evaluate the relationship between physicochemical parameters of groundwater. In addition, groundwater chemical types were classified using the Shukarev classification method and Piper trilinear diagram, while the main controlling factors of groundwater chemistry in the study area were determined using the Gibbs diagram. The main ion sources of Sr-rich groundwater were studied using ionic ratio diagrams. The formation mechanism of Sr in groundwater was investigated based on the geological and hydrogeological conditions as well as hydrogeochemical theory.

3. Results and Discussion
3.1. Hydrochemical Characteristics
3.1.1. Descriptive Analysis of Hydrochemical Characteristics

As reported in Table 1, the Sr^{2+} content in Sr-rich groundwater in the study area ranged from 0.85 to 2.99 mg·L^{-1}, with a mean value of 1.45 mg·L^{-1}, which is 7.25 times higher than the national limit value of Sr^{2+} content in natural mineral water for drinking (0.20 mg·L^{-1}). Compared with Sr^{2+} content in Sr-rich groundwater in other areas of China, the Sr^{2+} content in groundwater in the study area is close to that in Xintian County of Hunan Province (1.57 mg·L^{-1}), and lower than that in Guanling area of Guizhou Province (2.43 mg·L^{-1}). The Sr^{2+} content in groundwater in the study area is higher than that in Changchun, Jilin province (0.29 mg·L^{-1}), Xianning, Hubei Province (0.63 mg·L^{-1}), Yutian county, Xinjiang Province (1.1 mg·L^{-1}) and Pingxiang city, Jiangxi Province (1.29 mg·L^{-1}). Sr-rich groundwater pH range was 7.50~8.30, with an average value of 7.92, indicating a slightly alkaline Sr-rich groundwater. Part of the water samples in the study had a high PH value of 8.3, mainly because there was a little amount of CO_3^{2-} in the groundwater. CO_3^{2-} will combine with H^+ in water, leading to the decrease of hydrogen ion concentration in groundwater. The pH value of Sr-rich groundwater meets the requirements of the Chinese Standards (6.5–8.5) and the WHO Guidelines (6.5–8.5). TDS (total dissolved solids) ranged from 332 to 718 mg·L^{-1}, with a mean value of 492.57 mg·L^{-1}, showing low TDS levels in Sr-rich groundwater. The total hardness ranged from 205.00 to 825.00 mg·L^{-1}, with a mean value of 324.90 mg·L^{-1}, suggesting medium-hard and ultra-hard water. The TDS of Sr-rich groundwater meets the requirements of the Chinese Standards (1000 mg·L^{-1}) and the WHO Guidelines (1000 mg·L^{-1}). The dominant cations in Sr-rich groundwater were Ca^{2+}, Na^+, and Mg^{2+}, with average contents of 71.09, 60.57, and 31.77 mg·L^{-1}, accounting for 42.73, 36.41, and 19.10% of total cation contents, respectively, whereas K^+ showed the lowest content in Sr-rich groundwater, with an average value of 2.92 mg·L^{-1}, accounting for only 1.76% of total cation contents. The Na^+ content of Sr-rich groundwater is lower than the maximum Na^+ content required by the Chinese Standards (200 mg·L^{-1}) and the WHO Guidelines (200 mg·L^{-1}). On the other hand, HCO_{3-} was the dominant anion in

Sr-rich groundwater, with mean values of 348.80 mg·L^{-1}, accounting for 73.52% of total anion contents, whereas NO$_3^-$ showed the lowest content in Sr-rich groundwater, with an average value of 7.76 mg·L^{-1}, accounting for only 1.63% of total anion contents. The Cl$^-$ content of Sr-rich groundwater is lower than the maximum Cl$^-$ content required by the Chinese Standards (250 mg·L^{-1}) and the WHO Guidelines (250 mg·L^{-1}). The SO$_4^{2-}$ content of Sr-rich groundwater is lower than the maximum SO$_4^{2-}$ content required by the Chinese Standards (250 mg·L^{-1}) and the WHO Guidelines (500 mg·L^{-1}). The NO$_3^-$ content of Sr-rich groundwater is lower than the maximum NO$_3^-$ content required by the WHO Guidelines (50 mg·L^{-1}) and higher than the maximum NO$_3^-$ content required by the Chinese Standards (20 mg·L^{-1}). This is due to the agricultural activities of local residents in the study area, resulting in the increase of NO$_3^-$ content. The coefficients of variation of the hydrochemical parameters of Sr-rich groundwater were all less than 1, indicating relatively stable contents of these hydrochemical components in Sr-rich groundwater. This finding may be due primarily to the fact that Sr-rich groundwater in the study area is strongly controlled by geological conditions and less influenced by human activities.

Table 1. Statistics on the hydrochemical characteristics of Sr-rich groundwater.

Projects	mg·L^{-1}											pH
	K$^+$	Na$^+$	Ca^{2+}	Mg^{2+}	Cl$^-$	SO$_4^{2-}$	HCO$_3^-$	NO$_3^-$	Sr^{2+}	TH *	TDS	
Min	1.63	33.6	39.2	23.1	17.7	33.6	272	1.98	0.85	205	332	7.5
Max	5.59	89.6	122	40.4	126	81.7	415	30.3	2.99	825	718	8.3
Mean	2.92	60.57	71.09	31.77	56.72	61.16	348.8	7.76	1.45	324.9	492.57	7.92
SD	1.01	15.36	22.27	4.02	36.79	13.08	40.94	5.74	0.51	115.25	112.14	0.21
CV/%	34.53	25.36	31.33	12.66	64.86	21.38	11.74	73.97	34.92	35.47	22.77	2.64
CS [1]	/	200	/	/	250	250	/	20	/	/	1000	6.5–8.5
WG [2]	/	200	/	/	250	500	/	50	/	/	1000	6.5–8.5

* TH is the total hardness, [1] CS is the Chinese Standards, [2] WG is the WHO Guidelines.

3.1.2. Hydrochemical Facies

Piper's trilinear diagram can be used to characterize the total chemical properties and major ionic composition changes of a water body [36,37]. As shown in Figure 2, the groundwater sample points were concentrated in zone 4 of the diamond-shaped domain of the Piper trilinear diagram, indicating that the chemistry of Sr-rich groundwater in the study area is dominated by alkaline earth metal elements (calcium and magnesium) rather than the alkali metal elements (sodium). On the other hand, the groundwater sample points were concentrated in the E region of the anion triangle of the Piper trilinear diagram, indicating the dominance of weak acidic anions (carbonic acid), whereas the observed cations were concentrated in the B region of the Piper trilinear diagram (cationic triangular domain), suggesting a non-dominant type. In general, the distribution of Sr-rich groundwater sampling sites in the study area was relatively concentrated, showing six main types, among which HCO$_3^-$ Ca·Mg·Na was the main groundwater facies type, accounting for 70% of the total groundwater samples, followed by HCO$_3^-$ Ca·Na·Mg and HCO$_3^-$ Ca·Mg groundwater facies types, both accounting for 20% of the total groundwater samples, while HCO$_3^-$ Na·Mg·Ca, HCO$_3^-$ Mg·Ca·Na and HCO$_3^-$ Mg·Na·Ca groundwater facies type were less abundant. In addition, TIS salinity diagram can further characterize the chemical composition of water [38]. As shown in Figure 3, (Cl$^-$ + HCO$_3^-$) content is 5.11~10.21 meq·L^{-1}, SO$_4^{2-}$ content is 1.4~3.6 meq·L^{-1}. Indeed, the total ionic salinity (TIS) of water ranges from 14.2 to 26.4 meq·L^{-1}. The strontium-rich groundwater has high TIS value, which indicates that the strontium-rich groundwater is the result of a long-term evolution of the groundwater system in the study area.

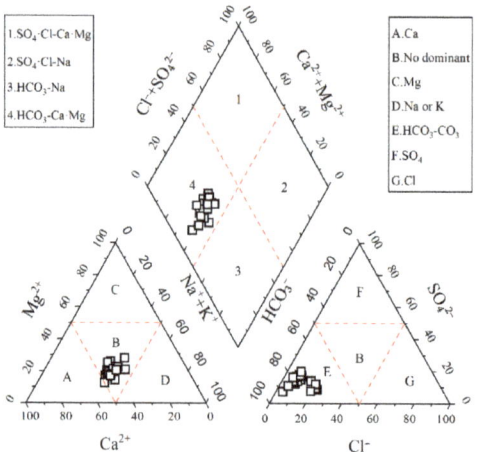

Figure 2. Piper trilinear diagram of Sr-rich groundwater.

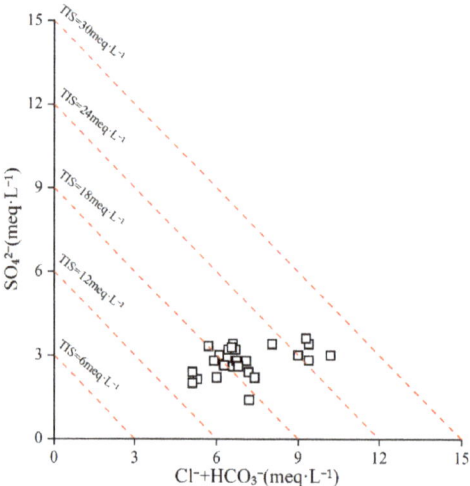

Figure 3. TIS salinity diagram of Sr-rich groundwater.

3.1.3. Correlation Analysis of Hydrochemical Compositions

Correlation analysis can reveal the consistency and variability of the sources of groundwater hydrochemical parameters. As shown in Figure 4, Cl^- showed a high positive correlation with Na^+ and Ca^{2+} and moderate correlation with Mg^{2+}, with correlation coefficients of 0.92, 0.94, and 0.59, respectively. HCO_3^- showed moderately positive correlations with Na^+ and Ca^{2+}, with correlation coefficients of 0.74 and 0.69, respectively, while SO_4^{2-} and NO_3^- revealed negative correlations with all cations in groundwater. NO_3^- showed a moderate negative correlation with K^+, with a correlation coefficient of −0.56. As for Cl^-, it showed a moderate positive correlation with HCO_3^-, with a correlation coefficient of 0.59. Sr^{2+} showed a high correlation with Na^+ and Cl^-, with correlation coefficients of 0.85 and 0.84, respectively, suggesting that the source of Sr in groundwater is closely related to the effect of atmospheric precipitation. In addition, Sr^{2+} revealed moderate positive correlations with Ca^{2+} and HCO_3^-, with correlation coefficients of 0.73 and 0.59, respectively, indicating that the source of Sr may be the same as these two ions. The correlation between TDS and ions can better reflect the genesis of groundwater. Indeed, TDS showed positive correlation coefficients with Cl$^-$ and Na+ of 0.81 and 0.76, indicating high and moderate correlation, respectively, and an atmospheric precipitation source of groundwater. The correlation coefficients of TDS with Ca^{2+} and HCO$_3^-$ were 0.84 and 0.60, showing high and moderate correlations, respectively, indicating that Ca^{2+} and HCO$_3^-$ contribute significantly to the formation of groundwater chemistry types in the carbonate-rich rock study area.

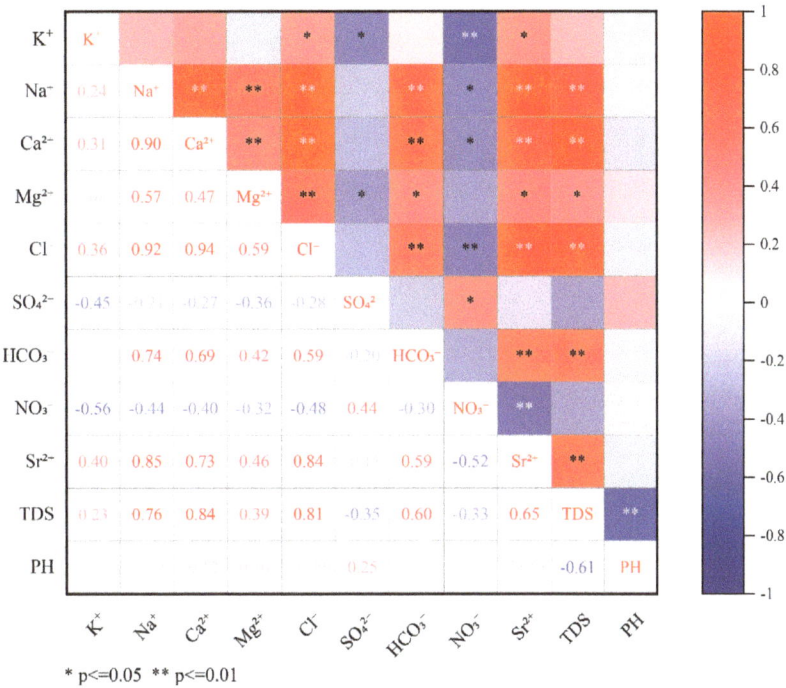

* $p<=0.05$ ** $p<=0.01$

Figure 4. Correlation coefficients between hydrochemical parameters of Sr-rich groundwater.

3.1.4. Controlling Factors and Natural Processes

A Gibbs diagram is an important tool for analyzing the chemical genesis of water and can be used to analyze the mechanism of ion formation in water bodies [39,40]. In the Gibbs diagram, the ion control mechanism is classified into three main effects, namely rock weathering, evaporation-concentration, and atmospheric precipitation. The variation of Cl$^-$/(Cl$^-$ + HCO$_3^-$) and Na$^+$/(Na$^+$ + Ca^{2+}) ratios in Sr-rich groundwater samples in the study area varied slightly from 0.04 to 0.25 and from 0.42 to 0.56, with mean values of 0.13 and 0.56, respectively. In addition, TDS values were at a moderate level, varying from 332 to 718 mg·L^{-1}, with a mean value of 493 mg·L^{-1}. As shown in Figure 5, the Sr-rich groundwater samples were mainly located in the rock weathering area of the Gibbs diagram, indicating that rock weathering was the most influential mechanism in the genesis of Sr-rich groundwater in the study area. In addition, some Sr-rich groundwater samples were close to the evaporation-concentration zone, indicating the slight influential effect of the evaporation-concentration process on the hydrochemical characteristics of Sr-rich groundwater in the study area. However, the groundwater samples were all far away from the atmospheric precipitation zone, suggesting the lack of any significant effect of atmospheric precipitation on the water chemistry of Sr-rich groundwater.

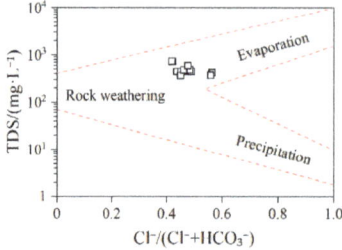
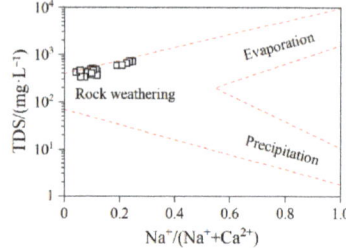

Figure 5. Gibbs plot of Sr-rich groundwater.

In this study, the direction and strength of the cation exchange interaction were assessed using the chloro-alkaline indices [41], namely CAI-I and CAI-II. These indices were calculated using the following formulas:

$$CAI-I = \frac{Cl^- - (Na^+ + K^+)}{Cl^-} \quad (1)$$

$$CAI-II = \frac{Cl^- - (Na^+ + K^+)}{HCO_3^- + SO_4^{2-} + CO_3^{2-} + NO_3^-} \quad (2)$$

When both CAI-I and CAI-II values are less than 0, it indicates that Ca^{2+} or Mg^{2+} in groundwater exchanged ions with Na^+ in aquifer minerals. When CAI-I and CAI-II values are greater than 0, it indicates that Na^+ in the groundwater exchanged ions with Ca^{2+} or Mg^{2+} in the aquifer minerals. In addition, the higher the absolute values of CAI-I and CAI-II the chloride index, the higher the degree of ion exchange.

The results showed that CAI-I and CAI-II varied from −0.23 to 0.01 and from −3.71 to 0.004, with a mean value of −0.128 and −1.155, respectively (Figure 6a). In addition, almost groundwater samples revealed negative values of chloro-alkaline indices, except for a few points. This result suggests that the cation exchange process generally occurs in groundwater in the study area, more specifically replacement of Na^+ with Ca^{2+} and Mg^{2+} from aquifer minerals to the groundwater. Except for a few samples, relatively high absolute values of the chloro-alkaline indices were observed in Sr-rich groundwater samples, suggesting a strong degree of cation exchange. A very small number of water samples do not undergo alternate cation adsorption, which may be due to the influence of some other sources. In addition, the relationship between $Na^+ + K^+$ and Cl^- can also be investigated to assess cation exchange. As shown in Figure 6b, the water samples were distributed above the $[(Na^+ + K^+) - Cl^-]$ ratio 1:1 line, indicating a significant enrichment of groundwater by $(Na^+ + K^+)$ compared to Cl^-. Indeed, except for rock salt and potassium salt dissolution, there are also other sources of Na^+ and K^+, mainly due to the presence of a certain degree of cation exchange to increase the Na^+ content.

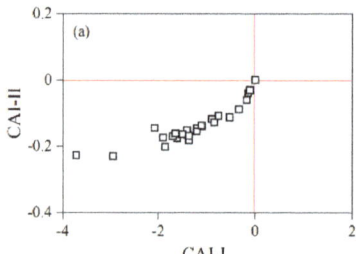
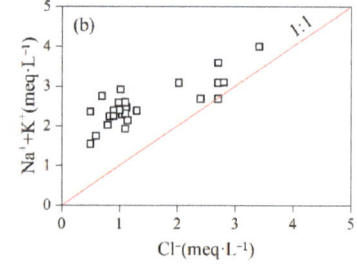

Figure 6. Chloro-alkaline indices (**a**) and the relationship between $Na^+ + K^+$ and Cl^- (**b**) in Sr-rich groundwater.

The mineral saturation index (SI) can be used to indicate the dissolved equilibrium state of groundwater components. The results showed positive values of saturation indices of dolomite, gypsum, and calcite (Table 2), indicating that all three minerals were in supersaturated states and were in sedimentation states during groundwater evolution. However, the saturation index range of rock salt in the study area was negative, indicating that the rock salt was in a dissolved state during groundwater evolution.

Table 2. Mineral saturation index in Sr-rich groundwater.

SI	Min	Max	Mean	SD
Dolomite	5.24	5.98	5.55	0.29
Gypsum	1.97	2.95	2.36	0.37
rock salt	−5.73	−3.92	−5.13	0.75
Calcite	4.53	4.92	4.71	0.14

3.1.5. Qualitative Analysis of Ion Sources

Water bodies are constantly interacting with the natural environment through various hydrogeochemical processes. In the process of groundwater circulation, hydrochemical ratios have a certain regularity, which can be used to assess the water chemistry and its main sources [42].

The Na^+/Cl^- ratio can characterize the degree of Na^+ enrichment in groundwater [43]. According to the correlation analysis (Figure 4), a high correlation was observed between Na^+ and Cl^- (r = 0.92). However, most of the water samples fall above the rock salt dissolution line (Figure 7a), with a Na^+/Cl^- ratio range of 0.96–4.58, with an average value of 2.09. The result showed a higher Na^+ enrichment compared to Cl^-, suggesting that Na^+ does not originate from the dissolution of rock salt only. Indeed, besides cation exchange that may increase the Na^+ content in Sr-rich groundwater in Tianjiazhai, the dissolution of silicate minerals (e.g., sodium feldspar) can contribute to the increase in Na^+ concentrations in groundwater.

Ca^{2+}/Mg^{2+}, $(Ca^{2+} + Mg^{2+})/HCO_3^-$, and $(Ca^{2+}+Mg^{2+})/(HCO_3^- + SO_4^{2-})$ ratios can be used to determine the sources of Ca^{2+} and Mg^{2+} in groundwater [44,45]. From Figure 7b, the results showed a Ca^{2+}/Mg^{2+} ratio range of 0.33–1.05, with a mean value of 0.67. Except for a few samples located on the 1:1 line, most groundwater samples fell below the 1:1 line, indicating that the sources of Ca^{2+} and Mg^{2+} in the study area are calcite and dolomite dissolution; this is mainly due to the fact that calcite reached saturation faster than dolomite. As shown in Figure 7c, all groundwater samples fall above the 1:1 line of the $Ca^{2+}+Mg^{2+}/HCO_3^-$ ratio. In addition, the results showed a $Ca^{2+}+Mg^{2+}/HCO_3^-$ ratio range of 1.18–1.99, with an average value of 1.55, indicating that Ca^{2+} and Mg^{2+} were more enriched than HCO_3^-. This finding suggests that Ca^{2+} and Mg^{2+} were derived from the dissolution of carbonate rocks as well as from other sources, such as the dissolution of gypsum. In addition, most groundwater samples fall at the 1:1 line of the $[(Ca^{2+}+Mg^{2+})/(HCO_3^-+SO_4^{2-})]$ ratio (Figure 7d), which indicates equilibrium states of these ions, while chloro-alkaline indices indicated the prevalence of cation exchange. Moreover, this result demonstrates that Ca^{2+} and Mg^{2+} were not only from the dissolution of calcite, dolomite, and gypsum but also from silicate dissolution. $(Ca^{2+}+Mg^{2+})$ and $(HCO_3^-+SO_4^{2-})$ were in equilibrium under the combined influence of carbonate rock, silicate rock, and cation exchange.

Most of the groundwater samples fell near the 1:1 line of the Ca^{2+}/SO_4^{2-} ratio (Figure 7e), while only a few samples fell below the 1:1 line, showing that most of the groundwater samples have relatively balanced Ca^{2+} and SO_4^{2-}, while a few samples are depleted in Ca^{2+} or enriched in SO_4^{2-}. This result may be due to the presence of a certain degree of cation exchange and sedimentation of carbonate rocks during the evolution of groundwater.

The relationship between the HCO_3^- and $(SO_4^{2-}+Cl^-)$ can further reflect the dissolved carbonate rock in groundwater. As shown in Figure 7f, the water samples fall above the 1:1 line of the $HCO_3^-/(SO_4^{2-}+Cl^-)$ ratio. Moreover, the $HCO_3^-/(SO_4^{2-}+Cl^-)$ ratios range from 1.02 to 3.53, with an average value of 1.45. The HCO_3^- enrichment compared to $(SO_4^{2-}+Cl^-)$ indicated the dominance of HCO_3^- in Sr-rich groundwater in Tianjiazhai, suggesting that the water chemistry is mainly influenced by carbonate rock dissolution, which is consistent with the groundwater facies type.

The Cl^-/Na^+ and NO_3^-/Na^+ ratios in groundwater are generally higher when the groundwater is affected by anthropogenic activities [46]. The variation in the Cl^-/Na^+ and NO_3^-/Na^+ ratios in the Sr-rich groundwater samples in Tianjiazhai ranged from 0.22 to 1.04 and from 0 to 0.22, with mean values of 0.56 and 0.05, respectively. The low Cl^-/Na^+ and NO_3^-/Na^+ ratios obtained suggested that Sr-rich groundwater in Tianjiazhai is less affected by anthropogenic activities.

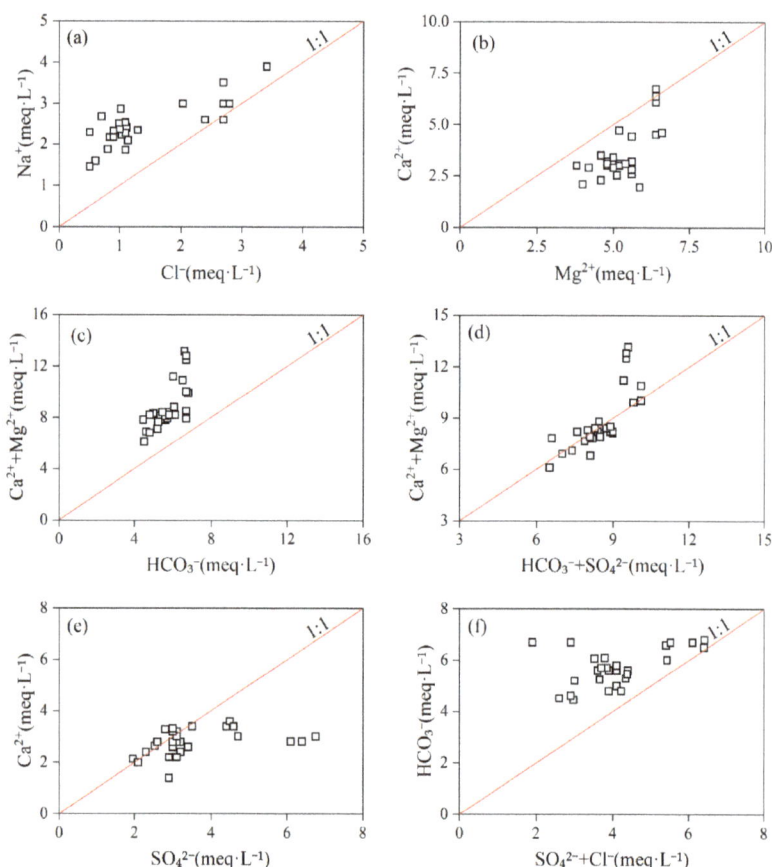

Figure 7. Relationship between the main ion ratios of Sr rich groundwater. (**a**) Cl^- and Na^+; (**b**) Mg^{2+} and Ca^{2+}; (**c**) HCO_3^- and $(Ca^{2+} + Mg^{2+})$; (**d**) $(HCO_3^- + SO_4^{2-})$ and $(Ca^{2+} + Mg^{2+})$; (**e**) SO_4^{2-} and Ca^{2+}; (**f**) $(SO_4^{2-} + Cl^-)$ and HCO_3^-.

3.2. Formation Mechanism of Sr in Groundwater

3.2.1. Source Conditions

The upper part of the lithosphere is rich in trace elements, such as Sr, which is abundantly contained in most rocks [47]. Indeed, Sr is mostly distributed in rock-forming minerals and is relatively concentrated in amphibolites, granites, and carbonates [48]. The Sr content in the clastic rocks of the Yanchang Formation in the study area is high. In addition, the lithology of this formation is a thick-bedded feldspathic sandstone of light gray-green medium to fine-grained feldspathic sandstone containing a large amount of potassium feldspar, which is rich in Sr. Strontium in groundwater is mainly derived from the dissolution of strontium in the sandstones of the Yanchang Formation. The Sr abundance in the surrounding rocks can reflect the Sr content in groundwater. Indeed, the presence of Sr-bearing minerals is the material basis for the formation of Sr-rich groundwater.

3.2.2. Hydrodynamic Conditions

The Sr-rich groundwater in Tianjiazhai is mainly stored in the fractured unconfined aquifer of the Yanchang Formation. The natural geographical, geological, and hydrogeological conditions in the area determine the alternating characteristics of the overall circulation of Sr-rich groundwater. During the rainy season, the main sources of recharge for weathered fractured unconfined water in the Yanchang Formation are atmospheric precipitation, Quaternary loose layer unconfined water, and surface water, while during the dry period, the weathered fractured unconfined water of the Yanchang Formation is mainly recharged from atmospheric precipitation and through transient runoff. Indeed, discharge occurs from high to low and from West to East, recharging surface water and pore unconfined water in the loose layer of the Quaternary, thus locally recharging the Shimachuan area through runoff, from upstream to downstream. The weathered fissure groundwater in the Yanchang Formation is located in different geomorphological units, resulting in different recharge modes, runoff directions, and discharge conditions. The Yanchang Formation of weathered fissure unconfined water, which is distributed in the Loess Mountains, mainly receives infiltration recharge from the overlying Quaternary aquifer and then flows from the Loess to the surrounding lowlands and discharges to the surface in the lowlands of the topographically cut gullies to recharge the weathered fissure groundwater of the Yanchang Formation in the river valley area. The weathered fractured groundwater of the Yanchang Formation, distributed in the valley area, mainly receives the recharge from the pore water of the overlying loose layer and the lateral recharge of the same aquifer. Besides the low recharge rates from the lower confined water, the exposed section of the aquifer is recharged directly from the atmospheric precipitation when the local section of the valley area is in the favorable part of the geomorphology before flowing along the river valley downstream from high to low elevations and from upstream to downstream areas. The discharge of the aquifer occurs in the form of submerged flow discharge in the riverbed, followed by downstream discharge. In addition, discharge by evaporation and artificial mining is also the main discharge process. There is no aquifuge between aquifers of the Quaternary loose layer pore unconfined water and the weathered fissure unconfined water of the Yanchang Formation, indicating a relatively close hydraulic connection. Surface water bodies and the Quaternary-Holocene alluvial unconfined water have a complementary relationship, resulting in the close relationship between weathered fissure unconfined water in the Yanchang Formation and surface water. According to the results of the dynamic surface water observation and groundwater level data observed from observation wells in Shimachuan, the dynamic changes of groundwater are closely related to atmospheric precipitation with the characteristics of synchronous changes. All this provides hydrodynamic conditions for the Sr enrichment in groundwater.

3.2.3. Hydrochemical Conditions

The Sr-rich groundwater in Tianjiazhai is mainly stored in the weathered fractured unconfined aquifer of the Yanchang Formation. Indeed, the rocks in the study area have a large weathering thickness and fracture development, mostly in the form of lines and veins and locally in the form of a network. Atmospheric precipitation recharges groundwater through fissure infiltration, resulting in CO_2 seepage into the ground with precipitation during the infiltration process (the free carbon dioxide in groundwater is determined by the volumetric method of standard base solution). Under certain temperature and pressure conditions, Sr-bearing minerals (rocks) undergo hydrolysis and dissolution filtration to form

Sr-rich groundwater. The thick layer of weathered rocks developed in the Tianjiazhai area, endowed with rich groundwater resources, provides good water conductivity channels and reaction space for groundwater circulation and water–rock interaction.

The Sr content in the upper and lower parts of Borehole G019 (Figure 8a) ranged from 0.14 to 0.20‰ and from 0.33 to 0.39‰, respectively. The Sr content in the upper and lower parts of borehole G020 (Figure 8b) ranged from 0.11 to 0.40‰ and from 0.51 to 0.62‰, respectively. The Sr contents of rock samples at different depths in the hydrogeological borehole were generally lower in the upper strata than in the lower strata. This result may be due to the higher weathering degree of the rocks in the upper strata than in the lower strata, resulting in high Sr element concentration, with high content, entering the groundwater. As can be seen from Table 3, Sr was prevalently present in the strata of the study area, ranging from 0.0741 to 0.147‰. By comparing the Sr content in the borehole and profile, lower Sr contents were observed in the profile than those in the borehole, which may be due to the longer leaching time in the area where the profile is located, allowing Sr to easily dissociate from rocks and enter groundwater. Therefore, the Sr enrichment in groundwater in Tianjiazhai is influenced by the degree of weathering and water–rock reaction time. The higher the degree of weathering and the longer the water–rock reaction time, the more easily Sr is leached from the rock into the groundwater system.

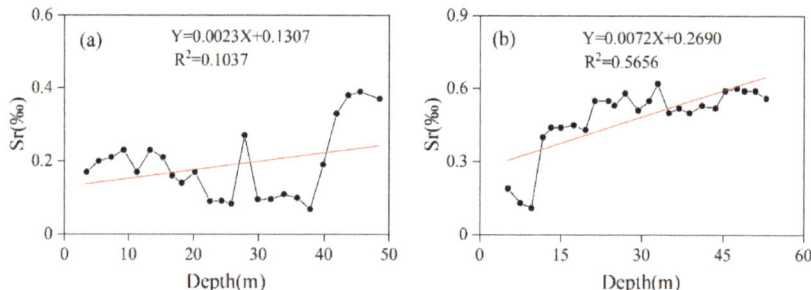

Figure 8. Sr content of strata in hydrogeological boreholes G019 (a) and G020 (b).

Table 3. Sr content in rock sample profiles.

Rock	Lithology	Strontium (‰)	Rock	Lithology	Strontium (‰)
pm-1	Medium-grained sandstone	0.126	pm-12	Fine-grained sandstone	0.080
pm-2	Conglomerate	0.103	pm-13	Mudstone	0.143
pm-3	Fine-grained sandstone	0.101	pm-14	Fine-grained sandstone	0.111
pm-4	Fine-grained sandstone	0.076	pm-15	Mudstone	0.110
pm-5	Coarse-grained sandstone	0.098	pm-16	Mudstone	0.074
pm-6	Coarse-grained sandstone	0.147	pm-17	Mudstone	0.128
pm-7	Coarse-grained sandstone	0.120	pm-18	Mudstone	0.119
pm-8	Fine-grained sandstone	0.116	pm-19	Fine-grained sandstone	0.126
pm-9	Fine-grained sandstone	0.145	pm-20	Fine-grained sandstone	0.121
pm-10	Fine-grained sandstone	0.111	pm-21	Loess	0.255
pm-11	Fine-grained sandstone	0.116	pm-22	Loess	0.266

Strontium enrichment in groundwater is related to groundwater circulation and transport characteristics. In groundwater flow systems, the farther the runoff distance, the longer the circulation path, the longer the retention time of groundwater in the aquifer, and the more favorable the dissolution and enrichment of strontium in groundwater [49–51]. The Sr-rich groundwater in Tianjiazhai is mainly stored in the weathered fractured aquifer of the Yanchang Formation, with a single water-bearing lithology and good groundwater circulation conditions. The Sr contents in the altered bedrock fracture water are relatively stable. In addition, the clastic rock layer is less permeable and provides a semi-closed/semi-open groundwater environment, with slow groundwater runoff, long retention time, and sufficient water–rock interaction, resulting in dissolution and enrichment of Sr in the groundwater. Long-term contact between groundwater and surrounding Sr-bearing minerals (rocks) is the fundamental condition for the formation of Sr-rich groundwater. The longer the contact time between groundwater and Sr minerals (rocks), the higher the Sr content in the groundwater.

4. Conclusions

In this study, the hydrochemical characteristics and sources of Sr-rich groundwater in the Tianjiazhai area were analyzed using descriptive analysis, correlation analysis, Piper trilinear diagrams, Gibbs plots and ion ratios. In addition, the formation mechanisms of Sr in groundwater were discussed, taking into account the specific geological and hydrogeological conditions of the study area, as well as the hydrogeochemical theory. The results showed that:

(1) Sr^{2+} content in Sr-rich groundwater ranged from 0.85 to 2.99 mg·L^{-1}, with an average value of 1.45 mg·L^{-1}, which was 7.25 times higher than the national limit value of Sr^{2+} content in natural mineral water for drinking (0.20 mg·L^{-1}). In addition, the pH and TDS values ranged from 7.50 to 8.30 and from 332.00 to 718.00 mg·L^{-1}, respectively, indicating that Sr-rich groundwater is generally slightly alkaline. The dominant cation and anion were Ca^{2+} and HCO_3^-, respectively. On the other hand, HCO_3-Ca·Mg·Na was the main facies type of Sr-rich groundwater in the Tianjiazhai area. The coefficients of variation of the hydrochemical parameters of Sr-rich groundwater were less than 1, indicating that the contents of all hydrochemical parameters indicators were relatively stable.

(2) The correlation coefficients of Sr^{2+} with Ca^{2+} and HCO_3^- were 0.73 and 0.59, respectively, indicating that the source of Sr may be the same as that of these two ions. The correlation coefficients of TDS with Cl^- and Na^+ were 0.81 and 0.76, respectively, mainly because that the groundwater in the Tianjiazhai area originated from atmospheric precipitation, whereas the correlation coefficients of TDS with Ca^{2+} and HCO_3^- were 0.84 and 0.60, respectively, indicating the important role of Ca^{2+} and HCO_3^- in the groundwater chemistry in the study area. The hydrochemical characteristics of Sr-rich groundwater in the Tianjiazhai area are controlled by water–rock interactions and influenced by cation exchange. In addition, water–rock interactions in Sr-rich groundwater are dominated by the dissolution of carbonate rocks.

(3) Weathering, dissolution, and leaching in the sandstone of the Yanchang Formation are the main processes controlling Sr content in groundwater. Indeed, the Sr-rich groundwater is mainly stored in the weathering fracture unconfined aquifer of the Yanchang Formation. In addition, the higher the degree of weathering of Sr-bearing minerals (rocks) and the longer the reaction time of water–rock interaction, the more favorable the dissolution and enrichment of Sr in groundwater. The hydrogeological conditions and geological and tectonic environment of the Tianjiazhai area provide favorable conditions for the formation of Sr-rich groundwater.

The sandstone of Yanchang Formation is widely distributed in the Shimachuan River basin, and the hydrodynamic and hydrochemical conditions of Shimachuan River basin are conducive to the enrichment of strontium in groundwater, and the strontium in groundwater flowing into Yellow River will be more enriched. Therefore, the development of this study is of typical significance and can provide a scientific basis for the study of strontium-rich groundwater in the whole Shimachuan River basin. However, due to the special geological environment of the study area, soil erosion is more serious, the groundwater quality is poor in the downstream region, and the sediment content in the groundwater injected into the Yellow River is especially large. Therefore, the above changes should be further explored in future research.

This study provides a deeper understanding of the hydrochemical characteristics and sources of Sr-rich groundwater in the Tianjiazhai area. The results obtained can also provide a scientific basis and theoretical guidance for the development, utilization, and protection of Sr-rich groundwater.

Author Contributions: Research conceptualization, C.L. and W.W.; data curation, D.W. and A.O.; methodology, C.L.; writing—original draft, C.L.; writing—review and editing, W.W. and X.K., supervision, W.W. All authors have read and agreed to the published version of the manuscript.

Funding: This research received no external funding.

Institutional Review Board Statement: Not applicable.

Informed Consent Statement: Not applicable.

Data Availability Statement: All processed data generated or used during the study appear in the submitted article. Raw data may be provided on reasonable request from the corresponding author.

Acknowledgments: We acknowledge the Shaanxi Province 139 Coal Geology and Hydrology Co. for the access to original information.

Conflicts of Interest: The authors declare no conflict of interest.

References

1. Karunanidhi, D.; Aravinthasamy, P.; Deepali, M.; Subramani, T.; Bellows, B.C.; Li, P. Groundwater quality evolution based on geochemical modeling and aptness testing for ingestion using entropy water quality and total hazard indexes in an urban-industrial area (Tiruppur) of Southern India. *Environ. Sci. Pollut. Res.* **2021**, *28*, 18523–18538. [CrossRef]
2. Li, P.; He, X.; Guo, W. Spatial groundwater quality and potential health risks due to nitrate ingestion through drinking water: A case study in Yan'an City on the Loess Plateau of northwest China. *Hum. Ecol. Risk Assess. Int. J.* **2019**, *25*, 11–31. [CrossRef]
3. Alvarez, M.d.P.; Funes, D.; Dapeña, C.; Bouza, P.J. Origin and hydrochemical characteristics of groundwater in the Northeastern Patagonia, Argentina: The relationship with geomorphology and soils. *Environ. Earth Sci.* **2020**, *79*, 503. [CrossRef]
4. Li, Y.; Li, P.; Cui, X.; He, S. Groundwater quality, health risk, and major influencing factors in the lower Beiluo River watershed of northwest China. *Hum. Ecol. Risk Assess. Int. J.* **2021**, *27*, 1987–2013. [CrossRef]
5. Srinivas, Y.; Aghil, T.B.; Hudson Oliver, D.; Nithya Nair, C.; Chandrasekar, N. Hydrochemical characteristics and quality assessment of groundwater along the Manavalakurichi coast, Tamil Nadu, India. *Appl. Water Sci.* **2017**, *7*, 1429–1438. [CrossRef]
6. Liu, J.; Gao, Z.; Wang, M.; Li, Y.; Shi, M.; Zhang, H.; Ma, Y. Hydrochemical characteristics and possible controls in the groundwater of the Yarlung Zangbo River Valley, China. *Environ. Earth Sci.* **2019**, *78*, 76. [CrossRef]
7. Chotpantarat, S.; Thamrongsrisakul, J. Natural and anthropogenic factors influencing hydrochemical characteristics and heavy metals in groundwater surrounding a gold mine, Thailand. *J. Asian Earth Sci.* **2021**, *211*, 104692. [CrossRef]
8. Feng, W.; Qian, H.; Xu, P.; Hou, K. Hydrochemical Characteristic of Groundwater and Its Impact on Crop Yields in the Baojixia Irrigation Area, China. *Water* **2020**, *12*, 1443. [CrossRef]
9. Bowman, R.S. Aqueous Environmental Geochemistry. *Eos Trans. Am. Geophys. Union* **1997**, *78*, 586. [CrossRef]
10. Fuoco, I.; De Rosa, R.; Barca, D.; Figoli, A.; Gabriele, B.; Apollaro, C. Arsenic polluted waters: Application of geochemical modelling as a tool to understand the release and fate of the pollutant in crystalline aquifers. *J. Environ. Manag.* **2022**, *301*, 113796. [CrossRef]
11. Fuoco, I.; Marini, L.; De Rosa, R.; Figoli, A.; Gabriele, B.; Apollaro, C. Use of reaction path modelling to investigate the evolution of water chemistry in shallow to deep crystalline aquifers with a special focus on fluoride. *Sci. Total Environ.* **2022**, *830*, 154566. [CrossRef]
12. Yang, F.; Liu, S.; Jia, C.; Gao, M.; Chang, W.; Wang, Y. Hydrochemical characteristics and functions of groundwater in southern Laizhou Bay based on the multivariate statistical analysis approach. *Estuar. Coast. Shelf Sci.* **2021**, *250*, 107153. [CrossRef]
13. Van Geldern, R.; Schulte, P.; Mader, M.; Baier, A.; Barth, J.A.C.; Juhlke, T.R.; Lee, K. Insights into agricultural influences and weathering processes from major ion patterns. *Hydrol. Processes* **2018**, *32*, 891–903. [CrossRef]
14. Ismail, E.; Abdelhalim, A.; Heleika, M.A. Hydrochemical characteristics and quality assessment of groundwater aquifers northwest of Assiut district, Egypt. *J. Afr. Earth Sci.* **2021**, *181*, 104260. [CrossRef]
15. Zhang, B.; Zhao, D.; Zhou, P.; Qu, S.; Liao, F.; Wang, G. Hydrochemical Characteristics of Groundwater and Dominant Water–Rock Interactions in the Delingha Area, Qaidam Basin, Northwest China. *Water* **2020**, *12*, 836. [CrossRef]
16. Dai, L.L.; Mei, M.L.; Chu, C.H.; Lo, E.C.M. Remineralizing effect of a new strontium-doped bioactive glass and fluoride on demineralized enamel and dentine. *J. Dent.* **2021**, *108*, 103633. [CrossRef]
17. Neves, N.; Campos, B.B.; Almeida, I.F.; Costa, P.C.; Cabral, A.T.; Barbosa, M.A.; Ribeiro, C.C. Strontium-rich injectable hybrid system for bone regeneration. *Mater. Sci. Eng. C* **2016**, *59*, 818–827. [CrossRef]
18. Haixue, L.; Xuxue, C.; Yuekun, M.; Weipo, L.; Bin, Z. Characteristics and formation mechanism of strontium-rich groundwater in Malian River drainage basin, Southern Ordos Basin. *Geoscience* **2021**, *35*, 682.
19. Wang, R.; Wu, X.; Zhai, Y.; Su, Y.; Liu, C. An Experimental Study on the Sources of Strontium in Mineral Water and General Rules of Its Dissolution—A Case Study of Chengde, Hebei. *Water* **2021**, *13*, 699. [CrossRef]

20. Cao, Y.; Song, L.; Liu, L.; Zhu, W.; Cui, S.; Wang, Y.; Guo, P. Preliminary study on strontium-rich characteristics of shallow groundwater in Dingtao area, China. *J. Groundw. Sci. Eng.* **2020**, *8*, 244–258.
21. Pit'eva, K.E.; Fortygina, M.A.; Mikheev, A.V. Formation of the composition and quality of carbonate mineral water of the Elburgan water bearing complex (Essentuki Town). *Mosc. Univ. Geol. Bull.* **2007**, *62*, 277–285. [CrossRef]
22. Lu, S.; Zhou, H.; Liu, W.; Chen, Q.; Yan, Z.; Chen, L. Distribution and enrichment of strontium in the Zigui karst watershed. *Geol. China* **2021**, *48*, 1865–1874.
23. Zuo, R.; Teng, Y.; Wang, J. Modeling migration of strontium in sand and gravel aquifer in the candidate VLLW disposal site. *J. Radioanal. Nucl. Chem.* **2009**, *281*, 653. [CrossRef]
24. Zhu, X.; Liu, W.; Li, Z.; Chen, T.; Ren, Y.; Shao, H.; Wang, L. Distribution and characterization analyses of strontium-bearing mineral spring water in the Chengde region. *Hydrogeol. Eng. Geol.* **2020**, *47*, 65–73.
25. Li, D.; Gan, S.; Li, J.; Dong, Z.; Long, Q.; Qiu, S.; Zhou, Y.; Lu, C. Hydrochemical characteristics and formation mechanism of strontium-rich groundwater in Shijiazhuang, north China plain. *J. Chem.* **2021**, *2021*, 5547924. [CrossRef]
26. Yang, Y.; Shao, C.; Lu, T.; Zhang, A. Discovery of Strontium-rich mineral water in Tingsiqiao and Guantangyi Town, Xianning. *Geol. China* **2021**, *48*, 1661–1662.
27. Li, P.; He, S.; Yang, N.; Xiang, G. Groundwater quality assessment for domestic and agricultural purposes in Yan'an City, northwest China: Implications to sustainable groundwater quality management on the Loess Plateau. *Environ. Earth Sci.* **2018**, *77*, 775. [CrossRef]
28. Xu, P.; Zhang, Q.; Qian, H.; Li, M.; Hou, K. Characterization of geothermal water in the piedmont region of Qinling Mountains and Lantian-Bahe Group in Guanzhong Basin, China. *Environ. Earth Sci.* **2019**, *78*, 442. [CrossRef]
29. Chung, S.Y.; Rajendran, R.; Senapathi, V.; Sekar, S.; Ranganathan, P.C.; Oh, Y.Y.; Elzain, H.E. Processes and characteristics of hydrogeochemical variations between unconfined and confined aquifer systems: A case study of the Nakdong River Basin in Busan City, Korea. *Environ. Sci. Pollut. Res.* **2020**, *27*, 10087–10102. [CrossRef]
30. Li, P.; He, S.; He, X.; Tian, R. Seasonal Hydrochemical Characterization and Groundwater Quality Delineation Based on Matter Element Extension Analysis in a Paper Wastewater Irrigation Area, Northwest China. *Expo. Health* **2018**, *10*, 241–258. [CrossRef]
31. Kwon, E.; Park, J.; Lee, J.M.; Kim, Y.-T.; Woo, N.C. Spatiotemporal changes in hydrogeochemistry of coastal groundwater through the construction of underground disposal facility for low and intermediate level radioactive wastes in Korea. *J. Hydrol.* **2020**, *584*, 124750. [CrossRef]
32. Su, C.; Huang, C.; Zou, S.; Xie, D.; Zhao, G.; Tang, J.; Luo, F.; Yang, Y. Enrichm entenvironment and sources of strontium of groundwater in Xintian county, Hunan Province. *Carsologica Sin.* **2017**, *36*, 678–683.
33. Zhou, C.; Zou, S.; Xia, R.; Xue, Q.; Zhu, D.; Li, L.; Cao, J.; Li, J.; Xie, H. Extreme enrichment of strontium discovered in surface water and groundwater in the Guanling area of Guizhou Province. *Geol. China* **2021**, *48*, 961–962.
34. Zhang, Q.; Qian, H.; Xu, P.; Li, W.; Feng, W.; Liu, R. Effect of hydrogeological conditions on groundwater nitrate pollution and human health risk assessment of nitrate in Jiaokou Irrigation District. *J. Clean. Prod.* **2021**, *298*, 126783. [CrossRef]
35. Zhang, J.; Shi, Z.; Wang, G.; Jiang, J.; Yang, B. Hydrochemical characteristics and evolution of groundwater in the Dachaidan area, Qaidam Basin. *Earth Sci. Front.* **2021**, *28*, 194–205.
36. Piper, A. A Graphic Procedure in the Geochemical Interpretation of Water-Analyses. *Eos Trans. Am. Geophys. Union* **1944**, *25*, 914–923. [CrossRef]
37. He, X.; Li, P.; Wu, J.; Wei, M.; Ren, X.; Wang, D. Poor groundwater quality and high potential health risks in the Datong Basin, northern China: Research from published data. *Environ. Geochem. Health* **2021**, *43*, 791–812. [CrossRef]
38. Apollaro, C.; Di Curzio, D.; Fuoco, I.; Buccianti, A.; Dinelli, E.; Vespasiano, G.; Castrignanò, A.; Rusi, S.; Barca, D.; Figoli, A.; et al. A multivariate non-parametric approach for estimating probability of exceeding the local natural background level of arsenic in the aquifers of Calabria region (Southern Italy). *Sci. Total Environ.* **2022**, *806*, 150345. [CrossRef]
39. Marghade, D.; Malpe, D.B.; Duraisamy, K.; Patil, P.D.; Li, P. Hydrogeochemical evaluation, suitability, and health risk assessment of groundwater in the watershed of Godavari basin, Maharashtra, Central India. *Environ. Sci. Pollut. Res.* **2021**, *28*, 18471–18494. [CrossRef]
40. Gibbs Ronald, J. Mechanisms Controlling World Water Chemistry. *Science* **1970**, *170*, 1088–1090. [CrossRef]
41. Schoeller, H. Qualitative Evaluation of Groundwater Resources. In *Methods and Techniques of Groundwater Investigation and Development*; UNESCO: Paris, France, 1967.
42. Xing, L.; Guo, H.; Zhan, Y. Groundwater hydrochemical characteristics and processes along flow paths in the North China Plain. *J. Asian Earth Sci.* **2013**, *70–71*, 250–264. [CrossRef]
43. Zhao, Q.; Su, X.; Kang, B.; Zhang, Y.; Wu, X.; Liu, M. A hydrogeochemistry and multi-isotope (Sr, O, H, and C) study of groundwater salinity origin and hydrogeochemcial processes in the shallow confined aquifer of northern Yangtze River downstream coastal plain, China. *Appl. Geochem.* **2017**, *86*, 49–58. [CrossRef]
44. Ma, R.; Wang, Y.; Sun, Z.; Zheng, C.; Ma, T.; Prommer, H. Geochemical evolution of groundwater in carbonate aquifers in Taiyuan, northern China. *Appl. Geochem.* **2011**, *26*, 884–897. [CrossRef]
45. Li, C.; Gao, X.; Wang, Y. Hydrochemistry of high-fluoride groundwater at Yuncheng Basin, northern China. *Sci. Total Environ.* **2015**, *508*, 155–165. [CrossRef]

46. Fan, B.-L.; Zhao, Z.-Q.; Tao, F.-X.; Liu, B.-J.; Tao, Z.-H.; Gao, S.; Zhang, L.-H. Characteristics of carbonate, evaporite and silicate weathering in Huanghe River basin: A comparison among the upstream, midstream and downstream. *J. Asian Earth Sci.* **2014**, *96*, 17–26. [CrossRef]
47. Ehya, F.; Shakouri, B.; Rafi, M. Geology, mineralogy, and isotope (Sr, S) geochemistry of the Likak celestite deposit, SW Iran. *Carbonates Evaporites* **2013**, *28*, 419–431. [CrossRef]
48. Parisi, S.; Paternoster, M.; Perri, F.; Mongelli, G. Source and mobility of minor and trace elements in a volcanic aquifer system: Mt. Vulture (southern Italy). *J. Geochem. Explor.* **2011**, *110*, 233–244. [CrossRef]
49. Zhang, Y.; Ding, H.; Fu, D.; Huang, Z.; Li, A.; Yan, C.; Jin, X. A study of enrichment environment and formation mechanism of strontium mineral water in Gansu Province. *Geol. China* **2020**, *47*, 1688–1701.
50. Su, C.; Nie, F.; Zou, S.; Zhao, G.; Luo, F.; Huang, Q.; Ba, J.; Li, X.; Liang, J.; Yang, Y. Hydrochemical Characteristics and Formation Mechanism of Strontium-rich Groundwater in Xintian County, Hunan Province. *Geoscience* **2018**, *32*, 554–564.
51. Sun, Q.; Sun, Z.; Jia, L.; Tian, H.; Guo, X.; Du, J.; Li, X.; Li, X.; Jia, L. Formation mechanism of the strontium-rich and metasilicic acid groundwater in the Lianhuashan area, Changchun, Jilin Province. *Geol. China* **2020**, *44*, 1031–1032.

Article

Groundwater Health Risk Assessment Based on Monte Carlo Model Sensitivity Analysis of Cr and As—A Case Study of Yinchuan City

Zhiyuan Ma [1], Junfeng Li [2], Man Zhang [3], Di You [4], Yahong Zhou [4] and Zhiqiang Gong [1,*]

[1] Hebei Key Laboratory of Environment Monitoring and Protection of Geological Resources, Hebei Geological Environment Monitoring Institute, Shijiazhuang 050021, China
[2] Hebei Institute of Hydrogeology and Engineering Geology, Shijiazhuang 050000, China
[3] Baoding Water and Soil Conservation Workstation, Baoding 071051, China
[4] School of Water Resources and Environment, Hebei GEO University, Shijiazhuang 050031, China
* Correspondence: 18801333516@163.com

Citation: Ma, Z.; Li, J.; Zhang, M.; You, D.; Zhou, Y.; Gong, Z. Groundwater Health Risk Assessment Based on Monte Carlo Model Sensitivity Analysis of Cr and As—A Case Study of Yinchuan City. *Water* **2022**, *14*, 2419. https://doi.org/10.3390/w14152419

Academic Editor: Adriana Bruggeman

Received: 24 June 2022
Accepted: 28 July 2022
Published: 4 August 2022

Publisher's Note: MDPI stays neutral with regard to jurisdictional claims in published maps and institutional affiliations.

Copyright: © 2022 by the authors. Licensee MDPI, Basel, Switzerland. This article is an open access article distributed under the terms and conditions of the Creative Commons Attribution (CC BY) license (https://creativecommons.org/licenses/by/4.0/).

Abstract: Groundwater is an important resource for domestic use and irrigation in the Yinchuan region of northwest China. However, the quality of groundwater in this region is declining due to human activities, with adverse effects on human health. In order to study the effects of chemical elements in groundwater on human health, the human health risk of drinking groundwater was calculated based on the actual situation in China and on the U.S. Environmental Protection Agency (USEPA) model. Moreover, the sensitivity of contaminant exposure in drinking water wells was quantified using Monte Carlo simulation to minimize uncertainty in conjunction with USEPA risk assessment techniques, with the aim to identify the major carcinogenic factors. In addition, Visual Minteq was used to analyze the possible ionic forms of the major factors in the hydrogeological environment of the study area. The results showed that the mean CR values for As were $2.94 \times 10^{-0.5}$ and $5.93 \times 10^{-0.5}$ for the dry and rainy seasons, respectively, while for 2018 they were $5.48 \times 10^{-0.5}$ and $3.59 \times 10^{-0.5}$, respectively. In parallel, the CR values for children for 2017 were $6.28 \times 10^{-0.5}$ and $1.27 \times 10^{-0.4}$, respectively, and $1.17 \times 10^{-0.4}$ and $7.67 \times 10^{-0.5}$, respectively, indicating a considerably higher carcinogenic risk for children than for adults. results of the sensitivity analysis of Cr^{6+} and As using Crystal Ball software showed association values of 0.9958 and 1 for As and 0.0948 and 0 for Cr in the dry and rainy seasons in 2017, and 0.7424 and 0.5759 for As and 0.6237 and 0.8128 for Cr in the dry and rainy seasons in 2018, respectively. Only in the rainy season of 2018, the association values for As were lower than those for Cr, indicating that As is more sensitive to total carcinogenic risk. The results of the visual coinage model analysis showed that among all the possible ionic forms of As, the activity of $HAsO_4^{2-}$ had the largest logarithmic value and that of H_3AsO_4 had the smallest value, regardless of pH changes. This indicates that $HAsO_4^{2-}$ is the ionic form of As with the main carcinogenic factor in the hydrogeological environment of the study area. Therefore, corresponding environmental control measures need to be taken in time to strengthen the monitoring and control of As, especially $HAsO_4^{2-}$, in the groundwater of the study area. This study is of great significance for Yinchuan city to formulate groundwater pollution risk management and recovery.

Keywords: groundwater; health risk assessment; Monte Carlo model (crystal ball software); visual minteq; yinchuan

1. Introduction

Water resources are one of the most important natural resources on earth and indispensable and irreplaceable for the survival of all living beings, as well as for human life and production activities [1–3]. Groundwater is an important component of the Earth's water resources; compared to other water sources, it has the advantages of good quality, lower investments required for its exploitation, lower susceptibility to pollution, and multi-year regulation [4]. These characteristics are an important guarantee to maintain the virtuous cycle of the water resources system and have a positive impact on national livelihood, people's health and safety, and sustainable social development [5,6]. In recent years, groundwater extraction in China reached about 2.7 billion cubic meters per year, a number that is currently increasing. Moreover, 400 Chinese cities and about 45% of agricultural irrigation rely on groundwater as a source of water supply. Agricultural, industrial, and domestic activities are the main sources of groundwater pollution [7], mainly consisting of heavy metals [8,9], organic matter [10,11], As [12–14], and chromium contamination [15].

At present, groundwater pollution is a serious problem in China. According to the results of a statistical analysis of groundwater quality in 118 cities and regions nationwide, the majority of urban groundwater suffers from a certain degree of point-source and surface pollution, with 64% of urban groundwater severely polluted, 33% mildly polluted [16], and only 3% of urban groundwater basically clean [17,18]. Due to groundwater pollution, an increasing number of cities and regions are suffering from water shortage, posing rising risks to human health. According to the World Health Organization (WHO), 80% of the diseases suffered by third-world citizens are caused by water pollution [19]. Most cities in northwestern China use groundwater as their only drinking water supply. In these areas, the contamination of groundwater is expected to directly affect the health of citizens who use it for drinking purposes. Especially, the large land development projects in the Loess region and the proposed revival of the Silk Road Economic Belt may have a significant impact on the quantity and quality of groundwater, and require urgent and comprehensive research [20–22].

The study area of the present research is adjacent to the upper reaches of the Yellow River; it includes Xingqing District, Xixia District, Jinfeng District, Yongning County, and Helan County. It borders the Helan Mountains to the west and the west bank of the Yellow Rivulet River to the east. The exploited groundwater resources in Yinchuan are equal to 4.8842×10^8 m^3/a, accounting for 28.8% of the exploitable groundwater resources, equal to 16.9317×10^8 m^3/a, indicating a serious over-exploitation of groundwater in local areas. Groundwater is used mainly for urban domestic purposes, and also for industrial, and rural human and animal drinking water purposes [23]. In general, the ecological environment in the study area is very fragile [24].

Chen et al. found different degrees of carcinogenic and non-carcinogenic health risks from drinking groundwater in Yinchuan Plain; moreover, the health risks from inorganic contaminants such as arsenic, lead, and Cr(V) were found to be greater than those from organic contaminants such as benzene, trichloroethylene, toluene, 1.2-dichloroethane, and 1.1-dichloroethylene. More in detail, the carcinogenic and non-carcinogenic risk rates of arsenic were found to be equal to 47.3% and 8.6% respectively, i.e., considerably greater than those from other protective layer, which are also considered priority control contaminants for groundwater in Yinchuan Plain [25]. Therefore, considering the importance of groundwater in this region, in this study a health risk assessment of groundwater was conducted for the study area.

Currently, most of the health risk assessment site methods are based on the EPA health risk assessment model [26], but due to the uncertainty of the input parameters, the resulting health risk calculation results are highly uncertain. However, the Monte Carlo model can solve this problem by probabilistic methods.

The main objectives of this study are as follows: (1) to analyze the current groundwater distribution in the Yinchuan Plain through a data survey; (2) Monte Carlo and visual coinage are used to determine the main influencing factors and possible ion forms in hydrogeology, and conduct a health risk assessment of groundwater in Yinchuan City, providing support for strengthening the supervision and control of groundwater in the region.

2. Study Area

2.1. Location and Climate

The study area of this research is the Yinchuan region in northwest China. Geographically, the study area encompasses the Xingqing District, Xixia District, Jinfeng District, Yongning County, and Helan County, and is located near the upper reaches of the Yellow River. It is bounded by the Helan Mountains in the west, and the Yellow River's western bank in the east. The Yinchuan region has a typical continental climate characterized by a long winter, a short summer, low levels of rain and frequent droughts, high wind and evaporation, and dramatic temperature changes [27,28]. The dry season runs from February to May, while the rainy season runs from June to September. The total precipitation in the rainy season accounts for about 70% of the total rainfall throughout the whole year.

2.2. Topography and Geomorphology

Geomorphologically, the Yinchuan region is higher in the west and lower in the east, and the southern part is higher than the northern one. The western part is occupied by the Helan Mountains, reaching an altitude of 1500~3200 m. The central part is characterized by a flat plain tilted from southwest to northeast, which is formed by flood, alluvial flood, river, and lake deposits. The terrain of the Taoling salt platform in the eastern part is undulating. There are several loess-like clay sand, silty sand, sand gravel, and other strata accumulated due to flood and aeolian deposits, with a small amount of Quaternary gravel residue between them. The topography and geomorphology of the Yinchuan area today were formed through the comprehensive influence of physical, chemical, and biological effects across a long geological period. Based on the sedimentary origin, the terrain of the study area can be divided into erosion, accumulation, and aeolian terrain. Based on the morphology, it can be divided into the piedmont alluvial plain and alluvial plain. The accumulation terrain includes first-order terrace, second-order terrace, low-level alkaline beach, and the Yellow River floodplain. The aeolian terrain includes fixed, semi-fixed, and active dunes.

The study area is located in the Yinchuan fault basin. To the east, the Yellow River fault is connected with the Ordos block. To the west, the eastern foot of the Helan Mountains is connected with the mountain transition. A series of faults developed in the plain, some controlling the boundary of the plain, and some hiding inside the plain. The fault dips steeply towards the center of the plain, resulting in a "ladder" fault depression. The east-west and northwest faults mainly control the north-south boundary of the plain, and some are hidden in the plain [29–32]. The distribution and properties of faults with different strikes vary greatly.

The exposed strata in the study area are mainly Quaternary strata, which can be divided into four series according to age: early, middle, late Pleistocene, and Holocene. On the cross-section, Yinchuan pool sinks towards the center. The middle part of the Quaternary stratum is the deepest, and the East and west gradually become shallower

2.3. Groundwater

The flow of groundwater follows the terrain trend from west to east, although the specific direction and conditions of runoff vary across the whole area. In the study area, the groundwater is pore-water, according to the formation conditions and distribution, it can be divided into phreatic and confined aquifers. Hydrogeological conditions in the study area are, of course, controlled by formation lithology. It can be divided into two areas. The west is a single river aquifer, and the East is a multi-layer structure. The thickness of a single aquifer is thickened from north to south, and the general thickness is 230–265 m. The thickness of the phreatic aquifer within the multilayer structure zone is only 20–60 m. This study targeted the Phreatic aquifer. There are two major recharge sources of groundwater in the Yinchuan area: natural and artificial recharge. Natural recharge occurs mainly through atmospheric precipitation, lateral runoff, and flood loss, while artificial recharge occurs mainly through irrigation infiltration, groundwater irrigation recharge, and drainage system leakage. Evaporative discharge, with a proportion of more than 45%, exceeds manual mining and lateral runoff, which are the two other main discharge modes in the region [33]. Lateral runoff mainly occurs toward the Yellow River and drainage ditches.

3. Materials and Methods

3.1. Sample Collection and Analysis

The study flowchart is shown below, which shows the overall idea of the study, from the background investigation, and data collection, to health risk assessment, using the Monte Carlo model and Visual Minteq, finally deriving the main ion morphology of carcinogenic factors.

In this study, 49 groups of monitoring points with relatively continuous and complete observations were selected as the analysis data (20–30 m below the ground surface); the location of all the monitoring points is shown in Figure 1. These data were collected by the Groundwater Dynamic Monitoring Network of the Ningxia Hui Autonomous Region in 2017–2018 (the dry season (March–April) and the rainy season (July–August)). The collection, sealing, and transportation of water samples were carried out in strict accordance with national technical regulations [31]. Detailed test methods, instrument specifications, and detection limits for each indicator are shown in Table 1. The pH and conductivity of groundwater samples were measured in the field, while other parameters were measured in the laboratory. Test parameters include Na^+, Ca^{2+}, Mg^{2+}, K^+, HCO_3^-, SO_4^{2-}, Cl^-, As, Cr^{6+}, total hardness (TH), and total dissolved solids (TDS), which were analyzed in the laboratory of the Hebei GEO University using standard methods recommended by technical specifications for environmental monitoring of groundwater. The accuracy of the test results was measured by the calculation of percent charge balance errors (%CBE), as follows:

Table 1. Analytical methods, instruments, and detection limits of physiochemical parameters.

Index	Method	Instrument	Model	Detection Limit
pH	Glass-electrode method	pH meter	PHSJ-4A	0–14.000
TH	EDTA titration			0.32 mg/L
EC	Conductivity analyzer analysis method	Conductivity Analyzer	DDS-11A	$0–1.999 \times 10^5$ μS/cm
TDS	Conductivity analyzer analysis method	Conductivity Analyzer	DDS-11A	$0–1.999 \times 10^5$ μS/cm
Na^+	Flame atomic absorption spectrophotometry	Flame photometer	6400A	0.01 mg/L
k^+	Flame atomic absorption spectrophotometry	Flame photometer	6400A	0.05 mg/L
Ga^{2+}	EDTA titration			0.2 mg/L
Mg^{2+}	EDTA titration			0.12 mg/L
SO_4^{2-}	Ion chromatography	Ion chromatograph	ICS-90A	0.09 mg/L
HCO_3^-	Titration			5 mg/L
Cl^-	Silver nitrate titration method			10–500 mg/L
As	Atomic fluorescence spectrometry	AFS-920	AFS-920	0.000046 mg/L
Cr^{6+}	Inductively coupled plasma atomic emission spectrometry	ICP-MS	ICAPQc	0.0006 mg/L

The study flowchart is shown below (Figure 2), which shows the overall idea of the study, from background investigation, data collection, to health risk assessment, using the Monte Carlo model and Visual Minteq, finally deriving the main ion morphology of carcinogenic factors.

The final detection results show that the relative error of all samples was <±5%, indicating that all 49 samples could be used for analysis.

$$\%CBE = \frac{\Sigma cations - \Sigma anions}{\Sigma cations + \Sigma anions} \times 100\% \tag{1}$$

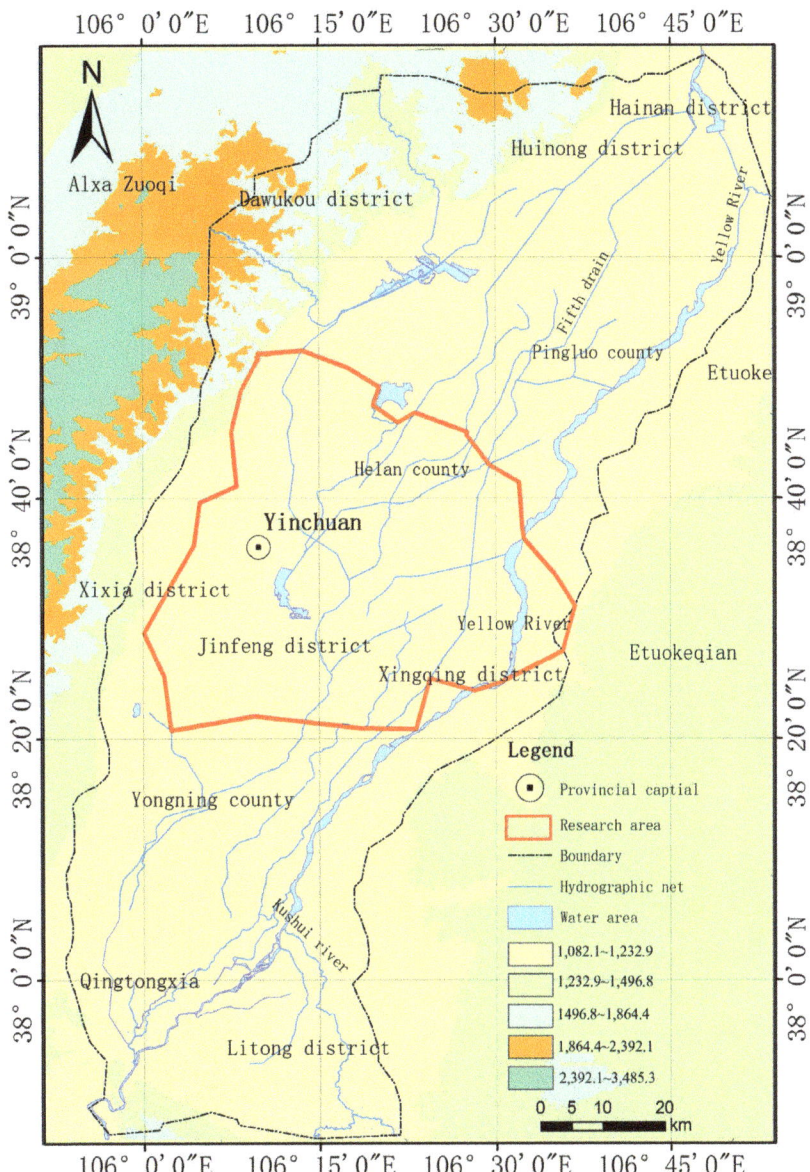

Figure 1. Study area, geological profile, and location of samplings.

3.2. Health Risk Assessment

The main pathways through which heavy metals pose human health risks are dermal contact and food chain. Human health risks resulting from direct or indirect exposure to groundwater can be assessed by the models recommended by the Ministry of Environmental Protection of Environmental Protection of China, based on models of the United States Environmental Protection Agency (USEPA). Data were analyzed according to the concentration period of precipitation (dry or rainy season) and human physiology (i.e., adults and children) [32,33]. Based on the actual situation in China, the Chinese model adopts specific parameters, which are presented in Table 2.

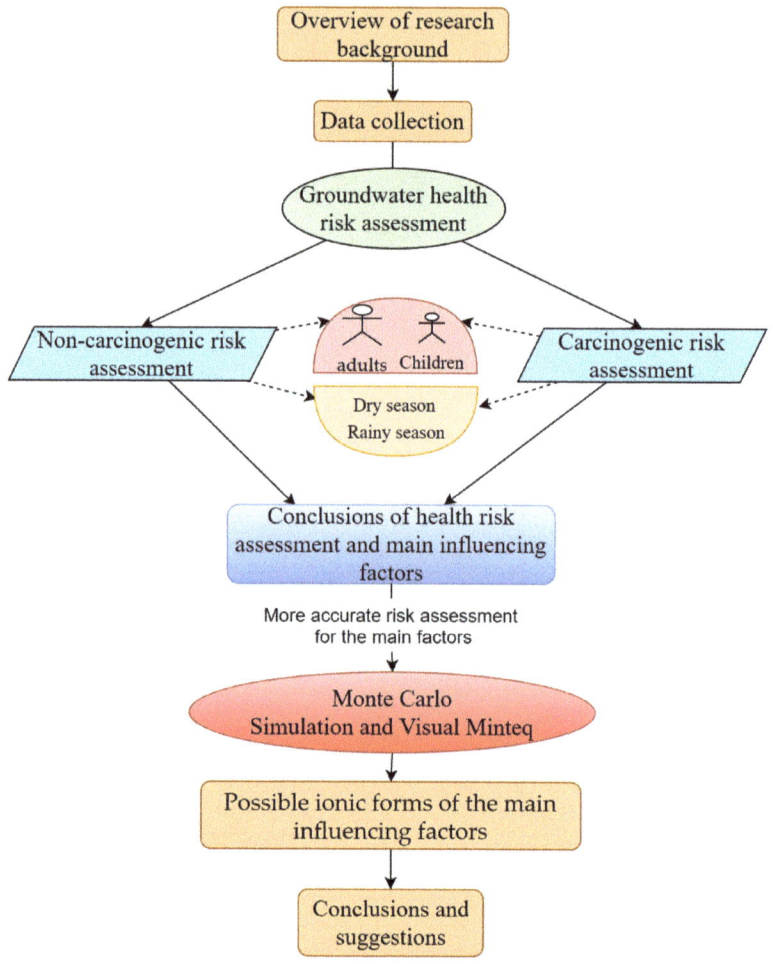

Figure 2. Research idea flowchart.

Table 2. Limit values of several parameters for oral intake and skin contact by type of sensitive group.

	TR (L/day)	EF (day/a)	ED (a)	BW (kg)	AT (day)	EV	K (cm/h)	t (h/day)	CF	H (cm)
Children	0.7	365	12	15	4380	1	0.001	0.4	0.001	99.4
Adults	1.5		30	70	10,950					165.3

The average daily dose for oral intake and dermal contact was calculated as follows:

$$Intake_{oral} = \frac{C \times IR \times EF \times ED}{BW \times AT} \quad (2)$$

$$Intake_{dermal} = \frac{DA \times EV \times SA \times EF \times ED}{BW \times AT} \quad (3)$$

$$DA = K \times C \times t \times CF \qquad (4)$$

$$SA = 239 \times H^{0.417} \times BW^{0.517} \qquad (5)$$

where $Intake_{oral}$ is the average daily dose of oral intake (mg (kg/d)$^{-1}$); DA and SA are defined as the exposure dose (mg/cm^2) and skin contact area (cm^2) of each event, respectively; and C indicates the concentration of pollutants in groundwater (mg/L^{-1}) obtained by laboratory tests. The limit values of oral intake and skin contact of several parameters for the two sensitive groups investigated are illustrated in Table 2.

The non-carcinogenic risk of oral intake and skin contact was calculated as follows:

$$RfD = RfD \times BAS_{gi} \qquad (6)$$

$$HQ_{oral} = \frac{Intake_{oral}}{RfD_{oral}} \qquad (7)$$

$$HQ_{dermal} = \frac{Intake_{dermal}}{RfD_{dermal}} \qquad (8)$$

where HQ and RfD indicate the non-carcinogenic hazard quotients and the reference doses, respectively [34]. In this study, the RfD values of Mn, NO$_2^-$, Fe, F, Pb, Cr^{6+}, Cd, As, and ammonia nitrogen (in terms of N), were found to be 0.14, 0.1, 0.3, 0.04, 0.0014, 0.0003, 0.003, 0.0003, and 0.97 mg (kg d) 1, respectively [35]. RfD_{dermal} indicates a gastrointestinal absorption factor that can be calculated by RfD_{oral}. Apart from the value of BAS_{gi} for Cr^{6+} which was 0.025, the other values BAS_{gi} were all equal to 1.

The non-carcinogenic risk of oral intake and skin contact absorption was calculated as the total risk, as follows:

$$HI_i = HQ_{oral} + HQ_{dermal} \qquad (9)$$

$$HI_{total} = \sum_{i=1}^{n} HI_i \qquad (10)$$

where HI is a health risk assessment index, which refers to the sum of multiple HQs of multiple substances through the two exposure pathways considered. HQ and HI values lower than 1 are considered safe for human health [36,37]. By contrast, residents may face non-carcinogenic risks when these values exceed 1. The standard value of HI was proposed by the Ministry of Environmental Protection of the China in 2014.

In this study, As and Cr^{6+} were considered the main carcinogenic factors; their carcinogenic health risk values were calculated as follows:

$$SF_{dermal} = \frac{SF_{oral}}{ABS_{gi}} \qquad (11)$$

$$CR_{oral} = Intake_{oral} \times SF_{oral} \qquad (12)$$

$$CR_{demal} = Intake_{demal} \times SF_{demal} \qquad (13)$$

$$CR_{total} = CR_{oral} + CR_{demal} \qquad (14)$$

where the SF_{oral} values of As and Cr^{6+} were set at 1.5 and 0.5 mg (kg/day)$^{-1}$, respectively; and CR indicates the risk of cancer. According to the regulations of the Ministry of Environmental Protection of China, the acceptable limit value for both parameters is 0.6 [38]. When the calculated value exceeds the limit value, this indicates that there is a risk of cancer in the region, and that it is necessary to take corresponding environmental control measures in time [39].

3.3. Monte Carlo Simulation and Visual Minteq

Due to individual differences and sampling limitations, the results of the health risk calculation were highly uncertain. In order to overcome this defect, this study used the Monte Carlo Simulation (MCS) method, which is a probabilistic statistical mathematical method to evaluate uncertainty by random sampling each variable value [40]. This simulation method was combined with the U.S. Environmental Protection Agency risk assessment technology to quantify and minimize the uncertainty, and analyze the sensitivity of pollutant exposure in drinking water wells. Different numbers related to each variable were sampled for a large number of digital repetition operations. The main steps are as follows: (1) The respective distribution functions were fitted to the data of each variable, and the hypothetical variable was defined; (2) The calculation formula was inputted and the decision variable was defined; (3) A random sampling was conducted from the distribution of the hypothetical variables; (4) The randomly selected parameter sequence was used for a large number of repeated operations, to obtain as output the probability distribution of the operation results (Figure 3). In this study, the Monte Carlo Model with 10,000 iterations was used to evaluate the carcinogenic risk of exposure to As and Cr6+ for children and adults during the rainy and dry seasons from 2017 to 2018. Then, sensitivity analysis was carried out using the MCS method with 10,000 repetitions using Oracle Crystal Ball®. Finally, the frequency diagram and sensitivity analysis diagram drawn by it were used to identify the input parameters with the greatest impact on the output of the risk assessment model. After using the sensitivity analysis to obtain the main influencing factors of health risk assessment results, Visual Minteq was used to analyze the possible ion forms of the main factors in the hydrogeological environment of the study area.

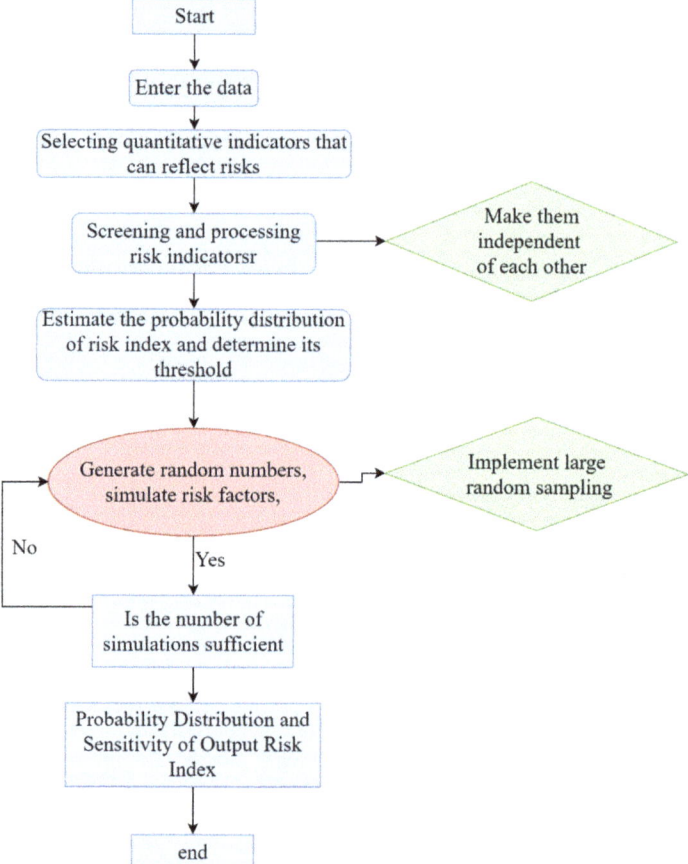

Figure 3. Monte Carlo model operation flowchart.

4. Results and Discussion

4.1. Health Risk Assessment

4.1.1. Health Risk Assessment

The concentrations of ions in water were used to assess the possible health risks from human exposure to drinking water through both oral intake and skin contact. This study considered two groups, namely adults and children [41], and evaluated the carcinogenic risk of eight ions and the non-carcinogenic risk of two heavy metal ions [42].

4.1.2. Non-Carcinogenic Risk Assessment

The Hazard Index (HI) was used to measure non-carcinogenic health risk as the sum of the hazard quotients (HQs) from both oral intake and dermal contact of several parameters including F^-, NH_4^+, NO_2^-, NO_3^-, Cr^{6+}, Mn, As, Fe (including Fe^{2+}, Fe^{3+}), and others, as shown in Tables 3 and 4 [43]. From 2017 to 2018, a total of 17 sampling points for adults and 67 for children had an HI index value higher than 1, with the latter far exceeding the number of the former.

Table 3. Non-carcinogenic risk assessment for adults.

	2017		2008	
Sample	Dry Season	Rainy Season	Dry Season	Rainy Season
W1	0.303	0.471	0.406	0.264
W2	0.382	0.401	0.332	0.461
W3	0.234	0.266	0.275	0.343
W4	1.442	1.820	2.847	0.859
W5	0.456	0.490	0.093	0.355
W6	0.776	1.056	0.638	0.797
W7	0.893	0.592	0.521	0.703
W8	0.442	0.352	0.265	0.452
W9	0.396	0.485	0.334	0.389
W10	0.567	0.410	0.279	0.352
W11	1.412	1.646	2.759	0.706
W12	0.492	1.114	0.991	0.789
W13	0.292	0.171	0.125	0.360
W14	0.439	0.259	0.431	0.653
W15	0.245	0.114	0.167	0.232
W16	0.423	0.420	0.193	0.448
W17	0.133	0.127	0.126	0.188
W18	0.458	0.572	0.588	0.428
W19	0.489	0.335	0.497	0.196
W20	0.465	0.436	0.260	0.611
W21	0.370	0.254	0.191	0.268
W22	0.234	0.269	0.227	0.388
W23	0.366	0.687	0.297	1.143
W24	0.385	0.515	0.118	0.483
W25	0.265	0.235	0.181	0.347
W26	0.977	2.064	1.495	0.583
W27	0.167	0.087	0.509	0.530
W28	0.678	0.593	0.649	0.706
W29	0.732	0.753	0.807	0.687
W30	0.373	0.421	0.419	0.500
W31	0.434	0.469	0.447	0.475
W32	0.238	0.496	0.270	0.072
W33	0.502	0.364	0.309	0.322
W34	0.298	0.229	0.201	0.286
W35	0.284	0.269	0.271	0.160
W36	0.241	0.234	0.478	0.386
W37	0.325	0.470	0.320	0.538

Table 3. Cont.

	2017		2008	
Sample	Dry Season	Rainy Season	Dry Season	Rainy Season
W38	0.663	1.700	0.738	0.621
W39	0.188	0.295	0.189	0.280
W40	0.314	0.192	0.193	0.235
W41	0.226	0.235	0.204	0.256
W42	2.575	2.688	2.601	2.530
W43	0.387	0.480	147.253	0.512
W44	0.131	0.150	0.108	0.384
W45	0.150	0.185	0.119	0.437
W46	0.187	0.184	0.197	0.390
W47	0.200	0.450	0.335	0.324
W48	0.369	0.455	0.361	0.351
W49	0.393	0.311	0.331	0.383
Num	3.000	7.000	5	2
Sum		17.000		

Table 4. Non-carcinogenic risk assessment for children.

	2017		2018	
Sample	Dry Season	Rainy Season	Dry Season	Rainy Season
W1	0.647	1.007	0.866	0.564
W2	0.817	0.857	0.7100	0.986
W3	0.500	0.568	0.587	0.732
W4	3.081	3.888	6.082	1.836
W5	0.974	1.047	0.198	0.759
W6	1.659	2.257	1.363	1.703
W7	1.907	1.265	1.114	1.502
W8	0.945	0.753	0.566	0.965
W9	0.846	1.036	0.714	0.832
W10	1.212	0.875	0.596	0.753
W11	3.017	3.517	5.894	1.509
W12	1.052	2.380	2.118	1.686
W13	0.624	0.366	0.267	0.770
W14	0.915	0.553	0.922	1.396
W15	0.523	0.243	0.358	0.496
W16	0.903	0.897	0.412	0.958
W17	0.285	0.271	0.270	0.401
W18	0.980	1.221	1.256	0.914
W19	1.045	0.715	1.061	0.419
W20	0.993	0.932	0.556	1.306
W21	0.790	0.542	0.408	0.573
W22	0.499	0.574	0.484	0.829
W23	0.781	1.469	0.635	2.443
W24	0.823	1.099	0.252	1.033
W25	0.474	0.502	0.340	0.694
W26	2.086	4.411	3.194	1.245
W27	0.357	0.185	1.087	1.133
W28	1.448	1.268	1.388	1.508
W29	1.564	1.609	1.723	1.467
W30	0.797	0.899	0.894	1.069
W31	0.928	1.002	0.955	1.015
W32	0.507	1.059	0.578	0.154
W33	1.072	0.777	0.660	0.688
W34	0.636	0.489	0.429	0.612
W35	0.607	0.574	0.580	0.342
W36	0.515	0.500	1.021	0.825

Table 4. Cont.

Sample	2017 Dry Season	2017 Rainy Season	2018 Dry Season	2018 Rainy Season
W37	0.694	1.004	0.683	1.148
W38	1.417	3.633	1.577	1.328
W39	0.401	0.631	0.403	0.597
W40	0.671	0.409	0.413	0.503
W41	0.482	0.502	0.435	0.547
W42	5.502	5.742	5.557	5.406
W43	0.827	1.025	314.623	1.094
W44	0.281	0.321	0.231	0.821
W45	0.321	0.395	0.253	0.933
W46	0.400	0.393	0.420	0.833
W47	0.428	0.961	0.716	0.693
W48	0.789	0.972	0.771	0.751
W49	0.840	0.665	0.708	0.819
Num	13	20	15	19
Sum		67		

4.1.3. Carcinogenic Risk Assessment

The International Agency for Cancer Research classified five elements (i.e., chromium, cadmium, arsenic, nickel, and cobalt) as possible carcinogens (IARC, 2013). Since cadmium and cobalt have no SF, the oral intake and skin contact for chromium, arsenic, and nickel were considered pathways of human contact, with a special focus on arsenic. In Tables 3 and 4, it can be seen that from 2017 to 2018 a total of 114 records of CR, which indicate the carcinogenic risk of adults and children, exceeded the value of $1 \times 10^{-0.6}$, demonstrating that the study area was seriously polluted, and that the main pollution factor was As. From 2017 to 2018, the average CR values for adults for As in the study area were $2.94 \times 10^{-0.5}$, $5.93 \times 10^{-0.5}$, $5.48 \times 10^{-0.5}$, and $3.59 \times 10^{-0.5}$ (Table 5), respectively, while those of children were $6.28 \times 10^{-0.5}$, $1.27 \times 10^{-0.4}$, $1.17 \times 10^{-0.4}$, and $7.67 \times 10^{-0.5}$, respectively, which were higher than those of adults (Table 6). Therefore, we mainly analyzed the influence of As ions on children's health risk and the possible ion forms under local hydrogeological conditions (Table 7).

Table 5. Summary of data for As.

As	2017 Dry Season	2017 Rainy Season	2018 Dry Season	2018 Rainy Season
Adults	$2.94 \times 10^{-0.5}$	$5.93 \times 10^{-0.5}$	$5.48 \times 10^{-0.5}$	$3.59 \times 10^{-0.5}$
Children	$6.28 \times 10^{-0.5}$	$1.27 \times 10^{-0.4}$	$1.17 \times 10^{-0.4}$	$7.67 \times 10^{-0.5}$

Table 6. Non-carcinogenic risk assessment for children.

Sample	2017 Dry Season	2017 Rainy Season	2018 Dry Season	2018 Rainy Season
W1	0	0	0	0.0000167
W2	0	0	0	0.00003
W3	0	0.0000167	0.0000133	0.00005
W4	33+F15:F53	0.00055	0.000967	0.0000733
W5	0	0.0000133	0	0.0000167
W6	0.0000433	0.000207	0.0000433	0.0000667
W7	0.0000333	0.0000167	0	0.0000567
W8	0	0	0	0.00003
W9	0	0.0000133	0	0.0000267

Table 6. *Cont.*

Sample	2017 Dry Season	2017 Rainy Season	2018 Dry Season	2018 Rainy Season
W10	0.0000333	0.0000233	0	0.0000233
W11	0.000327	0.000483	0.00096	0.0000733
W12	0.0000267	0.0000833	0.0000133	0
W13	0	0	0	0.0000333
W14	0.0000532	0.0000133	0.00000667	0.00016
W15	0	0	0	0.0000233
W16	0	0	0	0.0000167
W17	0	0	0	0.0000133
W18	0.0000233	0.0000333	0.00002	0
W19	0	0	0	0.0000133
W20	0	0.00002	0	0.0000567
W21	0.0000267	0.0000233	0	0.0000133
W22	0	0.00002	0	0.00006
W23	0	0.0000233	0	0.00006
W24	0	0.0000167	0.00000667	0.0000467
W25	0.000213	0.0000167	0.000106	0.000176
W26	0.000137	0.000527	0.00036	0
W27	0	0	0.00001	0.0000533
W28	0.00001	0	0	0.00002
W29	0.0000567	0.0000833	0.0000633	0
W30	0	0	0	0.00003
W31	0	0.0000167	0.00000667	0.0000233
W32	0	0.0000167	0.00001	0
W33	0.0000733	0.0000167	0	0
W34	0.00003	0.0000167	0.0000133	0.0000267
W35	0.0000533	0.0000767	0.0000867	0.0000267
W36	0	0	0	0.0000567
W37	0	0.0000133	0	0.0000533
W38	0.00008	0.000417	0.0000467	0.0000233
W39	0	0	0	0.00002
W40	0.0000533	0	0	0
W41	0	0.0000167	0	0.0000267
W42	0	0.0000167	0	0.0000367
W43	0	0.0000167	0	0.0000267
W44	0	0.0000133	0	0.00007
W45	0	0.0000167	0.00000333	0.0000667
W46	0	0	0	0.0000633
W47	0	0.0000133	0	0.0000467
W48	0.00002	0.0000533	0.0000567	0.00003
W49	0.00003	0	0	0.00003
Num	20	33	19	42
Sum		114		
W1	0	0	0	0.0000167
W2	0	0	0	0.00003
W3	0	0.0000167	0.0000133	0.00005
W4	33+F15:F53	0.00055	0.000967	0.0000733
W5	0	0.0000133	0	0.0000167
W6	0.0000433	0.000207	0.0000433	0.0000667
W7	0.0000333	0.0000167	0	0.0000567
W8	0	0	0	0.00003
W9	0	0.0000133	0	0.0000267
W10	0.0000333	0.0000233	0	0.0000233
W11	0.000327	0.000483	0.00096	0.0000733
W12	0.0000267	0.0000833	0.0000133	0
W13	0	0	0	0.0000333

Table 6. *Cont.*

	2017		2018	
Sample	Dry Season	Rainy Season	Dry Season	Rainy Season
W14	0.0000532	0.0000133	0.00000667	0.00016
W15	0	0	0	0.0000233
W16	0	0	0	0.0000167
W17	0	0	0	0.0000133
W18	0.0000233	0.0000333	0.00002	0
W19	0	0	0	0.0000133
W20	0	0.00002	0	0.0000567
W21	0.0000267	0.0000233	0	0.0000133
W22	0	0.00002	0	0.00006
W23	0	0.0000233	0	0.00006
W24	0	0.0000167	0.00000667	0.0000467
W25	0.000213	0.0000167	0.000106	0.000176
W26	0.000137	0.000527	0.00036	0
W27	0	0	0.00001	0.0000533
W28	0.00001	0	0	0.00002
W29	0.0000567	0.0000833	0.0000633	0
W30	0	0	0	0.00003
W31	0	0.0000167	0.00000667	0.0000233
W32	0	0.0000167	0.00001	0
W33	0.0000733	0.0000167	0	0
W34	0.00003	0.0000167	0.0000133	0.0000267
W35	0.0000533	0.0000767	0.0000867	0.0000267
W36	0	0	0	0.0000567
W37	0	0.0000133	0	0.0000533
W38	0.00008	0.000417	0.0000467	0.0000233
W39	0	0	0	0.00002
W40	0.0000533	0	0	0
W41	0	0.0000167	0	0.0000267
W42	0	0.0000167	0	0.0000367
W43	0	0.0000167	0	0.0000267
W44	0	0.0000133	0	0.00007
W45	0	0.0000167	0.00000333	0.0000667
W46	0	0	0	0.0000633
W47	0	0.0000133	0	0.0000467
W48	0.00002	0.0000533	0.0000567	0.00003
W49	0.00003	0	0	0.00003
Num	20	33	19	42
Sum		114		

Table 7. Carcinogenic risk assessment for children.

	2017		2018	
Sample	Dry Season	Rainy Season	Dry Season	Rainy Season
W1	0.303	0.471	0.406	0.264
W2	0.382	0.401	0.332	0.461
W3	0.234	0.266	0.275	0.343
W4	1.442	1.820	2.847	0.859
W5	0.456	0.490	0.093	0.355
W6	0.776	1.056	0.638	0.797
W7	0.893	0.592	0.521	0.703
W8	0.442	0.352	0.265	0.452
W9	0.396	0.485	0.334	0.389
W10	0.567	0.410	0.279	0.352
W11	1.412	1.646	2.759	0.706

Table 7. Cont.

Sample	2017 Dry Season	2017 Rainy Season	2018 Dry Season	2018 Rainy Season
W12	0.492	1.114	0.991	0.789
W13	0.292	0.171	0.125	0.360
W14	0.439	0.259	0.431	0.653
W15	0.245	0.114	0.167	0.232
W16	0.423	0.420	0.193	0.448
W17	0.133	0.127	0.126	0.188
W18	0.458	0.572	0.588	0.428
W19	0.489	0.335	0.497	0.196
W20	0.465	0.436	0.260	0.611
W21	0.370	0.254	0.191	0.268
W22	0.234	0.269	0.227	0.388
W23	0.366	0.687	0.297	1.143
W24	0.385	0.515	0.118	0.483
W25	0.265	0.235	0.181	0.347
W26	0.977	2.064	1.495	0.583
W27	0.167	0.087	0.509	0.530
W28	0.678	0.593	0.649	0.706
W29	0.732	0.753	0.807	0.687
W30	0.373	0.421	0.419	0.500
W31	0.434	0.469	0.447	0.475
W32	0.238	0.496	0.270	0.072
W33	0.502	0.364	0.309	0.322
W34	0.298	0.229	0.201	0.286
W35	0.284	0.269	0.271	0.160
W36	0.241	0.234	0.478	0.386
W37	0.325	0.470	0.320	0.538
W38	0.663	1.700	0.738	0.621
W39	0.188	0.295	0.189	0.280
W40	0.314	0.192	0.193	0.235
W41	0.226	0.235	0.204	0.256
W42	2.575	2.688	2.601	2.530
W43	0.387	0.480	147.253	0.512
W44	0.131	0.150	0.108	0.384
W45	0.150	0.185	0.119	0.437
W46	0.187	0.184	0.197	0.39
W47	0.200	0.450	0.335	0.324
W48	0.369	0.455	0.361	0.351
W49	0.393	0.311	0.331	0.383
Num	3	7	5	2
Sum		17		

4.2. Risk Characterization Based on MCS

MCS allows to iteratively generate time series by setting up a stochastic process, calculating parameter estimates and statistics, and studying the characteristics of their distribution. Because As and Cr^{6+} concentrations vary under local hydrogeological conditions and population properties, this study used a Crystal Ball model based on MCS to produce an assessment of the possible carcinogenic health risks to children.

As shown in Figure 4, The forecast for the 2017 dry season ranged from 0–$1.80 \times 10^{-0.5}$, with a mean of $3.26 \times 10^{-0.6}$ and a standard deviation of $5.75 \times 10^{-0.6}$, while the forecast for the rainy season of the same year ranged from 0–$1.00 \times 10^{-0.3}$, with a mean of $1.39 \times 10^{-0.4}$ and a standard deviation of $2.92 \times 10^{-0.4}$. The forecast for the 2018 dry season ranged from 0–$6.00 \times 10^{-0.4}$, with a mean of $2.06 \times 10^{-0.4}$ and a standard deviation of $3.47 \times 10^{-0.4}$, while the forecast for the rainy season of the same year ranged from 0–$5.00 \times 10^{-0.4}$, with a mean of $1.79 \times 10^{-0.4}$ and a standard deviation of 2.65 c. The simulations yielded a CR for children between 0 and $1 \times 10^{-0.6}$ with probabilities in 2017 of 16.59% and 0%, and in 2018 of 0% and 0% for the dry and rainy season, respectively, indicating that children are at a higher health risk and prompt action on this respect should be considered. MCS can also be used to determine the sensitivity of the chosen parameters depending on their uncertainty [44–46]. In fact, different parameters have different sensitivities to carcinogenic risk values. Accordingly, in this study, the analysis of the effect of As and Cr on carcinogenic risk was performed.

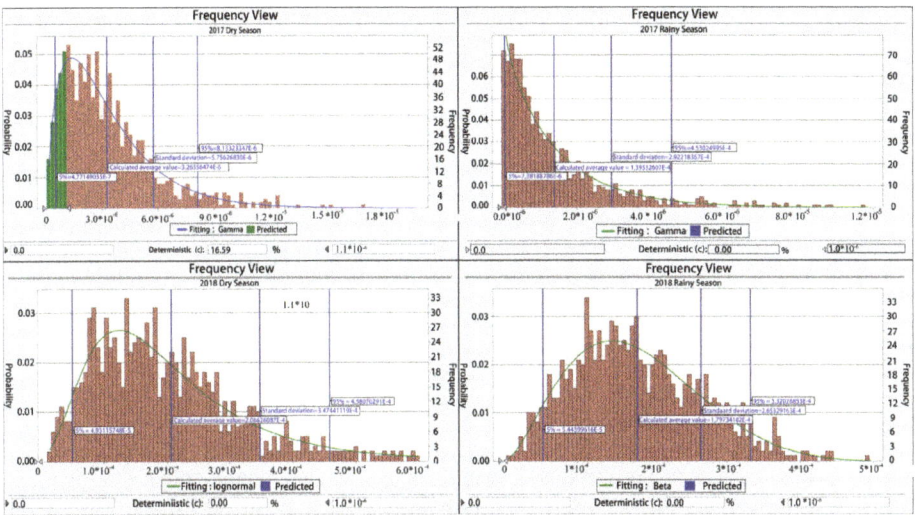

Figure 4. Assessment of the possible carcinogenic health risks to children.

As shown in Figure 5, the association values for As in the dry and rainy seasons for 2017 were 0.9958 and 1, respectively, and Cr was 0, while the association values for As in the 2018 dry and rainy seasons were 0.7424 and 0.5759, respectively, and Cr was 0.6237 and 0.8182, respectively. As you can see, during the 2018 rainy season, the association value of As was lower than the association value of Cr. In the 2017 dry rainy season and the 2018 dry season, the association value of As was higher than the correlation value of Cr.

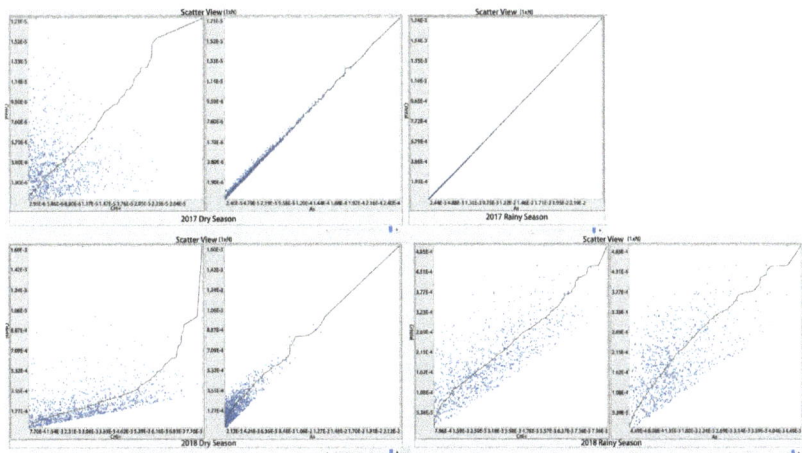

Figure 5. Association values for As and Cr.

Moreover, as can be seen in Figure 6, in 2017 the sensitivity of As was 100%, while that of Cr was 3% and 0% in the dry and rainy seasons, respectively. In 2018, the sensitivity of As was 54.1% and 57.8% and that of Cr was 45.9% and 42.2% in the dry and rainy seasons, respectively. This indicates that As was the most sensitive to total carcinogenic risk [47].

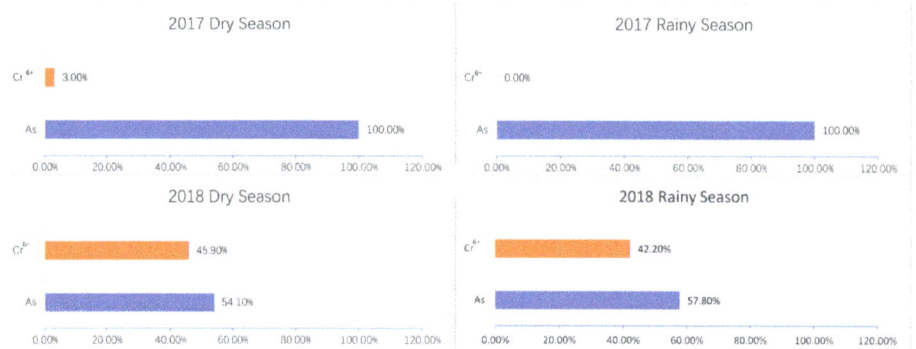

Figure 6. Sensitivity of As and Cr.

Based on these results, the analysis of the possible patterns of As present in the water environment of the study area was conducted.

In summary, in this study, limited data were employed through the use of MCS to determine probability density functions and confidence intervals for carcinogenic risks, and an uncertainty analysis was performed to reveal the possible influence of various parameters on human health risks [48].

4.3. Visual Minteq Model Analysis

In order to further analyze the possible ionic forms that could be formed as a result of chemical reactions between As ions and other ions in the water environment of the study area, Visual Minteq was used to simulate the pH, ion concentrations, and possible products of the water environment. The data to be entered included pH, temperature, K^+, Na^+, Ca^{2+}, Mg^{2+}, Fe^{2+}, and As. The possible compounds of As produced by the simulations in the study area context are depicted in Figure 7 and include AsO_4^{3-}, $H_2AsO_4^-$, H_3AsO_4 and $HAsO_4^{2-}$. As can be seen from Figure 7, among all the possible ionic forms of As formed during the wet and dry seasons of 2017 and 2018, the activity of $HAsO_4^{2-}$ had the largest logarithmic value while that of H_3AsO_4 had the smallest value, regardless of pH change. The minimum and maximum values for $HAsO_4^{2-}$ were −7.331 and −5.784 in the 2017 dry season and −7.274 and −5.645 in the wet season, respectively. The minimum and maximum values for $HAsO_4^{2-}$ were −7.85 and −5.371 in the 2018 dry season and −7.342 and −6.212 in the wet season, respectively. The minimum and maximum values for H_3AsO_4 were −15.142 and −12.222 in the dry season and −14.648 and −11.895 in the rainy season, respectively. The minimum and maximum values of H_3AsO_4 were −14.75 and −12.129 in the 2018 dry season and −14.666 and −12.509 in the rainy season, respectively.

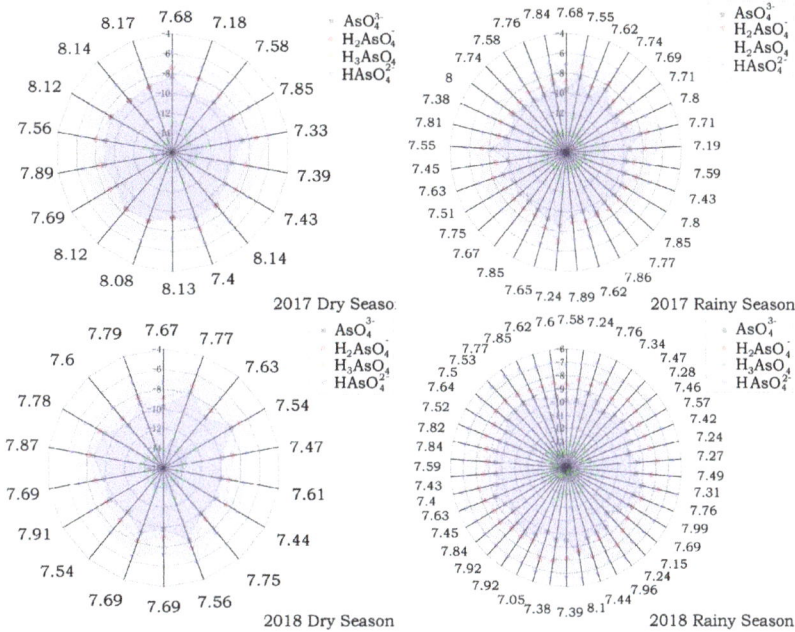

Figure 7. Logarithmic values of the activities of the possible compounds of As generated by the simulations.

5. Conclusions and Recommendations

In this study, 49 sets of observations from relatively continuous and complete monitoring sites were used to compare the health risks posed to adults and children by oral intake and dermal contact with drinking water in Yinchuan during the dry and rainy seasons in 2017 and 2018, assessing the health risks of Mn, NO_2^-, Fe, F, Pb, Cr^{6+}, Cd, As, and ammonia nitrogen. According to these data, the HI index exceeded the value of 1 at 17 sampling points for adults and 67 for children. The average CR values for adults for As in the study area for 2017 were $2.94 \times 10^{-0.5}$ and $5.93 \times 10^{-0.5}$ for the dry and rainy seasons, respectively, while for 2018 they were $5.48 \times 10^{-0.5}$ and $3.59 \times 10^{-0.5}$, respectively. In parallel, the CR values for children for 2017 were $6.28 \times 10^{-0.5}$ and $1.27 \times 10^{-0.4}$, respectively, and $1.17 \times 10^{-0.4}$ and $7.67 \times 10^{-0.5}$, respectively, indicating a considerably higher carcinogenic risk for children than for adults.

The results of the sensitivity analysis of Cr^{6+} and As using Crystal Ball software showed that in 2017 the association values were 0.9958 and 1 for As and 0.0948 and 0 for Cr for the dry and rainy seasons, respectively, while in 2018 they were 0.7424 and 0.5759 for As and 0.6237 and 0.8182 for Cr for the dry and rainy seasons, respectively. This means that the association values for As were lower than those for Cr only in the 2018 rainy season, indicating that As is most sensitive to overall carcinogenic risk.

The results of the visual coinage model analysis allowed us to conclude that the activity of $HAsO_4^{2-}$ had the largest logarithmic value, while that of H_3AsO_4 had the smallest logarithmic value among all the possible ionic forms of As, regardless of the change in pH. This indicates that $HAsO_4^2$ is the ionic form of As with the main carcinogenic factor in the hydrogeological environment of the study area.

Based on these results, it is recommended to strengthen the monitoring and control of As, especially of $HAsO_4^{2-}$, levels in groundwater in the study area.

Author Contributions: Conceptualization, Z.M. and J.L.; methodology, M.Z.; software, D.Y.; validation, Z.M., J.L. and M.Z.; formal analysis, Y.Z.; investigation, Z.M.; resources, Y.Z.; data curation, M.Z.; writing—original draft preparation, M.Z.; writing—review and editing, D.Y.; visualization, Z.G.; supervision, Z.G.; project administration, Y.Z.; funding acquisition, Z.G. All authors have read and agreed to the published version of the manuscript.

Funding: This research is funded by Open project of Hebei key laboratory of geological resources and environmental monitoring and protection (JCYKT202101); Natural Science Foundation of Hebei Province of China (D2022403016); Hebei water conservancy science and technology plan project (2021-45). Hebei University Science and technology research project (ZD2022119); Science and technology innovation team project of Hebei GEO University (KJCXTD-2021-14); Introduction of foreign intelligence project in Hebei province in 2021 (2021ZLYJ-1).

Institutional Review Board Statement: Not applicable.

Informed Consent Statement: Not applicable.

Data Availability Statement: Data availability statement can be said to be the monitoring data of Hebei Institute of hydrogeology and engineering geology, real and reliable.

Acknowledgments: The author thanks Hebei Geological Environment Monitoring Institute, and Hebei Institute of hydrogeology and engineering geology providing real and reliable data.

Conflicts of Interest: The authors declare no conflict of interest.

References

1. Li, P.; Li, X.; Meng, X.; Li, M.; Zhang, Y. Appraising Groundwater Quality and Health Risks from Contamination in a Semiarid Region of Northwest China. *Expo. Health* **2016**, *8*, 361–379. [CrossRef]
2. Li, P.; Qian, H. Human health risk assessment for chemical pollutants in drinking water source in Shizuishan city-northwest china. *Iran. J. Environ. Health Sci. Eng.* **2011**, *8*, 41–48.
3. Wang, X. Groundwater resources development and utilization in China. In *Inner Mongolia*; Inner Mongolia People's Publishing House: Hohhot, China, 1992.
4. Zivanovic, V.; Jemcov, I.; Dragisic, V.; Atanackovic, N.; Magazinovic, S. Karst groundwater source protection based on the time-dependent vulnerability assessment model: Crnica springs case study, eastern Serbia. *Environ. Earth Sci.* **2016**, *75*, 1224. [CrossRef]
5. Tai, T.; Wang, J.; Wang, Y.; Bai, L. Groundwater pollution risk evaluation method research progress in our country. *J. Beijing Norm. Univ. Nat. Sci.* **2012**, *6*, 648–653. (In Chinese)
6. Li, P.; Qian, H.; Wu, J. Accelerate research on land creation. *Nature* **2014**, *510*, 29–31. [CrossRef]
7. Wang, Q.; Wang, J. Research on groundwater pollution and its prevention-control policy in China. *J. China Univ. Geosci.* **2008**, *8*, 72–75.
8. Anokhin, Y. Toxic aluminium and heavy metals in groundwater of middle Russia: Health risk assessment. *Int. J. Environ. Res. Public Health* **2005**, *2*, 214–218.
9. Wongsasuluk, P.; Chotpantarat, S.; Siriwong, W.; Robson, M. Heavy metal contamination and human health risk assessment in drinking water from shallow groundwater wells in an agricultural area in Ubon Ratchathani province, Thailand. *Environ. Geochem. Health* **2014**, *36*, 169–182. [CrossRef]
10. Manecki, P.; Gałuszka, A. Groundwater quality as a geoindicator of organochlorine pesticide contamination after pesticide tomb reclamation: A case study of Franciszkowo, northwestern Poland. *Environ. Earth Sci.* **2012**, *67*, 2441–2447. [CrossRef]
11. Han, D.; Tong, X.; Jin, M.; Hepburn, E.; Tong, C.; Song, X. Evaluation of organic contamination in urban groundwater surrounding a municipal landfill, Zhoukou, China. *Environ. Monit. Assess.* **2013**, *185*, 3413–3444. [CrossRef]
12. Akter, A.; Ali, M. Arsenic contamination in groundwater and its proposed remedial measures. *Int. J. Environ. Sci. Technol.* **2011**, *8*, 433–443. [CrossRef]
13. Nasrabadi, T.; Bidabadi, N. Evaluating the spatial distribution of quantitative risk and hazard level of arsenic exposure in groundwater, case study of Qorveh County, Kurdistan Iran. *Iran J. Environ. Health Sci. Eng.* **2013**, *10*, 30. [CrossRef]
14. Uddin, M.; Harun-Ar-Rashid, A.; Hossain, S.; Hafiz, M.; Nahar, K.; Mubin, S. Slowarsenic poisoning of the contaminated groundwater users. *Int. J. Environ. Sci. Technol.* **2006**, *3*, 447–453. [CrossRef]
15. Song, G. *Study on the Mechanism of Chromium Pollution in Groundwater*; Xi'an College of Engineering: Xi'An, China, 1997.
16. Jing, S.; Wang, L.; Chen, Q.; Wu, B. Research on evaluation and countermeasures to present status of groud water environment in Tianjin. *North. Environ.* **2012**, *6*, 40–42.
17. Deepesh, M.; Kumar, J.; Singh, V.; Chinchu, M. Assessment and mapping of groundwater vulnerability to pollution: Current status and challenges. *Earth-Sci. Rev.* **2018**, *185*, 901–927.
18. Wang, G.; Liu, Z.; Gao, Y.; Fan, Q. Study on groundwater resources and the protective strategy in shiyang river basin. *J. Arid. Land Resour. Environ.* **2007**, *21*, 48–51.

19. World Health Organization. *Guidelines for Drinking Water Quality*, 4th ed.; WHO: Geneva, Switzerland, 2011.
20. Zhou, Y.; Wei, A.; Li, J.; Li, Y.; Li, J. Groundwater Quality Evaluation and Health Risk Assessment in the Yinchuan Region, Northwest China. *Expo. Health* **2016**, *8*, 443–456. [CrossRef]
21. Teng, Y.; Jie, S.; Zhai, Y.; Wang, J. A review on the overlay and index method for groundwater pollution risk assessment. *Adv. Earth Sci.* **2012**, *27*, 1140–1147.
22. Liu, S.; Wang, H. Review on Groundwater Pollution Risk Assessment in China. In Proceedings of the 2012 4th International Conference on Environmental Science and Information Application Technology (ESIAT 2012), Shanghai, China, 12 November 2012.
23. Lai, Y.; Li, Y.; Song, W.; Zhou, L.; Song, Y. Current situation and countermeasures of exploitation and utilization of groundwater resources in Shenzhen. *Environ. Sci. Survey* **2018**, *37*, 25–27.
24. Qian, H.; Li, P.; Howard, K.; Yang, C.; Zhang, X. Assessment of groundwater vulnerability in the Yinchuan Plain, Northwest China using OREADIC. *Environ. Monit. Assess.* **2012**, *6*, 3613–3628. [CrossRef]
25. Chen, X.; Zhu, L.; Liu, J. Study on health risk evaluation and risk control of drinking groundwater in Yinchuan Plain. *Anhui Agric. Sci.* **2019**, *47*, 7.
26. Mohammadpour, A.; Gharehchahi, E.; Badeenezhad, A.; Parseh, I.; Khaksefidi, R.; Golaki, M.; Dehbandi, R.; Azhdarpoor, A.; Derakhshan, Z.; Rodriguez Chueca, J.; et al. Nitrate in Groundwater Resources of Hormozgan Province, Southern Iran: Concentration Estimation, Distribution and Probabilistic Health Risk Assessment Using Monte Carlo Simulation. *Water* **2022**, *14*, 564. [CrossRef]
27. Zhang, H.; Qian, H.; Chen, J.; Qiao, L. Assessment of Groundwater Chemistry and Status in a Heavily Used Semi-Arid Region with Multivariate Statistical Analysis. *Water* **2014**, *6*, 2212–2232. [CrossRef]
28. Zhang, X.; Qian, H.; Wu, H.; Chen, J.; Qiao, L. Multivariate Analysis of Confined Groundwater Hydrochemistry of a Long-Exploited Sedimentary Basin in Northwest China. *J. Chem.* **2016**, *2016*, 3812125. [CrossRef]
29. Li, P.; Wu, Q.; Wu, J. Groundwater suitability for drinking and agricultural usage in Yinchuan area, China. *Int. J. Environ. Sci.* **2010**, *1*, 1241–1249.
30. Qian, H.; Li, P. Hydrochemical characteristics of groundwater in Yinchuan plain and their control factors. *Asian J. Chem.* **2011**, *23*, 2927–2938.
31. Paul, R.; Brindha, K.; Gowrisankar, G.; Mou, L.; Mahesh, K. Identification of hydrogeochemical processes controlling groundwater quality in Tripura, Northeast India using evaluation indices, GIS, and multivariate statistical methods. *Environ. Earth Sci.* **2019**, *78*, 470. [CrossRef]
32. Yin, Z.; Duan, R.; Li, P.; Li, W. Water quality characteristics and health risk assessment of main water supply reservoirs in Taizhou City, East China. *Hum. Ecol. Risk Assess.* **2021**, *27*, 2142–2160. [CrossRef]
33. Mustafa, K.G.; Nadia, R.M.; Zafar, I.S.; Jalal, U.; Saima, N.; Ur, R.J.; Ijaz, H.A.; Amin, M.K. The determination of potentially toxic elements (PTEs) in milk from the Southern Cities of Punjab, Pakistan: A health risk assessment study. *J. Food Compos. Anal.* **2022**, *108*, 104446. [CrossRef]
34. Kumar, D.L.; Kumar, P.K.; Ratnakar, D.; Vurimindi, H. Health risk assessment of nitrate and fluoride toxicity in groundwater contamination in the semi-arid area of Medchal, South India. *Appl. Water Sci.* **2022**, *12*, 11. [CrossRef]
35. Chen, G.; Wang, X.; Wang, R. Health risk assessment of potentially harmful elements in subsidence water bodies using a Monte Carlo approach: An example from the Huainan coal mining area, China. *Ecotoxicol. Environ. Saf.* **2019**, *171*, 737–745. [CrossRef]
36. Shukla, S.; Saxena, A.; Khan, R.; Li, P. Spatial analysis of groundwater quality and human health risk assessment in parts of Raebareli district, India. *Environ. Earth Sci.* **2021**, *80*, 800. [CrossRef]
37. Junjie, B.; Fufang, G.; Cong, P.; Jun, L. Characteristics of nitrate and heavy metals pollution in Huixian Wetland and its health risk assessment. *Alex. Eng. J.* **2022**, *61*, 9031–9042. [CrossRef]
38. Nhu, H.M.; Le, V.P.; Vinh, B.T.; Pham, H.; Khai, H.Q. Health risk assessment of arsenic in drinking groundwater: A case study in a central high land area of Vietnam. In Proceedings of the IOP Conference Series: Earth and Environmental Science, Al-Qadisiyah, Iraq, 19–20 January 2022. [CrossRef]
39. Nsabimana, A.; Li, P.; He, S.; He, X.; Alam, S.; Fida, M. Health Risk of the Shallow Groundwater and Its Suitability for Drinking Purpose in Tongchuan, China. *Water* **2021**, *13*, 3256. [CrossRef]
40. Ahmad, B.; Majid, R.; Hassan, P.; Iman, P.; Fariba, A.; Saeid, R. Factors affecting the nitrate concentration and its health risk assessment in drinking groundwater by application of Monte Carlo simulation and geographic information system. *Hum. Ecol. Risk Assess. Int. J.* **2019**, *27*, 1458–1471.
41. Mthembu, P.; Elumalai, V.; Li, P.; Uthandi, S.; Rajmohan, N.; Chidambaram, S. Integration of heavy metal pollution indices and health risk assessment of groundwater in semi-arid coastal aquifers, South Africa. *Expo. Health* **2022**, *14*, 487–502. [CrossRef]
42. Geng, M.; Qi, H.; Liu, X.; Gao, B.; Yang, Z.; Lu, W.; Sun, R. Occurrence and health risk assessment of selected metals in drinking water from two typical remote areas in China. *Environ. Sci. Pollut. Res.* **2016**, *23*, 8462–8469. [CrossRef]
43. Qu, L.; Huang, H.; Xia, F.; Liu, Y.; Dahlgren, R.A.; Zhang, M.; Mei, K. Risk analysis of heavy metal concentration in surface waters across the rural-urban interface of the Wen-Rui Tang River, China. *Environ. Pollut.* **2018**, *237*, 639–649. [CrossRef]
44. Sy, A.; Jian, Z.; Sxc, B.; Cc, C.; Jx, A.; Xl, A. Status assessment and probabilistic health risk modeling of metals accumulation in agriculture soils across China: A synthesis. *Environ. Int.* **2019**, *128*, 165–174.
45. Ns, A.; Msr, B.; Mba, C.; Jlz, C.; Hhn, C.; Wg, C. Industrial metal pollution in water and probabilistic assessment of human health risk. *J. Environ. Manag.* **2017**, *185*, 70–78.

46. Ghuniem, M.; Khorshed, M.; El-Safty, S.; Souaya, E.; Khalil, M. Potential human health risk assessment of potentially toxic elements intake via consumption of soft drinks purchased from different Egyptian markets. *Int. J. Environ. Anal. Chem.* **2020**, 1–23. [CrossRef]
47. Wang, L.; Li, P.; Duan, R.; He, X. Occurrence, Controlling Factors and Health Risks of Cr^{6+} in Groundwater in the Guanzhong Basin of China. *Expo. Health* **2022**, *14*, 239–251. [CrossRef]
48. Wang, Y.; Li, P. Appraisal of shallow groundwater quality with human health risk assessment in different seasons in rural areas of the Guanzhong Plain (China). *Environ. Res.* **2022**, *207*, 112210. [CrossRef] [PubMed]

MDPI
St. Alban-Anlage 66
4052 Basel
Switzerland
Tel. +41 61 683 77 34
Fax +41 61 302 89 18
www.mdpi.com

Water Editorial Office
E-mail: water@mdpi.com
www.mdpi.com/journal/water

www.ingramcontent.com/pod-product-compliance
Lightning Source LLC
LaVergne TN
LVHW070412100526
838202LV00014B/1442